Fuel Cell-Based and Hybrid Power Generation Systems Modeling

Fuel Cell-Based and Hybrid Power Generation Systems Modeling

Editor

Orazio Barbera

MDPI

Basel • Beijing • Wuhan • Barcelona • Belgrade • Novi Sad • Cluj • Manchester

Editor
Orazio Barbera
Italian National Research
Council (CNR)
Messina
Italy

Editorial Office
MDPI AG
Grosspeteranlage 5
4052 Basel, Switzerland

This is a reprint of articles from the Special Issue published online in the open access journal *Energies* (ISSN 1996-1073) (available at: https://www.mdpi.com/journal/energies/special_issues/Fuel_Cell_Based_and_Hybrid_Power_Generation_Systems_Modeling).

For citation purposes, cite each article independently as indicated on the article page online and as indicated below:

Lastname, A.A.; Lastname, B.B. Article Title. *Journal Name* **Year**, *Volume Number*, Page Range.

ISBN 978-3-7258-0353-8 (Hbk)
ISBN 978-3-7258-0354-5 (PDF)
doi.org/10.3390/books978-3-7258-0354-5

Contents

About the Editor

Orazio Barbera

Dr. Barbera has been a permanent researcher at CNR-ITAE (National Research Council of Italy – Institute for Advanced Energy Technologies) since 2009, graduated from "Politecnico di Milano" with a degree in mechanical engineering, and received a PhD in "Materials for Environment and Energy" from the University of Tor Vergata, Rome.

He focuses his research activity on technologies for generating and storing energy based on fuel cells and batteries. In particular, he focuses on engineering methods for low-temperature hydrogen/methanol fuel cells, metal–air batteries, and photoelectrochemical cells. He studied these design methodologies in national as well as international projects and private partnership contracts.

Prototyping is the direct expression of this research activity, allowing technology to be transferred from the lab scale to actual implementation; Dr. Barbera has designed numerous prototypes of fuel cell stacks for stationary, marine, space, and portable applications, which have been manufactured and tested. He also developed and tested lab-scale metal–air batteries as well as prototypes for photoelectrochemical water splitting. All of this allowed for studying the behaviour of the devices and validating the methodology, which has a general character. Moreover, the influence of the design methodology on performance has been a further important aspect of the investigation. Lately, he has focused on system integration in order to complete the research activity on device design.

Teaching and dissemination activities are further important aspects of Dr. Barbera's science curriculum. He has held seminars on hydrogen economy and engineering applied to electrochemical devices at universities, schools, and public institutions.

He is the author of numerous peer-reviewed scientific articles and book chapters, as well as the editor of journal Special Issues and books.

Preface

The Earth's climate has changed throughout history; over the past 650,000 years, seven cycles of glaciation have occurred. However, the current warming trend is particularly significant because it is most likely the result of the use of fossil fuels since the mid-20th century. In this context, near-zero-emissions systems based on fuel cells are a potential key factor in the green energy transition.

This Special Issue contains ten papers ranging from the generation of green hydrogen to the modeling of fuel-cell-based systems of different natures and for various applications. The breadth of the topics covered is its greatest strength and offers the reader a comprehensive view of how complex and exciting research's efforts for an advanced but environmentally friendly society are.

I have to thank all of the experienced and highly qualified papers' authors and my Assistant Editor, Ms Angela Zang; she was a generous and professional person who guided me to making this book come true.

Orazio Barbera
Editor

energies

MDPI

Editorial

Fuel Cell-Based and Hybrid Power Generation Systems Modelling

Orazio Barbera

Consiglio Nazionale delle Ricerche—Istituto di Tecnologie Avanzate per l'Energia "Nicola Giordano", Via Salita S. Lucia Sopra Contesse 5, 98126 Messina, Italy; orazio.barbera@itae.cnr.it

Citation: Barbera, O. Fuel Cell-Based and Hybrid Power Generation Systems Modelling. *Energies* **2024**, *17*, 3340. https://doi.org/10.3390/en17133340

Received: 8 May 2024
Accepted: 17 June 2024
Published: 8 July 2024

The World Economic Forum's Global Risks Report 2022 identifies climate change as a paramount threat to humanity. Global temperatures have risen by a concerning 1.2 °C since pre-industrial times, and this alarming acceleration is projected to continue, with estimates suggesting a further increase of 5 °C by the end of the century. The repercussions of climate change are far-reaching, impacting sectors like agriculture, energy, economics, finance, social dynamics, political landscapes, and public health [1], and achieving sustainable and inclusive development hinges on effectively addressing environmental degradation, necessitating a resolute political commitment from the international community. While this commitment is widely acknowledged as essential, there remains a significant disparity in the level of action taken by individual nations [2]. For 35 years, the scientific community has been acutely aware of the threat posed by global warming, a direct consequence of humanity's escalating anthropogenic carbon dioxide emissions. This period has been marked by sustained international efforts to mitigate this existential challenge. The Intergovernmental Panel on Climate Change (IPCC) was established in 1988 and produced its first science report in 1990. Two years later, the landmark Rio Earth Summit yielded the Rio Declaration and Agenda 21, a comprehensive framework for achieving sustainable development. This pivotal event coincided with the adoption of the "United Nations Framework Convention on Climate Change" (UNFCCC), a critical forum for negotiating global greenhouse gas (GHG) emission limitations. Building upon the UNFCCC, the first international agreement to mandate GHG emission reductions, the Kyoto Protocol, was adopted at COP3 (Conference of the Parties) in 1997. The fight against climate change gained further traction with the birth of the Paris Agreement at COP21 in 2015. This landmark accord garnered commitments from 196 nations to limit global temperature rise to well below 2 °C above pre-industrial levels, with an aspirational goal of 1.5 °C. The most recent COP26, held in Glasgow, witnessed significant development and numerous countries pledging to achieve net-zero emissions. COP27 in Egypt further solidified these commitments while establishing an innovative loss and damage fund to assist developing nations disproportionately impacted by climate change [3]. COP28 in Dubai marked a pivotal shift in climate action. The final agreement enshrined a move away from fossil fuels, potentially representing a turning point. It also significantly bolstered renewables and energy efficiency, with ambitious plans to expand their impact in the rest of this decade. Notably, COP28 finalised the loss and damage fund for developing countries, and fostered a greater sense of inclusivity in the discussion on solutions, including the return of nuclear power to the climate debate [4].

The preceding statements underscore the paramount significance of the battle against global warming in the context of international politics. It is a battle that can only be won through the application of cutting-edge scientific research, towards the development of advanced technologies for non-pollutant GHG-free power generation systems.

Pursuing a successful energy transition demands a multifaceted scientific strategy anchored by two pillars: fundamental research, to elucidate underlying scientific principles, and technological research, to develop practical solutions. The latter focuses on mitigating

GHG emissions through two key strategies: (i) the adoption of alternative energy sources and (ii) the use of hybrid and fuel cell-based power systems. Hybrid power systems can be defined as energy systems that integrate multiple energy sources, for more efficient, reliable, and cost-effective electricity generation, compared to single-source systems. By capitalising on the complementary nature of various energy sources, hybrid systems offer a sophisticated approach to maximising energy output and enhancing efficiency. This is achieved, through the strategic integration of diverse generation technologies; allowing the system to adapt to fluctuating environmental conditions and optimise energy production based on the most readily available resource. Hybridisation tackles energy demands in remote areas and existing facilities. Additionally, hybrid systems can incorporate energy storage or conventional backup sources, like diesel generators, for further enhanced system reliability. On the other hand, hybrid configuration implies a further system complication due to the request for efficient and smart management of different power sources as a function, in principle, of loads and costs. System sizing and optimisation are, generally, the focus of research in hybrid power sources. Consequently, understanding the functioning of hybrid energy systems is fundamental and studies cannot ignore their numerical modelling, generally based on two main approaches: analytical modelling and simulation. The first technique employs mathematical equations and established scientific principles to represent the behaviour of each component within the system. While well-suited for simpler systems, it can become intricate for larger systems with numerous components, making analysis challenging. The second leverages computer software to create a virtual representation of the system, and to simulate its behaviour. The model is over-time evolving, based on pre-defined rules or relationships between variables. Offering versatility, it applies to simple and complex systems. Popular software tools used for hybrid power system modelling include HOMER (HOMER Energy LLC), MATLAB/Simulink (MathWorks), and PSYCADE (OPAL-RT Technologies). The choice of modelling technique depends on several factors, including the system's complexity, the desired level of accuracy, and the available resources. Numerical modelling including energy and exergy analysis of hybrid power systems is also considered by scholars as well as cost modelling [5,6].

Feasibility studies for hybrid power systems provide a holistic view, encompassing the electrical system, financials, and site-specific characteristics. They offer a scientific approach to optimising new hybrid power systems, reconfiguring existing ones for efficiency, and identifying new renewable assets. For example, the analysis of scenarios, which involves implementing a hybrid power system composed of photovoltaic, wind, and battery energy storage systems, is fundamental to determining the ideal proportion of energy supply through an analysis of technical, financial, and social aspects [7]. Hybrid systems can integrate heterogeneous technologies, as in [8], where anaerobic digestion, pyrolysis and solar PV were integrated forming a plant which can produce energy and fuels such as green hydrogen, refuse-derived fuels, bio-compressed natural gas and compost. The study compared the financial feasibility of three scenarios—generating electricity and fuels, generating electricity alone and generating fuels alone—by modelling their levels of energy output and financial performance. Hybrid power systems can be conceived without renewable energy sources and studied using energy, exergy, economic, and life cycle environmental analyses. A biogas power generation and hydrogen generation system can be integrated with a solar thermal energy storage unit, a SOFC-Micro Gas Turbine unit, and a waste heat utilisation unit. In this case, higher efficiency could be achieved concerning conventional power systems [9]. Another innovative configuration of a power plant was modelled in [10] where, alkaline fuel cells, which generate a significant amount of waste heat, were coupled with thermogalvanic cells for electric power generation by harvesting the low-grade exhaust heat produced.

Hydrogen and battery technologies, which have a primary role in the energy transition, due to their ability to mitigate the fluctuation in energy production from renewable sources and to be applied to stationary and automotive applications, are developed from electrochemical and engineering points of view. The research activity on active materials is supported by technological research addressing the methodologies for devices (fuel cell stack and battery) and power unit construction. The latter results from assembling the device and ancillaries, for gas and temperature management, current and voltage management, digital control units, and so on. Fuel cell (FC) technology offers a promising path towards a clean and sustainable energy future. However, optimising their design, operation, and integration with other energy sources necessitates robust mathematical modelling techniques. These models act as virtual representations of the complex electrochemical and physical processes occurring within an FC system, enabling researchers and engineers to calculate performance, and models can predict the FC system's electrical output, efficiency, and response to varying operating conditions. This is crucial to design control strategies that maximise efficiency and system stability, optimise the design by simulating different design parameters within the model before building physical prototypes, and reduce development time and cost while achieving desired performance characteristics. Various mathematical modelling approaches are used for FC power systems; each with its strengths and limitations: Empirical Models, which rely on experimental data to establish relationships between input variables (e.g., temperature, pressure, fuel flow) and output variables (e.g., voltage, current). They are simple to implement but may not be accurate under conditions outside the range of the experimental data used for model creation. Semi-empirical models, combining empirical relationships with fundamental physical principles governing the electrochemical processes within the FC. They offer improved accuracy compared to purely empirical models but may still require some experimental data for parameter estimation. Physics-based models, representing the most complex approach, account for the fundamental governing equations that describe mass transport, heat transfer, and electrochemical reactions within the FC. They offer the highest level of accuracy but require a deep understanding of the underlying physical and chemical phenomena and significant computational resources. More detailed information can be found in [11]. In this context, the industry and the scientific community need advanced FC systems models that can replicate real-world operating conditions. In [12], a comprehensive methodology to calibrate and validate multi-physics dynamic FC systems models was developed to accurately describe the behaviour of the FC stack and components of the "balance of plant". The model was calibrated using experimental data from a Toyota Mirai FC electric vehicle's controller area network (CAN) bus system. An FC system for aircraft application was modelled in [13] to evaluate the applicability of this technology to an existing regional aircraft and assess its electrification. Specifically, a general mathematical model was developed, involving multiple scales, ranging from individual cells to aircraft scales. Then, different sizing approaches were used to compute the overall weight of the hydrogen-based propulsion system, to optimise the system and minimise its weight. In [14], a novel combined heat and power system, integrating solid oxide FC and organic Rankine cycle, was studied by mathematical modelling to achieve the efficient and clean utilisation of poultry litter. In [15], a model of a SOFC system with a recycling of the anode off-gas was proposed to improve the system fuel utilisation under a low rate of stack fuel utilisation.

These systems/power units are often integrated into hybrid power systems formed by heterogeneous power sources such as photovoltaic, wind, internal combustion engines, batteries, turbines and others. In the following paragraphs, there will be illustrated some examples from the scientific literature; focusing on hybrid power systems using fuel cells and batteries.

In [16], fuel cells are combined with an internal combustion engine, utilising ammonia decomposition as the primary fuel source. The system comprises an ammonia decomposition subsystem, a fuel cell, an internal combustion engine, a heat exchanger, and a power electronic converter. A hybrid system combining solar-assisted reforming of methanol and FC power generation was modelled in [17], where methanol is used as a coolant for the FC subsystem to take away the waste heat, and reformed for hydrogen production with the assistance of the solar energy subsystem. An artificial intelligence-based methodology was employed for analysing and optimising a grid-tied solar photovoltaic-FC hybrid power system. Demonstrating how AI techniques can assist in decision-making, improving system performance, achieving higher energy efficiency levels, and financial viability [18]. A biomass-based power generation system which includes a system to produce syngas, a fuel cell, a gas turbine, and an organic Rankine cycle was modelled in [19]. An optimisation of a multisource hybrid photovoltaic/wind/diesel/FC system was performed in [20]. The study also examines, and compares, the techno-economic viability of an off-grid hybrid photovoltaic/wind/diesel/FC, photovoltaic/diesel/FC, and wind/diesel/FC systems. Finally, in transportation applications, the optimal design of a renewable energy hybrid power system (photovoltaic, FC, and battery) for a tugboat was presented in [21].

This Special Issue aims to illustrate the most research advances in modelling fuel cell-based and hybrid power systems. It focuses on the mathematical modelling of FC and hybrid systems, and shows different approaches, considering FC technology (PEFC, SOFC, DMFC), system architecture, hybridisation level, application, and power management. It consists of ten contributions spanning from modelling high- and low-temperature fuel cell-based and hybrid power systems for stationary and automotive applications, to hydrogen production modelling from reforming and photoelectrolysis. It furnishes a comprehensive overview of the applicability of modelling techniques across diverse domains relevant to low-GHG or GHG-free power systems (see list of contributions).

Hydrogen production from reforming and photoelectrochemical devices was the focus of Contribution 1, which describes a design-for-optimization study, aimed at unifying temperature distribution by novel arrangements of catalyst segments in the model of a biogas reforming reactor, and Contribution 10, proposing a multi-physics-based numerical model of a tandem photoelectrochemical cell, an innovative concept for solar-to-hydrogen water conversion. A fuel cell/battery hybrid power system for a UAV was modelled in Contribution 2, and SOFC systems were modelled in Contribution 3 and Contribution 4; in particular, the first considered an anode-off recirculation subsystem, the second the presence of an additively manufactured, high-temperature, heat exchanger in an LNG-fuelled SOFC system. Micro-combined heat and power cogeneration for residential applications was the object of Contribution 5, wherein a hybrid PV/SOFC power system was modelled. A numerical, dynamic model of a low-temperature FC stationary power system, for supporting and adding flexibility to a renewable energy-sourced grid, was proposed in Contribution 6. Low-temperature FC power systems of commercial hydrogen-fed vehicles (Toyota Mirai I and II) were modelled and analysed in Contribution 7; the models' validation was carried out using experimental data, acquired during use under road driving conditions. Contribution 8 presented a three-dimensional, steady-state, two-phase, multi-component and non-isothermal DMFC model. The DMFC model was validated against experimental measurements. Finally, a hybrid power system power train for application in an autonomous marine vehicle (low-temperature FC/battery) was modelled and investigated in Contribution 9.

Mathematical modelling plays a critical role in unlocking the full potential of new power generation technologies. By continually refining and advancing these models, researchers and engineers can design, optimise, and integrate FC-based and hybrid power systems for a cleaner and more sustainable energy future, paving the way for wider adoption of this promising technology.

Conflicts of Interest: The author declares no conflicts of interest.

List of Contributions:

1. Pajak, M.; Brus, G.; Szmyd, J. Catalyst Distribution Optimization Scheme for Effective Green Hydrogen Production from Biogas Reforming. *Energies* **2021**, *14*, 5558. https://doi.org/10.3390/en14175558.
2. Vasyukov, I.; Pavlenko, A.; Puzin, V.; Batishchev, D.; Bolshenko, I. Mathematical Modeling of an Electrotechnical Complex of a Power Unit Based on Hydrogen Fuel Cells for Unmanned Aerial Vehicles. *Energies* **2021**, *14*, 6974. https://doi.org/10.3390/en14216974.

3. Choi, E.; Yu, S.; Lee, S. Optimization of Operating Conditions of a Solid Oxide Fuel Cell System with Anode Off-Gas Recirculation Using the Model-Based Sensitivity Analysis. *Energies* **2022**, *15*, 644. https://doi.org/10.3390/en15020644.
4. Hollmann, J.; Fuchs, M.; Spieker, C.; Gardemann, U.; Steffen, M.; Luo, X.; Kabelac, S. System Simulation and Analysis of an LNG-Fueled SOFC System Using Additively Manufactured High Temperature Heat Exchangers. *Energies* **2022**, *15*, 941. https://doi.org/10.3390/en15030941.
5. Ghanem, R.; Nousch, L.; Richter, M. Modeling of a Grid-Independent Set-Up of a PV/SOFC Micro-CHP System Combined with a Seasonal Energy Storage for Residential Applications. *Energies* **2022**, *15*, 1388. https://doi.org/10.3390/en15041388.
6. Crespi, E.; Guandalini, G.; Nieto Cantero, G.; Campanari, S. Dynamic Modeling of a PEM Fuel Cell Power Plant for Flexibility Optimization and Grid Support. *Energies* **2022**, *15*, 4801. https://doi.org/10.3390/en15134801.
7. Pielecha, I. Modeling of Fuel Cells Characteristics in Relation to Real Driving Conditions of FCHEV Vehicles. *Energies* **2022**, *15*, 6753. https://doi.org/10.3390/en15186753.
8. Olesen, A.; Kær, S.; Berning, T. A Multi-Fluid Model for Water and Methanol Transport in a Direct Methanol Fuel Cell. *Energies* **2022**, *15*, 6869. https://doi.org/10.3390/en15196869.
9. De Lorenzo, G.; Piraino, F.; Longo, F.; Tinè, G.; Boscaino, V.; Panzavecchia, N.; Caccia, M.; Fragiacomo, P. Modelling and Performance Analysis of an Autonomous Marine Vehicle Powered by a Fuel Cell Hybrid Powertrain. *Energies* **2022**, *15*, 6926. https://doi.org/10.3390/en15196926.
10. Giacoppo, G.; Trocino, S.; Lo Vecchio, C.; Baglio, V.; Díez-García, M.; Aricò, A.; Barbera, O. Numerical 3D Model of a Novel Photoelectrolysis Tandem Cell with Solid Electrolyte for Green Hydrogen Production. *Energies* **2023**, *16*, 1953; https://doi.org/10.3390/en16041953.

References

1. Zhang, Y.; Liu, L.; Lan, M.; Su, Z.; Wang, K. Climate Change and Economic Policy Uncertainty: Evidence from Major Countries around the World. *Econ. Anal. Policy* **2024**, *81*, 1045–1060. [CrossRef]
2. Tadadjeu, S.; Njangang, H.; Woldemichael, A. Are Resource-Rich Countries Less Responsive to Global Warming? Oil Wealth and Climate Change Policy. *Energy Policy* **2023**, *182*, 113774. [CrossRef]
3. Maslin, M.A.; Lang, J.; Harvey, F. A Short History of the Successes and Failures of the International Climate Change Negotiations. *UCL Open Environ.* **2023**, *5*. [CrossRef] [PubMed]
4. Henderson, J. Ten Key Conclusions from COP28: A Farewell to Fossil Fuels? In *Energy Insight*; The Oxford Institute for Energy Studies: Oxford, UK, 2024; Volume 143.
5. Sinha, A.A.; Sanjay; Ansari, M.Z.; Shukla, A.K.; Choudhary, T. Comprehensive Review on Integration Strategies and Numerical Modeling of Fuel Cell Hybrid System for Power & Heat Production. *Int. J. Hydrogen Energy* **2023**, *48*, 33669–33704. [CrossRef]
6. Lee, J.Y.; Tseng, L.H.; Chen, C.L. *A Mathematical Technique for Optimal Design of Hybrid Power Systems Considering Demand-Side Management*; Elsevier Masson: Issy les Moulineaux, France, 2017; Volume 40, ISBN 9780444639653.
7. Borba, A.T.A.; Simões, L.J.M.; de Melo, T.R.; Santos, A.Á.B. Techno-Economic Assessment of a Hybrid Renewable Energy System for a County in the State of Bahia. *Energies* **2024**, *17*, 572. [CrossRef]
8. Armoo, E.A.; Mohammed, M.; Narra, S.; Beguedou, E.; Agyenim, F.B.; Kemausuor, F. Achieving Techno-Economic Feasibility for Hybrid Renewable Energy Systems through the Production of Energy and Alternative Fuels. *Energies* **2024**, *17*, 735. [CrossRef]
9. Ran, P.; Ou, Y.F.; Zhang, C.Y.; Chen, Y.T. Energy, Exergy, Economic, and Life Cycle Environmental Analysis of a Novel Biogas-Fueled Solid Oxide Fuel Cell Hybrid Power Generation System Assisted with Solar Thermal Energy Storage Unit. *Appl. Energy* **2024**, *358*, 122618. [CrossRef]
10. Wang, M.; Ruan, J.; Zhang, J.; Jiang, Y.; Gao, F.; Zhang, X.; Rahman, E.; Guo, J. Modeling, Thermodynamic Performance Analysis, and Parameter Optimization of a Hybrid Power Generation System Coupling Thermogalvanic Cells with Alkaline Fuel Cells. *Energy* **2024**, *292*, 130557. [CrossRef]
11. Wang, Y.; Yang, X.; Sun, Z.; Chen, Z. A Systematic Review of System Modeling and Control Strategy of Proton Exchange Membrane Fuel Cell. *Energy Rev.* **2024**, *3*, 100054. [CrossRef]
12. Lopez-Juarez, M.; Rockstroh, T.; Novella, R.; Vijayagopal, R. A Methodology to Develop Multi-Physics Dynamic Fuel Cell System Models Validated with Vehicle Realistic Drive Cycle Data. *Appl. Energy* **2024**, *358*, 122568. [CrossRef]
13. Chiara, M.; Pramotton, S.; Marocco, P.; Hugo, A.; Monteverde, A.; Santarelli, M. Optimal Design of a Hydrogen-Powered Fuel Cell System for Aircraft Applications. *Energy Convers. Manag.* **2024**, *306*, 118266. [CrossRef]
14. Yang, S.; Sun, H.; Liu, Z.; Deng, C.; Xie, N. Process Modeling and Analysis of a Combined Heat and Power System Integrating Solid Oxide Fuel Cell and Organic Rankine Cycle for Poultry Litter Utilization. *Appl. Therm. Eng.* **2024**, *236*, 121897. [CrossRef]
15. Ma, L.; Ru, X.; Wang, J.; Lin, Z. System Model and Performance Analysis of a Solid Oxide Fuel Cell System Self-Humidified with Anode off-Gas Recycling. *Int. J. Hydrogen Energy* **2024**, *57*, 1164–1173. [CrossRef]

16. Xu, J.H.; Zhang, B.X.; Yan, H.Z.; Ding, Q.; Zhu, K.Q.; Yang, Y.R.; Huang, T.M.; Li, S.; Wan, Z.M.; Wang, X.D. A Comprehensive Assessment of the Hybrid Power Generation System of PEMFC and Internal Combustion Engine Based on Ammonia Decomposition. *Energy* **2023**, *285*, 129559. [CrossRef]
17. Zhao, K.; Kong, H.; Tan, S.; Yang, X.G.; Zheng, H.; Yang, T.; Wang, H. Analysis of a Hybrid System Combining Solar-Assisted Methanol Reforming and Fuel Cell Power Generation. *Energy Convers. Manag.* **2023**, *297*, 117664. [CrossRef]
18. Soni, P.; Naveena Bhargavi, R.; Dave, V.; Paliwal, H. Artificial Intelligence-Enabled Techno-Economic Analysis and Optimization of Grid-Tied Solar PV-Fuel Cell Hybrid Power Systems for Enhanced Performance. *E3S Web Conf.* **2024**, *472*, 03012. [CrossRef]
19. Hai, T.; Alenizi, F.A.; Flaih, L.R.; Singh Chauhan, B.; Metwally, A.S.M. Dual-Objective Optimization of a Novel Hybrid Power Generation System Based on Hydrogen Production Unit for Emission Reduction. *Int. J. Hydrogen Energy* **2024**, *52*, 916–928. [CrossRef]
20. Koholé, Y.W.; Wankouo Ngouleu, C.A.; Fohagui, F.C.V.; Tchuen, G. Optimization of an Off-Grid Hybrid Photovoltaic/Wind/Diesel/Fuel Cell System for Residential Applications Power Generation Employing Evolutionary Algorithms. *Renew. Energy* **2024**, *224*, 120131. [CrossRef]
21. Xin, R.; Wang, Z.; Zhai, J.; Zhang, J.; Cui, D.; Ji, Y. Simulation of Design and Operation of a Hybrid PV (Photovoltaic)/PEMFC (Proton Exchange Membrane Fuel Cell)/Battery Power System for a Tugboat. *Adv. Transdiscipl. Eng.* **2023**, *38*, 250–262. [CrossRef]

energies

Article

Catalyst Distribution Optimization Scheme for Effective Green Hydrogen Production from Biogas Reforming

Marcin Pajak *, Grzegorz Brus and Janusz S. Szmyd

Department of Fundamental Research in Energy Engineering, AGH University of Science and Technology, 30-059 Krakow, Poland; brus@agh.edu.pl (G.B.); janusz.szmyd@agh.edu.pl (J.S.S.)
* Correspondence: mpajak@agh.edu.pl

Abstract: Green hydrogen technology has recently gained in popularity due to the current economic and ecological trends that aim to remove the fossil fuels share in the energy mix. Among various alternatives, biogas reforming is an attractive choice for hydrogen production. To meet the authorities' requirements, reforming biogas-enriched natural gas and sole biogas is tempting. Highly effective process conditions of biogas reforming are yet to be designed. The current state of the art lacks proper optimization of the process conditions. The optimization should aim to allow for maximization of the process effectiveness and limitation of the phenomena having an adverse influence on the process itself. One of the issues that should be addressed in optimization is the uniformity of temperature inside a reactor. Here we show an optimization design study that aims to unify temperature distribution by novel arrangements of catalysts segments in the model biogas reforming reactor. The acquired numerical results confirm the possibility of the enhancement of reaction effectiveness, coming from improving the thermal conditions. The used amount of catalytic material is remarkably reduced as a side effect of the presented optimization. To ensure an unhindered perception of the reaction improvement, the authors proposed a ratio of the hydrogen output and the amount of used catalyst as a measure.

Keywords: numerical optimization; genetic algorithm; green hydrogen production; catalyst distribution; biogas reforming; artificial intelligence methods; fuel cells

Citation: Pajak, M.; Brus, G.; Szmyd, J.S. Catalyst Distribution Optimization Scheme for Effective Green Hydrogen Production from Biogas Reforming. *Energies* **2021**, *14*, 5558. https://doi.org/10.3390/en14175558

Academic Editor: Muhammad Aziz

Received: 31 July 2021
Accepted: 1 September 2021
Published: 6 September 2021

1. Introduction

The hydrogen industry is currently at the beginning of an undeniable peak. The energy industry has started to pursue technology related to the production and consumption of hydrogen, to meet the ecological regulations introduced by the governmental organizations [1]. The trend gains its strength not only due to a pure interest in the fuel cells technology itself but also because of a rapidly growing necessity of energy storage. The demand for energy storage comes from its periodic oversupply from renewable sources [2]. The leading technologies used for production of hydrogen are hydrocarbons reforming and water electrolysis [3,4]. Water electrolysis from renewables would be the first choice, considering the ecological aspect. However, the technology is still under development and requires further research before its use would be economically encouraged [5]. The present state of the art makes the use of the water electrolysis justified only in applications implementing renewable energy sources, in which no other possibility of the surplus energy consumption is available [6]. Therefore, the majority of the hydrogen produced worldwide comes from the reforming of hydrocarbons [7]. The reforming sovereignty was established due to a relatively low cost of hydrogen production and the reaction placing second, considering the lowest environmental impact [8]. Production of hydrogen on the way of the reforming reaction is assured in the nearest future due to a lack of better alternatives. Furthermore, the size of the natural gas establishment induces no possibility of its immediate shutdown [9,10]. Even though the governmental regulations call for a termination of fossil fuel exploitation, it is impossible to achieve instantly. Thus, natural gas production is inevitable in the nearest future. Although, emissions coming from the energy production using natural gas may be heavily limited, if the natural gas would be converted into hydrogen, on the way of the steam reforming reaction. Considering the current economic sentiments, the reform of hydrocarbons remains the most feasible way of hydrogen production. To decrease the conversion process' environmental impact, a carbon-capture system can be introduced [11,12]. Furthermore, reform of biogas-enriched natural gas may be a reasonable solution for even better addressing of the ecological issue of the process [13], as biogas is qualified as a renewable energy source [14]. However, biogas production may often require additional energy input, limiting its feasibility. An interesting take on the clean biomass transformation is presented by Gonclaves et al. [15]. Processing of sole biogas, and its integration with a carbon capture system would reduce the reforming's environmental impact to a minimum [16]. Based on the presented literature review, scientific efforts on the development of the reforming process can not yet be forfeited. Despite its almost century-old history, the reforming technology remains flawed. Carbon deposition and degradation of the catalytic material, due to the thermal stress induced by temperature gradients, are the most remarkable issues [17,18]. Proper maintenance of the thermal conditions during the reaction has been already proven to alleviate the catalyst degradation [19]. Moreover, proper moderation of the temperature field inside the reactor has been reported to increase the overall process' efficiency [20]. Our team has developed an original strategy to address the issue of temperature gradients occurring in the reactor [21]. The presented article proposes an application of the macro-patterning concept. The concept predicts the division of the reactor's volume into separate segments, and their further filling with alternating catalytic and non-catalytic materials [21]. The main goal of macro-pattering is to allow for the moderation of the thermal conditions inside the reforming reactor. The strategy is combined with an evolutionary optimization to find the most optimal design of the catalytic insert for the reforming of biogas. The application of macro-patterning is predicted to improve the thermal conditions during the process. The conditions improvement would result in enhancement of the reaction thermodynamics, effecting in reduction of the carbon deposition and premature catalyst degradation [19,22]. The model biogas being the subject of this analysis is assumed to consist only of CH_4 and CO_2. The influence of other biogas components is out of the scope of the presented research [23]. The analysis prepared for the needs of the presented research included:

- Preparation of the numerical simulation considering reforming of model biogas
- Application of the macro-patterning concept for the reactor's geometry
- Combining the prepared numerical model with a genetic algorithm to find the most optimal catalyst insert design

2. Mathematical Model

2.1. Chemical Reactions

The presented research included formulation of an adequate mathematical model. The analysis of the biogas reforming process can be considered as a quasi three-dimensional, due to the reactor's geometry and its axial symmetry. Thanks to the axial symmetry, the computational domains may be simplified to two-dimensions, when proper conditions are constituted within the model (Section 3.1) [24]. The geometry implemented into the model of biogas reforming process is presented in Figure 1.

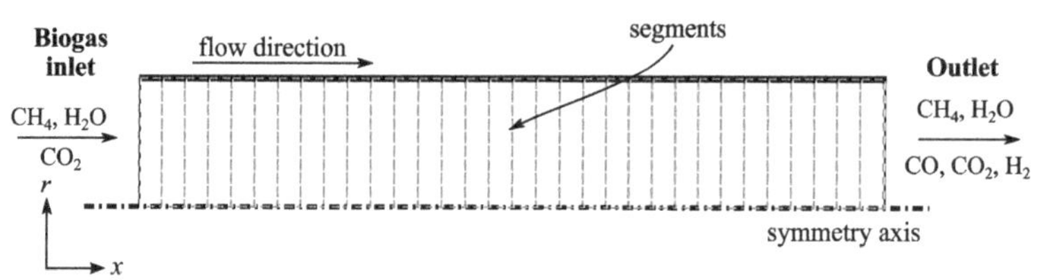

Figure 1. Domain defined for a biogas reforming reactor.

Due to the process character, two main chemical reactions are implemented in the mathematical model [25,26]. According to the previously conducted research, inclusion of the methane-steam reforming (MSR) (Equation (1)), dry reforming (DRY) (Equation (2) and water–gas-shift reaction (WGS) (Equation (3)) is sufficient for acquiring relevant results [27,28]. The reactions are described using the following equations:

$$CH_4 + H_2O \rightarrow 3H_2 + CO, \qquad \Delta H_{MSR} = 206.1\frac{kJ}{mol}, \tag{1}$$

$$CH_4 + CO_2 \rightleftharpoons 2H_2 + 2CO, \qquad \Delta H_{DRY} = 247\frac{kJ}{mol}, \tag{2}$$

$$CO + H_2O \rightleftharpoons H_2 + CO_2, \qquad \Delta H_{WGS} = -41.15\frac{kJ}{mol}. \tag{3}$$

The enthalpy change ΔH are taken from literature [21,29]. To allow the inclusion of the reactions into the model, knowledge of their rates is essential. According to the research conducted by Brus et al. [30], the effective rate of MSR and DRY reactions can be expressed with a common equation:

$$R_{\mathrm{eff}} = \dot{w}_{\mathrm{cat}} A_{\mathrm{MSR}} \exp\left(-\frac{E_{\mathrm{a}}}{RT}\right) p_{\mathrm{CH_4}}^{\alpha} \left(p_{\mathrm{H_2O}} + p_{\mathrm{CO_2}}\right)^{\beta}. \tag{4}$$

The individual reaction rates for the MSR and DRY reactions can be distinguished as follows:

$$R_{\mathrm{MSR}} = R_{\mathrm{eff}} \frac{p_{\mathrm{H_2O}}}{p_{\mathrm{CO_2}} + p_{\mathrm{H_2O}}}, \tag{5}$$

$$R_{\mathrm{DRY}} = R_{\mathrm{eff}} \frac{p_{\mathrm{CO_2}}}{p_{\mathrm{CO_2}} + p_{\mathrm{H_2O}}}. \tag{6}$$

The WGS reaction has a more unpredictable nature. Thus, preparation of a formula, returning proper values regardless of the process conditions, is not plausible. However, according to Ahmed and Föger, the WGS reaction can be assumed to maintain equilibrium under specific conditions [31]. The correctness of the described approach is confirmed by other research [32–34]. The boundary conditions applied in the presented research, described in Section 3.1, are designed to satisfy the equilibrium assumption. Therefore, the WGS rate is calculated using the following procedure:

$$K_{\mathrm{WGS}} = \frac{k_{\mathrm{WGS}}^{+}}{k_{\mathrm{WGS}}^{-}} = \frac{p_{\mathrm{CO_2}} p_{\mathrm{H_2}}}{p_{\mathrm{CO}} p_{\mathrm{H_2O}}} = \exp\left(-\frac{\Delta G_{\mathrm{WGS}}^{0}}{RT}\right). \tag{7}$$

Combining of the Equations (4) and (7) allows for formulation of the WGS reaction rate equation, as follows:

$$R_{\mathrm{WGS}} = k_{\mathrm{WGS}}^{+} p_{\mathrm{CO}} p_{\mathrm{H_2O}} + k_{\mathrm{WGS}}^{-} p_{\mathrm{H_2}} p_{\mathrm{CO_2}}. \tag{8}$$

The values of partial pressures, included in the Equations (4)–(8), are derived from an analysis of the reactions' stoichiometry [30].

2.2. Heat and Mass Transfer

The prepared mathematical model has the fundamental transport equations incorporated. Considering the defined computational domain, the equations are formulated for a two-dimensional case. Therefore, the model's equations are calculated for the longitudinal and radial directions only. The fluids taking part in the reaction are considered Newtonian and incompressible. Thus, the continuity equation (Equation (9)) takes the following form [35]:

$$\frac{\partial(\rho_0 U_x)}{\partial x} + \frac{1}{r}\frac{\partial(r\rho_0 U_r)}{\partial r} = 0. \tag{9}$$

The materials serving as the catalytic insert are porous. Therefore, the terms of the momentum conservation equation have to contain parameters depending on the structure morphology [36]. A separate equation for each dimension is formulated (Equations (10) and (11)):

$$\begin{aligned} \frac{\rho_0}{\varepsilon_0^2}\left(U_x\frac{\partial U_x}{\partial x}+U_r\frac{\partial U_x}{\partial r}\right) = \\ -\frac{\partial P}{\partial x}+\frac{\mu}{\varepsilon_0}\left[\frac{\partial^2 U_x}{\partial x^2}+\frac{1}{r}\frac{\partial}{\partial r}\left(r\frac{\partial U_x}{\partial r}\right)\right] \\ -\frac{\mu}{K_p}U_x-\frac{\rho_0 c_{\text{ine}}}{\sqrt{K_p}}U_x\sqrt{U_x^2+U_r^2}, \end{aligned} \tag{10}$$

$$\begin{aligned} \frac{\rho_0}{\varepsilon_0^2}\left(U_x\frac{\partial U_r}{\partial x}+U_r\frac{\partial U_r}{\partial r}\right) = \\ -\frac{\partial P}{\partial r}+\frac{\mu}{\varepsilon_0}\left[\frac{\partial^2 U_r}{\partial x^2}+\frac{1}{r}\frac{\partial}{\partial r}\left(r\frac{\partial U_r}{\partial r}\right)-\frac{U_r}{r^2}\right] \\ -\frac{\mu}{K_p}U_r-\frac{\rho_0 c_{\text{ine}}}{\sqrt{K_p}}U_r\sqrt{U_x^2+U_r^2}, \end{aligned} \tag{11}$$

The use of metallic foams implies a necessity of application of a relevant model, allowing us to calculate the effective thermal conductivity coefficient of the material λ_{eff} [37,38]. Knowledge of a proper λ_{eff} value is vital for the calculation of the energy conservation equation (Equation (12)) [36]. An adequate model is prepared, described in detail within our previous article [39].

$$\begin{aligned} \rho_0 C_p\left(U_x\frac{\partial T_{\text{loc}}}{\partial x}+U_r\frac{\partial T_{\text{loc}}}{\partial r}\right) = \frac{\partial}{\partial x}\left(\lambda_{\text{eff}}\frac{\partial T_{\text{loc}}}{\partial x}\right) \\ +\frac{1}{r}\frac{\partial}{\partial r}\left(r\lambda_{\text{eff}}\frac{\partial T_{\text{loc}}}{\partial r}\right)+Q_s, \end{aligned} \tag{12}$$

where the heat sources/sinks Q_s depend on enthalpy changes ΔH and rates of the MSR, DRY and WGS [25,30], and are defined using following equations:

$$Q_{\text{MSR}} = -\Delta H_{\text{MSR}} R_{\text{MSR}}, \tag{13}$$

$$Q_{\text{DRY}} = -\Delta H_{\text{DRY}} R_{\text{DRY}}, \tag{14}$$

$$Q_{\text{WGS}} = -\Delta H_{\text{WGS}} R_{\text{WGS}}. \tag{15}$$

The species conservation is calculated using molar fractions of species taking part in the reaction (Equation (16)). The formulated equation was derived from the Fick's law of diffusion [40]:

$$\begin{aligned} \rho_0\left(U_x\frac{\partial Y_j}{\partial x}+U_r\frac{\partial Y_j}{\partial r}\right) = \frac{\partial}{\partial x}\left(\rho_0 D_{j,\text{eff}}\frac{\partial Y_j}{\partial x}\right) \\ +\frac{1}{r}\frac{\partial}{\partial r}\left(r\rho_0 D_{j,\text{eff}}\frac{\partial Y_j}{\partial r}\right)+S_j. \end{aligned} \tag{16}$$

The mass sources and sinks S_j depend on the MSR, DRY and WGS rates and molar masses of the species taking part in the reaction [41,42]. The exact equations defining the values of S_j are described in the Table 1.

Table 1. Mass generation.

Species	Mass Generation MSR	Mass Generation WGS	Mass Generation DRY	Summarized Generation
H_2	$3R_{MSR}M_{H_2}$	$R_{WGS}M_{H_2}$	$2R_{DRY}M_{H_2}$	$3R_{MSR}M_{H_2}$ $+ R_{WGS}M_{H_2} + 2R_{DRY}M_{H_2}$
CO	$R_{MSR}M_{CO}$	$-R_{WGS}M_{CO}$	$2R_{DRY}M_{CO}$	$R_{MSR}M_{CO}$ $- R_{WGS}M_{CO} + 2R_{DRY}M_{CO}$
CO_2	0	$R_{WGS}M_{CO_2}$	$-R_{DRY}M_{CO_2}$	$R_{WGS}M_{CO_2}$ $-2R_{DRY}M_{H_2}$
CH_4	$-R_{MSR}M_{CH_4}$	0	$-R_{DRY}M_{H_2}$	$-R_{MSR}M_{CH_4}$ $-R_{DRY}M_{H_2}$
H_2O	$-R_{MSR}M_{H_2O}$	$-R_{WGS}M_{H_2O}$	0	$-R_{MSR}M_{H_2O}$ $-R_{WGS}M_{H_2O}$

3. Numerical Model

Proceeding with computations of the prepared mathematical model, requires its implementation within a relevant numerical procedure. To allow doing so, a discretization of the computational domain is necessary. The governing equations are applied into the numerical model using the Finite Volume Method [24,43]. The generalized transport equation, representing the partial differential Equations (9)–(12) and (16), takes the following form [24,44]:

$$\Psi_x \frac{\partial \phi}{\partial x} + \Psi_r \frac{\partial \phi}{\partial r} = \frac{\partial}{\partial x}\left(\Gamma \frac{\partial \phi}{\partial x}\right) + \frac{1}{r}\frac{\partial}{\partial r}\left(r\Gamma \frac{\partial \phi}{\partial r}\right) + \overline{S}. \tag{17}$$

The source terms $\overline{S}$ in Equation (17) are gathered in the Table 2. The values presented in the Table 2 are relevant for segments filled with the catalytic material. When a non-catalytic segment is considered, the source terms values are set to 0. The chemical reactions are assumed to be suppressed, just after the gases mixture enters the non-catalytic region of the reformer [21]. To acquire the discrete transport equation (Equation (18)), integration of the Equation (17) is conducted. After the application of simple, mathematical transformations, the discrete transport equation is expressed as follows:

$$\begin{aligned} &[(\Psi_x\phi)_e - (\Psi_x\phi)_w] r_m \Delta r + [(r\Psi_r\phi)_n - (r\Psi_r\phi)_s]\Delta x = \\ &\left(\Gamma_e \frac{\phi_E - \phi_P}{\delta x} - \Gamma_w \frac{\phi_P - \phi_W}{\delta x}\right) r_m \Delta r \\ &+ \left(r\Gamma_n \frac{\phi_N - \phi_P}{\delta r} - r\Gamma_s \frac{\phi_P - \phi_S}{\delta r}\right)\Delta x + \overline{S} r_m \Delta r \Delta x, \end{aligned} \tag{18}$$

$$a_P\phi_P = a_E\phi_E + a_W\phi_W + a_N\phi_N + a_S\phi_S + b, \tag{19}$$

$$a_P = a_E + a_W + a_N + a_S, \tag{20}$$

$$b = \overline{S} r_m \Delta r \Delta x. \tag{21}$$

The coefficients a_j represent the fluxes crossing the faces of the control volumes. The fluxes are calculated using the Power Law scheme [24]. The subscripts N, E, S, W correspond to the compass directions, and represent the faces above, right, below and left to the currently analyzed node.

The SIMPLE algorithm is implemented into the numerical model, for calculation of the pressure corrections. The use of the SIMPLE algorithm is necessary to calculate fluid velocity in the specific regions of the reactor [24,45]. The systems of equations created during the discretization processes are further applied to the numerical procedure, and solved using the Gauss-Seidl iterative method [46].

Table 2. Source terms in Equation (17).

Equation	$\overline{S}$
(10)	$-\frac{\partial P}{\partial x} - \frac{\mu}{K_p} U_x - \frac{\rho_0 c_{ine}}{\sqrt{K_p}} U_x \sqrt{U_x^2 + U_r^2}$
(11)	$\frac{\partial P}{\partial r} - \frac{\mu}{K_p} U_r - \frac{\rho_0 c_{ine}}{\sqrt{K_p}} U_r \sqrt{U_x^2 + U_r^2} - \frac{\mu U_r}{\varepsilon r^2}$
(12)	Q_s
(16)	S_j

3.1. Boundary Conditions

The presented numerical procedure is considered a quasi three-dimensional case. Thanks to the axial symmetry of the reactor, the computational domain is reduced to two dimensions [47]. The symmetry boundary conditions are applied at the axis of the reactor, simultaneously allowing for an acceleration of the computations. Afterward, thirty separate segments are designated. The segments divide the reactor into separate zones in the longitudinal direction. Each of the segments has th exact same dimensions and is predicted to be filled with a catalytic nickel/yttria-stabilized-zirconia (Ni/YSZ) composite or with a stainless steel metallic foam serving as a non-catalytic segment. The presence of the metallic foam segments serves to reheat the gases mixture before entering the subsequent catalytic segment. Therefore, the gases are carrying more energy, allowing for an intensification of the reforming process inside the proceeding catalytic region of the reactor. The prepared numerical procedure requires the definition of x adequate values of the thermal conductivity coefficients λ for the two materials. Considering the process temperature taking values between 800 and 900 K, the λ value for the Ni/YSZ is set at 22 W m^{-1} K^{-1} [48] and for the metallic foam at 30 W m^{-1} K^{-1} [49]. The reactors length L is equal to 0.3 m and the inlet velocity u_{in} is set at 0.15 m s^{-1}. The reactor's wall is assumed to be made of stainless steel and to have a thickness of 2×10^{-3} m. The non-slip and Neumann boundary condition, with the heat flux equal to 0, are set at the reactor's wall. The feedstock entering the reactor is considered to acquire the reactor's temperature immediately. The properties of specimens being the subject of the analysis are taken from the literature [50]. The numerical grid's dimensions are set at 150 elements in the longitudinal direction, and 25 elements in the radial direction. A sensitivity analysis of the grid resolution and model's falsification procedures are described in detail in our previous work [21]. All of the described boundary conditions are summarized in Figure 2.

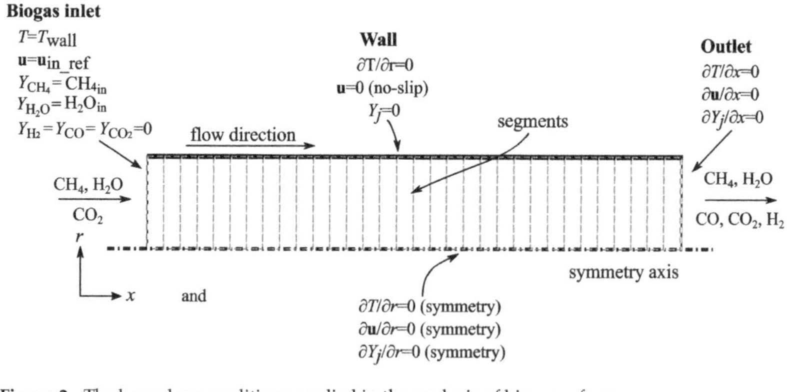

Figure 2. The boundary conditions applied in the analysis of biogas reformer.

4. Optimization Procedure

The optimization conducted in the presented analysis predicts the finding of the most optimal alignment of the catalytic and non-catalytic materials. The pursued optimal alignment has to lead to a unified distribution of temperature field without any significant losses in the produced hydrogen yield. The defined optimization problem is an arduous task due to the amount of possible segments configurations. Each segment can be distinguished as a catalytic or non-catalytic. Moreover, every single segment may vary in the values of porosity ε and average pore size d_p. The large variety of possible solutions induces the presence of numerous local extrema. Due to the problem characteristics, the genetic algorithm (GA) is chosen as the optimization technique. The genetic algorithm is a mature optimization concept, with a stochastic nature [51]. The GA is reported to return satisfying results for optimizations regarding adversely conditioned search spaces [52–54]. The genetic algorithm is successfully applied in researches considering the optimization of the steam reforming reaction [39,55,56]. Due to the algorithm having its roots in the process of natural selection, vocabulary originating from genetics is introduced to the nomenclature. A single solution is referred to as a specimen. Each specimen consists of segments, going by the name of chromosomes. A single segment has an individual set of parameters, called genes. In the case of the presented research, genes represent if a single segment has catalytic properties and its values of ε and d_p. A single iteration of the genetic algorithm includes the calculation of the reforming simulation for a set of specimens. The set is referred to as a population [57]. A numerical procedure is prepared for the needs of the presented research. The procedure starts with the preparation of the initial population. The chromosomes for the first set of solutions are initialized randomly. The algorithm is coupled with an in-house solver of the reforming process [39]. After randomization of the initial parameters, their values are loaded into instances of the reforming simulation. Afterward, the computation is commenced and proceeds until the convergence criteria for each specimen are met. The numerical procedure is assumed to converge after differences in values of the heat sources, acquired for two subsequent iterations, are lower than 10^{-5}. The sufficiency of the given threshold is proven in our previous research [21,39]. After converging each specimen, evaluation of the results is performed, based on user-defined fitness functions [51]. The higher the overall fitness f of a specific reactor, the higher are its chances to pass its chromosomes to the proceeding population. Each subsequent population is prepared using the crossover procedure. First, a set of parent specimens is chosen, basing on the roulette rule [39,51]. Afterward, the algorithm randomly selects a pair of reactors from the parents' set and translates their chromosomes' values into a single binary string. Then, the strings are split at the randomly selected crossover point, and their corresponding parts are interchanged. The operation results in the creation of a pair of specimens, inheriting the chromosomes from two different parents. The crossover procedure is repeated until a population of the same size is created. After each crossover, a mutation procedure is performed [58]. The procedure predicts flipping of a single bit in the binary string, occurring with a probability specified by the user. The mutation is introduced to prevent the GA from finding a local extremum instead of the global one. Another benefit of the mutation is the possibility of introducing new genes, which acquiring would be impossible in the way of the sole crossover. After having the new population defined, the GA calls the reforming simulation over the created specimens and its procedure is repeated. The presented analysis predicted thirty subsequent iterations of the genetic algorithm. Each population is constrained to include thirty specimens. The number of reactors in a single generation was reduced to be only thirty, due to an enormous computation cost of the numerical simulation of the reforming process [39]. The goal of the presented optimization procedure is to define catalyst distribution, allowing for unification of the temperature field with simultaneous maintenance of the methane conversion, at levels similar to the conventional reforming reactors. To define the robustness of the acquired optimization results, two different fitness functions are defined. The first function analyzes the amount of methane converted during the process. The amount of methane processed during the reaction is calculated basing on the methane (CH_4) fraction at the inlet and the outlet of the reformer, according to the following formula:

$$f_{CH_4} = \frac{frac_{CH_4in} - frac_{CH_4out}}{frac_{CH_4in}}, \tag{22}$$

where $frac_{CH_4in}$ and $frac_{CH_4out}$ stand for the methane fraction at the inlet and the outlet of the reactor, respectively. The second function considers minimization of the temperature gradients occurring inside the reformer. The algorithm analyzes the local temperature differences for each of the created control volumes (Section 3). The procedure iterates over each of the nodes and computes values of the temperature T differences between the analyzed node P and its neighbors: N, E, S, W. When all the differences are calculated, the highest difference is set as $\Delta T_{\text{loc}_{i,j}}$ of an analyzed control volume, where i and j represent the column and row number of the numerical grid (Equation (23)).

$$\Delta T_{\text{loc}_{i,j}} = \max\left\{ |T_{\text{P}_{i,j}} - T_{\text{N}_{i,j}}|, |T_{\text{P}_{i,j}} - T_{\text{E}_{i,j}}|, |T_{\text{P}_{i,j}} - T_{\text{S}_{i,j}}|, |T_{\text{P}_{i,j}} - T_{\text{W}_{i,j}}| \right\}. \tag{23}$$

After having calculated $\Delta T_{\text{loc}_{i,j}}$ values for each of the control volumes, the highest found is set as the temperature difference representing the whole reactor ΔT, according to the following formula:

$$\Delta T = \max\left\{ T_{\text{loc}_{1,1}}, T_{\text{loc}_{1,2}}, \ldots, T_{\text{loc}_{i,j}} \right\}. \tag{24}$$

The overall temperature fitness f_T is calculated using the following equation:

$$f_T = 1 - \frac{\Delta T}{\Delta T_{\text{max}}}, \tag{25}$$

where ΔT_{max} represents the maximal temperature difference, occurring in the reactor with continuous and homogeneous catalyst. The ΔT_{max} for the needs of the presented research is set at 25 K. After having computed the f_{CH_4} and f_T values, the overall fitness f is calculated using Equation (26)

$$f = \omega_1 \cdot f_{CH_4} + \omega_2 \cdot f_T, \tag{26}$$

where ω_1 and ω_2 represent the weights. The weights values are 0.6 for ω_1 and 0.4 for ω_2. The summarized procedure of the genetic algorithm is presented as a block schema in Figure 3.

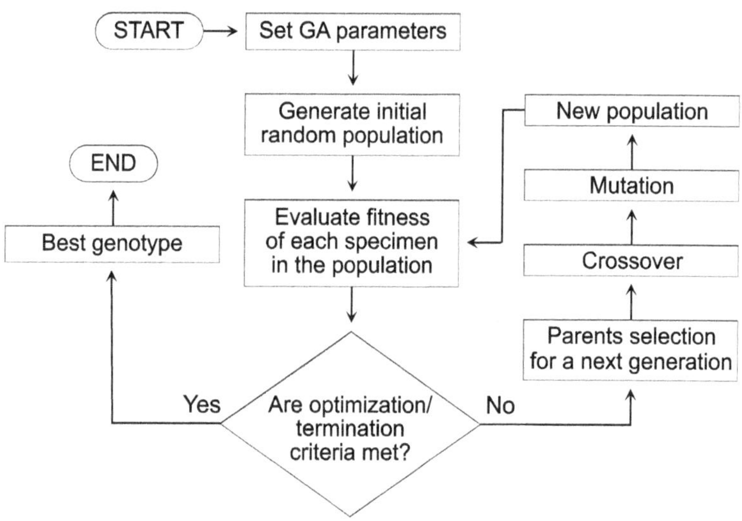

Figure 3. The block schema of the genetic algorithm procedure.

5. Numerical Results

The presented analysis included computation of the genetic algorithm procedure, for three different compositions of the inlet gas. The compositions are described in detail in Table 3. The different ratios of chemical specimens are defined in the numerical code by setting corresponding values of steam-to-carbon (SC) and carbon-to-carbon (CC) ratios [23]. The reference cases (Figure 4a–c) are reactors with homogenous and continuous catalytic material. The porosity of the catalytic material is set at 50% for each of the reference cases. The temperature field for the cases has a noticeable decrease in the temperature values at the inlet of the reactor (Figure 4a–c). The decrease occurs due to the reaction's activation. The temperature distribution becomes more uniform, closing to the end of the reactor. The unification is induced by the reaction progress. As there is less methane for conversion left, the reaction decreases in its intensity, consuming lower amounts of energy locally. The main goal of the carried optimization, is to unify the temperature field in the whole reactor, simultaneously preventing significant reduction of the methane conversion rate. The comparison of the temperature fields for subsequent generations in presented in Figures 5–7. To provide insight into the whole GA progress, the temperature fields of example reactors for different biogas compositions, from the initial generation are presented in Figures 5a1, 6b1 and 7c1. A single reactor is chosen for each composition. The optimization process is carried out for thirty subsequent generations. Temperature fields for the most optimal solutions acquired for the 30th, 30th and 30th generations are presented in Figure 5a2–a4 (composition (1)), Figure 6b2–b4 (composition (2)) and Figure 7c2–c4 (composition (3)). The information on fitness values is presented in Table 4. According to the data presented in Figure 5–7 and in Table 4, the algorithm manages to enhance the thermodynamic conditions inside the reformer. A visible improvement is reported for each of the inlet gas compositions. Detailed analysis of Table 4 gives information that methane conversion is decreased for each biogas compositions, when comparing with the reference cases. The methane conversion's decrease is directly caused by the reduction of the amount of the catalyst used for the most optimal solutions found (Table 5). The exact values of chemical specimens conversion are presented in Table 6. Considering the scale of the catalyst use limitation, forces to analyze the results from a different perspective. The use of the catalyst is reduced by circa 90% for each case, while the methane conversion fell only by circa 40%. A green hydrogen productivity ζ is introduced for a better insight on the increase of the effectiveness of the biogas reforming. The ζ parameter is an exact ratio of the hydrogen output and amount of the used catalyst ι for a specific reactor (Equation (27)).

$$\zeta = \frac{H2_{output}}{\iota} \tag{27}$$

The observed phenomenon clearly indicates improvement of the overall process effectiveness (green hydrogen productivity). Simultaneously confirming observations reported by Ricca et al. and Settar et al. [59,60]. Therefore, the presented results prove the macro-patterning concept, to be a valid strategy for an optimization of the biogas steam reforming. The acquired results indicate the necessity of defining of the minimal catalyst amount in the algorithm to prevent the limitation of the catalyst use of the reported magnitude. The methane conversion could be maintained by changing of the reactor dimensions [21]. However, an additional optimization procedure for defining the optimal radius and length of the reactor would be necessary.

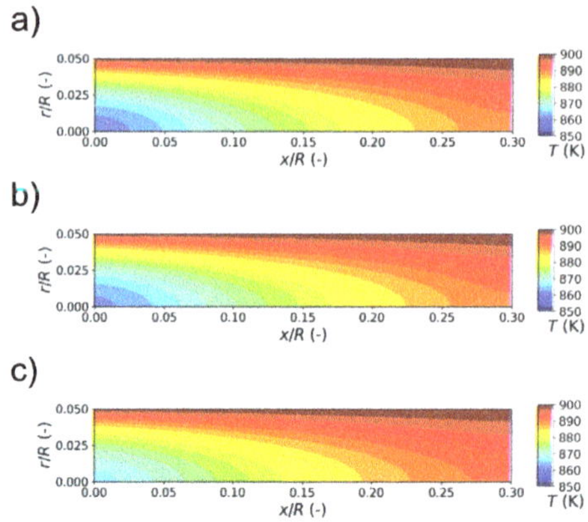

Figure 4. The temperature distribution for different biogas compositions (uniform catalyst distribution—reference case): (**a**) composition (1), (**b**) composition (2), (**c**) composition (3).

Table 3. Variants of the inlet gas compositions.

No.	CH_4	CO_2	H_2O	*SC*	*CC*
(1)	23%	30%	46%	2.0	1.3
(2)	20%	40%	40%	2.0	2.0
(3)	18%	50%	32%	2.0	2.9

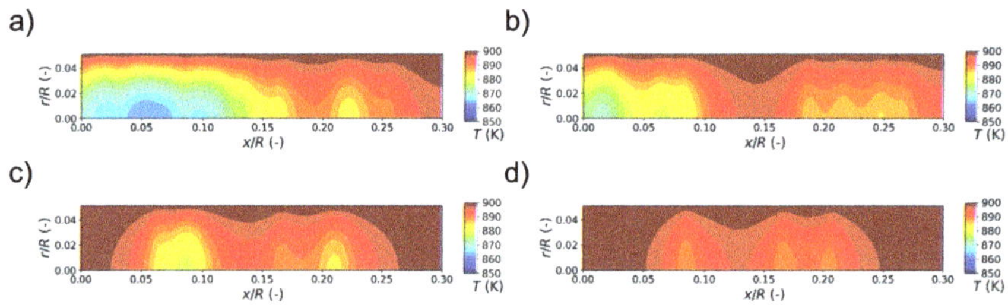

Figure 5. The temperature distribution for biogas composition (1): (**a**) initial, composition (1), (**b**) 10th gen., (**c**) 20th gen., (**d**) 30th gen.

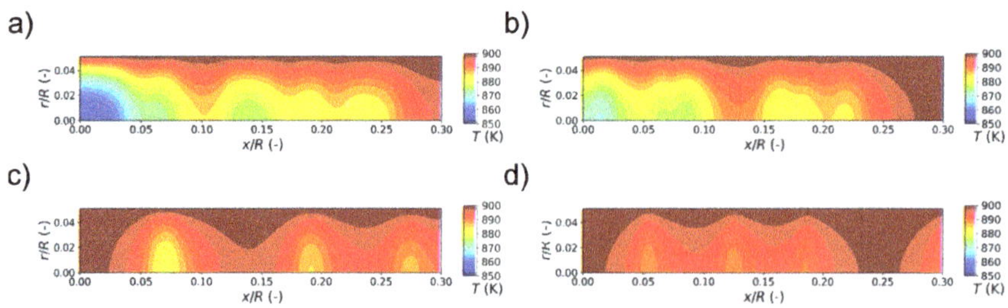

Figure 6. The temperature distribution for biogas composition (2): (**a**) initial, (**b**) 10th gen., (**c**) 20th gen., (**d**) 30th gen.

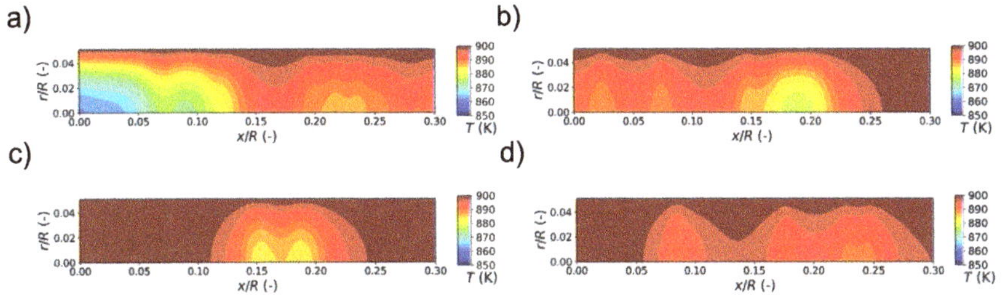

Figure 7. The temperature distribution for biogas composition (3): (**a**) initial, (**b**) 10th gen., (**c**) 20th gen., (**d**) (3) 30th gen.

Table 4. Fitness values and conversion acquired for the most optimal specimens in the subsequent generations and for the reference cases.

Gen.	Composition (1)			Composition (2)			Composition (3)		
	f_T	f_{CH_4}	f	f_T	f_{CH_4}	f	f_T	f_{CH_4}	f
REF	0.06	0.84	0.53	0.07	0.86	0.54	0.10	0.88	0.56
INIT	0.01	0.66	0.40	0.01	0.72	0.43	0.02	0.62	0.38
10th	0.17	0.57	0.41	0.04	0.61	0.38	0.29	0.52	0.42
20th	0.33	0.55	0.46	0.43	0.61	0.54	0.44	0.49	0.47
30th	0.61	0.43	0.51	0.67	0.50	0.57	0.64	0.52	0.58

Table 5. The amount of catalyst used in comparison with the reference cases.

Gen.	Composition (1)		Composition (2)		Composition (3)	
	ι	ζ	ι	ζ	ι	ζ
REF	100%	0.40	100%	0.38	100%	0.29
30th	17%	0.94	10%	1.8	6%	2.17

ζ—green hydrogen productivity (-); ι—amount of the catalyst used.

Table 6. Conversion of methane and carbon dioxide.

Gen.	Composition (1) CH_4	CO_2	Composition (2) CH_4	CO_2	Composition (3) CH_4	CO_2
REF	84%	19%	86%	21%	88%	23%
30th	43%	2%	50%	4%	52%	8%

ζ—green hydrogen productivity (-); ι—amount of the catalyst used.

6. Conslusions

The presented paper focuses on the optimization of the temperature distribution inside a biogas reforming reactor. The research pursues optimal distribution of the catalytic material to ensure enhancement of the reaction effectiveness. The optimization procedure succeeds in modification the reactors' design for the creation of a more unified thermal field. The reported results indicate a decrease in the hydrogen yield. However, the amount of the catalytic material used for the most optimal cases was reduced by: 83% for composition (1), 90% for composition (2) and 94% for composition (3). Considering the ratio of the hydrogen yield and the reduction of the used catalyst amount, the research confirms the improvement of the thermodynamic conditions inside the reactor, to be a valid strategy for increasing the biogas reforming effectiveness. The capabilities of the presented optimization strategy are most visible in the case of the composition (3). The optimization carried out for the composition (3) not only leads to an improvement in the thermal conditions but also manages to maintain conversion rate at a similar level as for the initial population. Following the presented results, the optimization procedure requires improvement before future calculations. For acquiring results of a higher quality, a sensitivity analysis of the evaluation procedure has to be conducted. The reaction effectiveness is ranked by evaluation of the biogas conversion rate. Hydrogen output at the end of the process should be considered instead of the methane fraction. The influence of magnitudes of the ω_1 and ω_2 values has to be investigated. Currently, the calculation of a thirty-generation algorithm run takes about 500 h. Therefore, the robustness of the numerical procedure has to be improved before the conduction of the sensitivity analysis. During the optimization, the overall amount of the catalyst used in the reactors is reduced significantly. A constrain defining a minimal amount of the catalyst used should be added to the algorithm's procedure to allow the unification of the temperature distribution, preventing losses in the amount of hydrogen produced.

Author Contributions: Conceptualization, M.P., G.B., J.S.S.; methodology, M.P., G.B., J.S.S.; software, M.P., G.B.; validation, M.P., G.B., J.S.S.; formal analysis, M.P., G.B., J.S.S.; numerical investigation M.P.; resources, G.B.; data curation, M.P.; writing—original draft preparation, M.P.; writing—review and editing, M.P., G.B., J.S.S.; visualization M.P.; supervision, G.B., J.S.S.; project administration, J.S.S.; funding acquisition, J.S.S. All authors have read and agreed to the published version of the manuscript.

Funding: The authors acknowledge the support of the Initiative for Excellence—Research University Project at AGH University of Science and Technology and AGH Grant No. 16.16.210.476.

Institutional Review Board Statement: Not applicable.

Informed Consent Statement: Not applicable.

Data Availability Statement: All data available on request.

Acknowledgments: The presented research was conducted using the computational power provided by PLGrid.

Conflicts of Interest: The authors declare no conflict of interest.The funders had no role in the design of the study; in the collection, analyses, or interpretation of data; in the writing of the manuscript, or in the decision to publish the results.

References

1. Sazali, N. Emerging technologies by hydrogen: A review. *Int. J. Hydrog. Energy* **2020**, *45*, 18753–18771. [CrossRef]
2. Widera, B. Renewable hydrogen implementations for combined energy storage, transportation and stationary applications. *Therm. Sci. Eng. Prog.* **2020**, *16*, 100460. [CrossRef]
3. Ayodele, F.O.; Mohammad, N.; Mustapa, S.I.; Ayodele, B.V. An overview of economic analysis and environmental impacts of natural gas conversion technologies. *Sustainability* **2020**, *12*, 148. [CrossRef]
4. Schmidt, O.; Gambhir, A.; Staffell, I.; Hawkes, A.; Nelson, J.; Few, S. Future cost and performance of water electrolysis: An expert elicitation study. *Int. J. Hydrog. Energy* **2017**, *42*, 30470–30492. [CrossRef]
5. Mutlu, R.N.; Kucukkara, I.; Gizir, A.M. Hydrogen generation by electrolysis under subcritical water condition and the effect of aluminium anode. *Int. J. Hydrog. Energy* **2020**, *45*, 12641–12652. [CrossRef]
6. Brauns, J.; Turek, T. Alkaline water electrolysis powered by renewable energy: A review. *Processes* **2020**, *8*, 248. [CrossRef]
7. Baykara, S.Z. Hydrogen: A brief overview on its sources, production and environmental impact. *Int. J. Hydrog. Energy* **2018**, *43*, 10605–10614. [CrossRef]
8. Kannah, R.Y.; Kavitha, S.; Karthikeyan, O.P.; Kumar, G.; Dai-Viet, N.V.; Banu, J.R. Techno-economic assessment of various hydrogen production methods—A review. *Bioresour. Technol.* **2021**, *319*, 124175. [CrossRef]
9. Pegram, J.; Falcone, G.; Kolios, A. A review of job role localization in the oil and gas industry. *Energies* **2018**, *11*, 2779. [CrossRef]
10. Tao, S.; Chen, S.; Pan, Z. Current status, challenges, and policy suggestions for coalbed methane industry development in China: A review. *Energy Sci. Eng.* **2019**, *7*, 1059–1074. [CrossRef]
11. Sharma, I.; Friedrich, D.; Golden, T.; Brandani, S. Exploring the opportunities for carbon capture in modular, small-scale steam methane reforming: An energetic perspective. *Int. J. Hydrog. Energy* **2019**, *44*, 14732–14743. [CrossRef]
12. Lee, H.; Jung, I.; Roh, G.; Na, Y.; Kang, H. Comparative analysis of on-board methane and methanol reforming systems combined with HT-PEM fuel cell and CO_2 capture/liquefaction system for hydrogen fueled ship application. *Energies* **2020**, *13*, 224. [CrossRef]
13. Alves, H.J.; Bley Junior, C.; Niklevicz, R.R.; Frigo, E.P.; Frigo, M.S.; Coimbra-Araújo, C.H. Overview of hydrogen production technologies from biogas and the applications in fuel cells. *Int. J. Hydrog. Energy* **2013**, *38*, 5215–5225. [CrossRef]
14. Gregorie, E.F.J.; Lamb, J.J.; Lien, K.M.; Pollet, B.G.; Burheim, O.S. Hydrogen and Biogas. In *Micro-Optics and Energy: Sensors for Energy Devices*; Lamb, J.J., Pollet, B.G., Eds.; Springer International Publishing: Cham, Switzerland, 2020; pp. 131–155. [CrossRef]

15. Gonçalves, A.; Puna, J.F.; Guerra, L.; Campos Rodrigues, J.; Gomes, J.F.; Santos, M.T.; Alves, D. Towards the Development of Syngas/Biomethane Electrolytic Production, Using Liquefied Biomass and Heterogeneous Catalyst. *Energies* **2019**, *12*, 3787. [CrossRef]
16. Zhao, X.; Joseph, B.; Kuhn, J.; Ozcan, S. Biogas Reforming to Syngas: A Review. *iScience* **2020**, *23*, 101082. 10.1016/j.isci.2020.101082. [CrossRef]
17. Meloni, E.; Martino, M.; Palma, V. A Short Review on Ni Based Catalysts and Related Engineering Issues for Methane Steam Reforming. *Catalysts* **2020**, *10*, 352. [CrossRef]
18. Buchireddy, P.R.; Peck, D.; Zappi, M.; Bricka, R.M. Catalytic hot gas cleanup of biomass gasification producer gas via steam reforming using nickel-supported clay minerals. *Energies* **2021**, *14*, 1875. [CrossRef]
19. Mozdzierz, M.; Brus, G.; Sciazko, A.; Komatsu, Y.; Kimijima, S.; Szmyd, J.S. Towards a Thermal Optimization of a Methane/Steam Reforming Reactor. *Flow Turbul. Combust.* **2016**, *97*, 171–189. [CrossRef]
20. Palma, V.; Ricca, A.; Martino, M.; Meloni, E. Innovative structured catalytic systems for methane steam reforming intensification. *Chem. Eng. Process. Process Intensif.* **2017**, *120*, 207–215. [CrossRef]
21. Pajak, M.; Mozdzierz, M.; Chalusiak, M.; Kimijima, S.; Szmyd, J.S.; Brus, G. A numerical analysis of heat and mass transfer processes in a macro-patterned methane/steam reforming reactor. *Int. J. Hydrog. Energy* **2018**, *43*, 20474–20487. [CrossRef]
22. Tomiczek, M.; Kaczmarczyk, R.; Mozdzierz, M.; Brus, G. A numerical analysis of heat and mass transfer during the steam reforming process of ethane. *Heat Mass Transf.* **2018**, *54*, 2305–2314. [CrossRef]
23. Nishino, T.; Szmyd, J.S. Numerical analysis of a cell-based indirect internal reforming tubular SOFC operating with biogas. *J. Fuel Cell Sci. Technol.* **2010**, *7*, 0510041–0510048. [CrossRef]
24. Patankar, S.V. *Numerical Heat Transfer and Fluid Flow*; Hemisphere: Washington, DC, USA, 1980.
25. Xu, J.; Froment, G.F. Methane steam reforming: II. Diffusional limitations and reactor simulation. *AIChE J.* **1989**, *35*, 97–103. [CrossRef]
26. Komatsu, Y.; Kimijima, S.; Szmyd, J.S. A Performance Analysis of a Solid Oxide Fuel Cell—Micro Gas Turbine Hybrid System Using Biogas. *ECS Trans.* **2009**, *25*, 1061–1070. [CrossRef]
27. Sciazko, A.; Komatsu, Y.; Brus, G.; Kimijima, S.; Szmyd, J.S. A novel approach to improve the mathematical modelling of the internal reforming process for solid oxide fuel cells using the orthogonal least squares method. *Int. J. Hydrog. Energy* **2014**, *39*, 16372–16389. [CrossRef]
28. Brus, G.; Nowak, R.; Szmyd, J.S.; Komatsu, Y.; Kimijima, S. An Experimental and Theoretical Approach for the Carbon Deposition Problem during Steam Reforming of Model Biogas. *J. Theor. Appl. Mech.* **2015**, *53*, 273–284. [CrossRef]
29. Mazhar, A.; Khoja, A.H.; Azad, A.K.; Mushtaq, F.; Naqvi, S.R.; Shakir, S.; Hassan, M.; Liaquat, R.; Anwar, M. Performance Analysis of TiO_2-Modified $Co/MgAl_2O_4$ Catalyst for Dry Reforming of Methane in a Fixed Bed Reactor for Syngas (H_2, CO) Production. *Energies* **2021**, *14*, 3347. [CrossRef]
30. Brus, G.; Komatsu, Y.; Kimijima, S.; Szmyd, J.S. An analysis of biogas reforming process on Ni/YSZ and Ni/SDC catalysts. *Int. J. Thermodyn.* **2012**, *15*, 43–51. [CrossRef]
31. Ahmed, K.; Föger, K. Approach to equilibrium of the water-gas shift reaction on a Ni/zirconia anode under solid oxide fuel-cell conditions. *J. Power Sources* **2001**, *103*, 150–153. [CrossRef]
32. Sciazko, A.; Komatsu, Y.; Brus, G.; Kimijima, S.; Szmyd, J.S. A novel approach to the experimental study on methane/steam reforming kinetics using the Orthogonal Least Squares method. *J. Power Sources* **2014**, *262*, 245–254. [CrossRef]
33. Iwai, H.; Yamamoto, Y.; Saito, M.; Yoshida, H. Numerical simulation of intermediate-temperature direct-internal-reforming planar solid oxide fuel cell. *Energy* **2011**, *36*, 2225–2234. [CrossRef]
34. Brus, G.; Kimijima, S.; Szmyd, J.S. Experimental and numerical analysis of transport phenomena in an internal indirect fuel reforming type Solid Oxide Fuel Cells using Ni/SDC as a catalyst. *J. Phys. Conf. Ser.* **2012**, *395*. [CrossRef]
35. Fanchi, J.R. Reservoir Simulation. In *Integrated Reservoir Asset Management*; Fanchi, J.R., Ed.; Gulf Professional Publishing: Boston, MA, USA, 2010; pp. 223–241.
36. Carbonell, R.G.; Whitaker, S. Heat and Mass Transfer in Porous Media. In *Fundamentals of Transport Phenomena in Porous Media*; Bear, J., Corapcioglu, M.Y., Eds.; Springer: Dordrecht, The Netherlands, 1984; pp. 121–198. [CrossRef]
37. Bhattacharya, A.; Calmidi, V.V.; Mahajan, R.L. Thermophysical properties of high porosity metal foams. *Int. J. Heat Mass Transf.* **2002**, *45*, 1017–1031. [CrossRef]
38. Dai, Z.; Nawaz, K.; Park, Y.G.; Bock, J.; Jacobi, A.M. Correcting and extending the Boomsma-Poulikakos effective thermal conductivity model for three-dimensional, fluid-saturated metal foams. *Int. Commun. Heat Mass Transf.* **2010**, *37*, 575–580. [CrossRef]
39. Pajak, M.; Buchaniec, S.; Kimijima, S.; Szmyd, J.S.; Brus, G. A multiobjective optimization of a catalyst distribution in a methane/steam reforming reactor using a genetic algorithm. *Int. J. Hydrog. Energy* **2021**, *46*, 20183–20197. [CrossRef]
40. Mozdzierz, M.; Chalusiak, M.; Kimijima, S.; Szmyd, J.S.; Brus, G. An afterburner-powered methane/steam reformer for a solid oxide fuel cells application. *Heat Mass Transf.* **2018**, *54*, 2331–2341. [CrossRef]
41. Tan, W.C.; Iwai, H.; Kishimoto, M.; Brus, G.; Szmyd, J.S.; Yoshida, H. Numerical analysis on effect of aspect ratio of planar solid oxide fuel cell fueled with decomposed ammonia. *J. Power Sources* **2018**, *384*, 367–378. [CrossRef]

42. Mozdzierz, M.; Brus, G.; Sciazko, A.; Komatsu, Y.; Kimijima, S.; Szmyd, J.S. An attempt to minimize the temperature gradient along a plug-flow methane/steam reforming reactor by adopting locally controlled heating zones. *J. Phys. Conf. Ser.* **2014**, *530*. [CrossRef]
43. Ferziger, J.H.; Peric, M. *Computational Methods for Fluid Dynamics*; Springer: New York, NY, USA, 2002. [CrossRef]
44. Chalusiak, M.; Wrobel, M.; Mozdzierz, M.; Berent, K.; Szmyd, J.S.; Brus, G. A numerical analysis of unsteady transport phenomena in a Direct Internal Reforming Solid Oxide Fuel Cell. *Int. J. Heat Mass Transf.* **2019**, *131*. [CrossRef]
45. Wang, H.; Wang, H.; Gao, F.; Zhou, P.; Zhai, Z.J. Literature review on pressure–velocity decoupling algorithms applied to built-environment CFD simulation. *Build. Environ.* **2018**, *143*, 671–678. [CrossRef]
46. MacCormack, R.W.; Candler, G.V. The solution of the Navier-Stokes equations using Gauss-Seidel line relaxation. *Comput. Fluids* **1989**, *17*, 135–150. [CrossRef]
47. Mozdzierz, M.; Brus, G.; Kimijima, S.; Szmyd, J.S. Numerical analysis of helium-heated methane/steam reformer. *J. Phys. Conf. Ser.* **2016**, *745*, 032081. [CrossRef]
48. Powell, R.; Tye, R.; Hickman, M. The thermal conductivity of nickel. *Int. J. Heat Mass Transf.* **1965**, *8*, 679–688. [CrossRef]
49. Zhao, C.Y.; Lu, T.J.; Hodson, H.P.; Jackson, J.D. The temperature dependence of effective thermal conductivity of open-celled steel alloy foams. *Mater. Sci. Eng. A* **2004**, *367*, 123–131. [CrossRef]
50. Todd, B.; Young, J.B. Thermodynamic and transport properties of gases for use in solid oxide fuel cell modelling. *J. Power Sources* **2002**, *110*, 186–200. [CrossRef]
51. Golberg, D.E. *Genetic Algorithms in Search Optimization & Machine Learning*; Addison-Wesley Longman Publishing Co., Inc.: Boston, MA, USA, 1989. [CrossRef]
52. Zhang, G.; Yu, L.; Shao, Q.; Feng, Y. A Clustering Based GA for Multimodal Optimization in Uneven Search Space. In Proceedings of the 2006 6th World Congress on Intelligent Control and Automation, Dalian, China, 21–23 June 2006; Volume 1, pp. 3134–3138.
53. Das, S.; Maity, S.; Qu, B.Y.; Suganthan, P. Real-parameter evolutionary multimodal optimization—A survey of the state-of-the-art. *Swarm Evol. Comput.* **2011**, *1*, 71–88. [CrossRef]
54. Zou, P.; Rajora, M.; Liang, S.Y. Multimodal Optimization of Permutation Flow-Shop Scheduling Problems Using a Clustering-Genetic-Algorithm-Based Approach. *Appl. Sci.* **2021**, *11*, 3388. [CrossRef]
55. Rajesh, J.K.; Gupta, S.K.; Rangaiah, G.P.; Ray, A.K. Multiobjective optimization of steam reformer performance using genetic algorithm. *Ind. Eng. Chem. Res.* **2000**, *39*, 706–717. [CrossRef]
56. Azarhoosh, M.J.; Ebrahim, H.A.; Pourtarah, S.H. Simulating and Optimizing Auto-Thermal Reforming of Methane to Synthesis Gas Using a Non-Dominated Sorting Genetic Algorithm II Method. *Chem. Eng. Commun.* **2016**, *203*, 53–63. [CrossRef]
57. Ye, G.Z.; Kang, D.K. Extended Evolutionary Algorithms with Stagnation-Based Extinction Protocol. *Appl. Sci.* **2021**, *11*, 3461. [CrossRef]
58. De Falco, I.; Della Cioppa, A.; Tarantino, E. Mutation-based genetic algorithm: Performance evaluation. *Appl. Soft Comput.* **2002**, *1*, 285–299. [CrossRef]
59. Ricca, A.; Palma, V.; Martino, M.; Meloni, E. Innovative catalyst design for methane steam reforming intensification. *Fuel* **2017**, *198*, 175–182. [CrossRef]
60. Settar, A.; Abboudi, S.; Lebaal, N. Effect of inert metal foam matrices on hydrogen production intensification of methane steam reforming process in wall-coated reformer. *Int. J. Hydrog. Energy* **2018**, *43*, 12386–12397. [CrossRef]

energies

MDPI

Article

Mathematical Modeling of an Electrotechnical Complex of a Power Unit Based on Hydrogen Fuel Cells for Unmanned Aerial Vehicles

Ivan V. Vasyukov, Alexander V. Pavlenko *, Vladimir S. Puzin, Denis V. Batishchev and Irina A. Bolshenko

Novocherkassk Polytechnic Institute, Platov South-Russian State Polytechnic University, 346428 Novocherkassk, Russia; vasuckov@gmail.com (I.V.V.); vspuzin@gmail.com (V.S.P.); batishchevd@gmail.com (D.V.B.); irenka84@mail.ru (I.A.B.)

* Correspondence: eea.srspu@gmail.com; Tel.: +7-(8635)-25-51-13

Abstract: The issues of mathematical and numerical simulation of an electrical complex of a power plant based on hydrogen fuel cells with a voltage step-down converter were considered. The work was aimed at developing a mathematical model that would provide for determining the most loaded operation mode of the complex components. The existing mathematical models do not consider the effect of such processes as the charge and discharge of the battery backup power supply on the power plant components. They often do not consider the nonlinearity of the fuel cell output voltage. This paper offers a mathematical model of an electrical complex based on the circuit analysis. The model combines a well-known physical model of a fuel cell based on a potential difference and a model of a step-down converter with a battery backup power supply developed by the authors. A method of configuring a fuel cell model based on the experimental current–voltage characteristic by the least-squares method has been proposed. The developed model provides for determining currents and voltages in all components of the power plant both in the nominal operating mode and in the mode of limiting the power consumed from the fuel cell when the battery backup power supply is being charged. The correctness of the calculated ratios and the mathematical model has been confirmed experimentally. Using the proposed model, a 1300 W power plant with a specific power of 529.3 W·h/kg was developed and tested.

Keywords: mathematical model; calculation method; step-down; converter; fuel cell; UAV; power unit

Citation: Vasyukov, I.V.; Pavlenko, A.V.; Puzin, V.S.; Batishchev, D.V.; Bolshenko, I.A. Mathematical Modeling of an Electrotechnical Complex of a Power Unit Based on Hydrogen Fuel Cells for Unmanned Aerial Vehicles. *Energies* **2021**, *14*, 6974. https://doi.org/10.3390/en14216974

Academic Editor: Orazio Barbera

Received: 21 September 2021
Accepted: 21 October 2021
Published: 24 October 2021

1. Introduction

There are a vast number of solutions for implementing an electrical part of an unmanned aerial vehicle (UAV). The simplest topology is to directly connect the load to the fuel cell outlet. The disadvantage of this solution is the increased voltage at the output in the no-load operation mode. Another drawback is the fact that during takeoff and landing, the UAV requires three to five times more energy in comparison with its consumption in the cruise mode. Moreover, the total time of such overloads accounts for 2% to 4% of the total flight time. To avoid using overweight fuel cells capable of withstanding such peak overloads, buffer energy storage devices are applied [1]. In the majority of cases, lithium polymer batteries act as storage devices. The solution to combine capacities from two energy sources is shown in Figure 1.

For unmanned aerial vehicles, the circuit in Figure 1a is more relevant, since it allows maintaining the voltage on the tire in a given narrow corridor, in contrast to the circuit in Figure 1b and, at the same time, it has a lower mass than the circuit in Figure 1c due to the absence of a DC–DC battery converter. However, no issues related to calculating the required capacity of such a battery are considered in the publications. Mathematical models of power plants considering the effect of the charging and discharging of a battery on the operating modes of the voltage converter components are not considered either.

The weight of the battery backup power supply represents a significant part of the weight of the UAV power plant electrics. The choice of its capacity (i.e., the weight) is one of the main tasks when designing such power plants. In this regard, it is appropriate to have a mathematical model of the electrical complex, which will allow obtaining data on the operating mode of the battery at a given power consumption of the UAV.

The complexity of the design process of the electrical part of the power unit lies in the fact that the voltage of the stack of fuel cells depends nonlinearly on the current consumption, which is why the known ratios for calculating converters operating from DC voltage sources need to be improved. The specific algorithm of the power plant control system, which limits the power consumption from the stack in the overload mode and stabilizes the battery charge current in the nominal mode, also does not allow using standard calculation methods.

In [2], some methods to design a power unit that comprises a fuel cell for an aircraft are presented. However, the present paper does not cover the issues of developing a DC–DC converter and its influence on the other system elements.

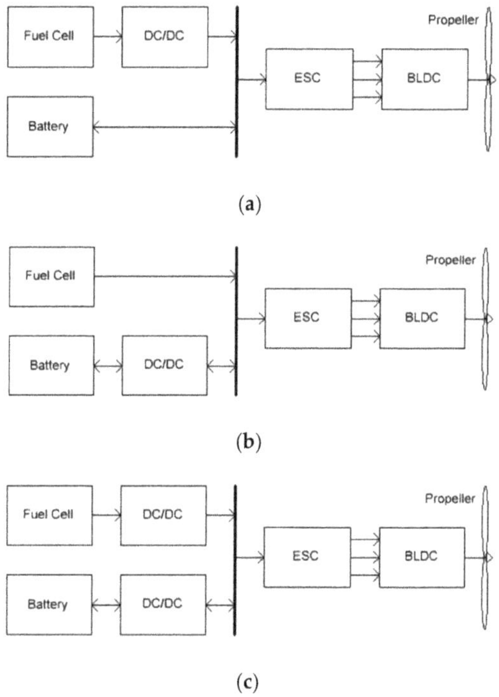

Figure 1. The solutions for a power unit with a buffer storage implementation: (**a**) with voltage stabilization of the fuel cell; (**b**) with stabilization of battery voltage; (**c**) with stabilization of the voltage of the fuel cell and battery.

In a great number of papers [3–7], such issues as a reduction in the current pulse of the fuel cell and an increase in the converter reliability by enhancing the number of parallel operating phases are studied. Such solutions to the power unit implementation are justified for land and water transport as well as for stationary consumers. When the power units are used as the main source of energy on aircraft, the requirements for specific energy intensity and surface power density dominate. The mass should be as small as possible. An increase in the number of converter phases and the use of supercapacitors lead to significant mass and dimension increase. UAV propulsion systems often use a low voltage, in the range of 12 to 36 V, while powerful fuel cells consist of several dozen interelectrode units (IEU). Thus, the voltage converter should be step down.

The work proposes a mathematical model of a power unit that is based on an iterative calculation of a step-down voltage converter operating from a fuel cell battery with a nonlinear current–voltage curve.

2. Method for Calculating the Converter

For UAV, the buck-converter circuit (step-down, copper, buck converter), which is shown in Figure 2, seems to be the most effective. When designing a converter, it is necessary to analyze the processes in the power supply system; calculate the voltages and currents in the elements; and determine the duty factor of the pulse-width modulation (PWM), the parameters of the choke, and the most-loaded modes of electronic components.

To derive the corresponding calculation formulas, the substitution circuit is used, shown in Figure 3. The transistor voltage was modeled by the electromotive force (EMF) source with voltage Uvt, the voltage drop on the diode is represented by the source Uvd, the output voltage is taken equal to Uout, and the input voltage is modeled by the source Uin. The active resistance of all parts of the circuit and the internal resistance of the EMF sources are taken to be zero.

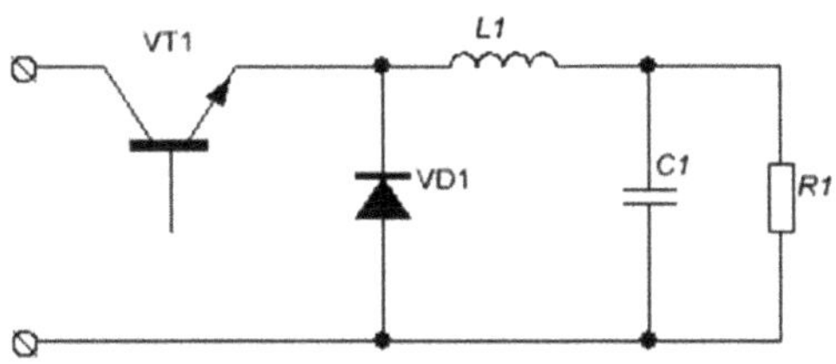

Figure 2. Buck-converter circuit.

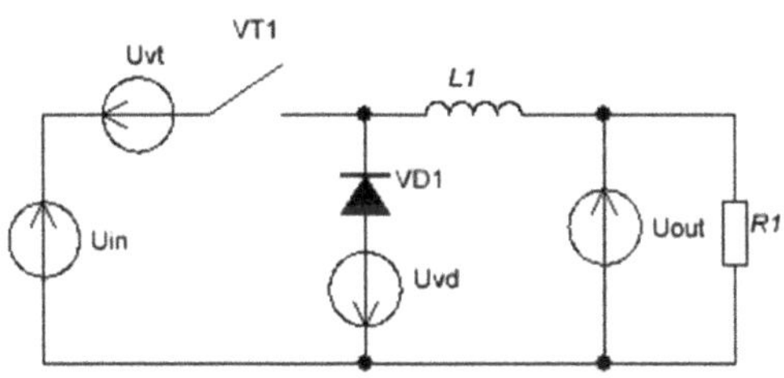

Figure 3. The substitute circuit of the converter.

The main dependencies on which the calculation is based are the formulas for determining the amplitude of the choke current in the discontinuous (DCM, Equation (1)) and continuous (CCM, Equation (2)) current modes [8], modified by the authors, by introducing voltages on the transistor and the diode. The discontinuous current mode is understood as a mode of operation of the converter when at the moment of the PWM pause, the choke current has the time to drop to zero. The continuous current mode is understood as the operating mode when at the moment of the PWM pause, the choke current decreases to some non-zero value. The authors have provided Equation (3) for determining the choke operation mode.

$$I_{Lmax\_DCM} = \sqrt{\frac{I_{OUT} \cdot 2 \cdot T \cdot (U_{IN} - U_{OUT} - U_{VT}) \cdot (U_{OUT} + U_{VD})}{L \cdot (U_{IN} + U_{VD} - U_{VT})}} \tag{1}$$

$$I_{Lmax\_CCM} = I_{OUT} + \frac{(U_{IN} - U_{OUT} - U_{VT}) \cdot (U_{OUT} + U_{VD}) \cdot T}{2 \cdot L \cdot (U_{IN} - U_{VT} + U_{VD})} \tag{2}$$

$$I_{L\max} = \begin{cases} I_{L\max\_DCM} \text{ if } I_{OUT} < \frac{I_{L\max\_DCM}}{2} \\ I_{L\max\_CCM} \text{ otherwise} \end{cases} \quad (3)$$

Having determined the mode of the choke current and its amplitude value, it is possible to calculate all the other electrical parameters of the circuit, including the duty cycle of the PWM. The resulting correlations are tested by computer simulation in the LTspice IV program. The converter model comprises ideal elements with lumped parameters that are placed in separate voltage sources. The simulation results confirm the correctness of the assumptions made to derive the formulas. The calculated data correspond to the characteristics and parameters obtained by simulating various operating modes of the circuit.

The input voltage, its source being the fuel cell stack, is nonlinearly dependent on the current draw. Therefore, in the process of calculation, it is necessary to consider a continuous dependence $U_{FC} = f(I_{FC})$. The abovementioned current–voltage characteristic (CVC) correlation can be obtained either experimentally or by tuning the mathematical model of the fuel cell battery. In the first case, a set of CVC points can be interpolated by cubic splines, for example, in the MathCAD program. The second solution for obtaining the CVC of a stack is discussed below.

Nowadays, the mathematical model of a fuel cell that comprises the Nernst potential difference and voltage drops in different parts of the CVC is the most popular one [9–15]. A similar model that additionally takes into account the transportation of liquid water is used in various multiphysics simulation environments, such as Amesim or Openmodelica [16]. When calculating the operating mode of the converter, it is sufficient to use a simpler model that simulates the CVC of a stack depending on its geometry, operating temperature, and gas pressures:

$$E_{FC} = E_{nernst} - E_{act} - E_{ohm} - E_{con}, \quad (4)$$

where E_{nernst} is the Nernst potential, E_{act} is a voltage drop in the initial part of the CVC (activation voltage), E_{ohm} is a voltage drop across active resistance in the working part of the CVC, E_{con} is a voltage drop in the CVC part where the mass transfer begins (concentration voltage).

The input parameters of the model are as follows:

$T_{meb} = 315.65$ K, operating temperature of the stack; $P_{Air} = 1$ atm, absolute partial air pressure; $P_{H2} = 2$ atm, absolute partial pressure of hydrogen; $\gamma_{O_2} = 20.94\%$, oxygen content in the air; $R = 8.314$ Дж/(mol·kg), universal gas constant; $F = 96485.33$ Кл/моль, Faraday constant; $E_0 = 1.229$ В, standard electrode potential for the reaction $O_2 + 4H_2 + 4e = 2H_2O$ under normal conditions; $\Delta S = -164.025$ enthropy; $T_{REF} = 298.15$ K, temperature of normal conditions; $n = 2$, number of electrons involved in the reaction; $S_{meb} = 83.22$ cm^2, area of the active part of the interelectrode unit (IEU); $L_{meb} = 0.018288$ cm, Nafion membrane thickness; ξ_1, ξ_2, ξ_3, ξ_4, b, and λ, empirical coefficients; I_{max} A, maximum current at which the CVC begins to bend down; R_C Ohm, electrode resistance; $P_{O2} = P_{Air} \cdot \gamma_{O_2} = 1 \cdot 0.2094 = 0.209$ Bar, absolute partial pressure of oxygen.

Nernst potential:

$$E_{nernst} = E_0 + \frac{\Delta S}{n \cdot F} \cdot (T_{meb} - T_{REF}) + \frac{R \cdot T_{meb}}{n \cdot F} \cdot \left(\ln(P_{H2}) + \frac{1}{2} \cdot \ln(P_{O2})\right). \quad (5)$$

Activation voltage:

$$E_{act} = -(\xi_1 + \xi_2 \cdot T_{meb} + \xi_3 \cdot T_{meb} \cdot \ln(C_{O_2}) + \xi_4 \cdot T_{meb} \cdot \ln(I)), \quad (6)$$

where $C_{O_2} = \frac{P_{O_2}}{5.08 \cdot 10^6 \cdot e^{-\left(\frac{498}{T_{meb}}\right)}}$ mol/cm^3 is the oxygen concentration in the gas/catalyst bed and I is the current through the fuel cell.

The coefficients $\xi_1 \ldots \xi_4$ can be either determined empirically or calculated by the formula given in [11]. They depend on protons, water, and hydrogen concentration.

Concentration tension $E_{con} = -b \cdot \left(1 - \frac{I}{I_{max}}\right)$ is the Tafel equation.

Internal resistance voltage $E_{ohm} = I \cdot (R_M + R_C)$, where $R_M = \frac{\rho_M \cdot L_{meb}}{S_{meb}}$ is the membrane resistance [17].

Here, the membrane resistivity is stated in the following way:

$$\rho_M = \frac{181.6 \cdot \left(1 + 0.03 \cdot \frac{I}{S_{meb}} + 0.062 \cdot \frac{T_{meb}}{303} \cdot \left(\frac{I}{S_{meb}}\right)^{2.5}\right)}{\left(\lambda - 0.634 - 3 \cdot \frac{I}{S_{meb}}\right) \cdot e^{4.18 \cdot \frac{T_{meb} - 303}{T_{meb}}}}. \tag{7}$$

The experimental CVC of the stack and the least-squares method were used to tune the model. For this, the correlation between the output voltage of the stack and the input parameters is written in the following form:

$$E_{FC} = f(I, \xi_1, \xi_2, \xi_3, \xi_4, \lambda, R_C, b). \tag{8}$$

Then, in a cyclic calculation, by enumerating the values of the parameter ξ_1 in a given range and determining the sum of the squared differences between the experimental CVC and the CVC of the model with the current value of ξ_1, an option with the highest index of coincidence of characteristics is found. The resulting parameter ξ_1 is used as the input when calculating the next parameter ξ_2, and so on, in the loop. After calculating the last parameter b, the sum of the squared differences between the experimental CVC and the CVC of the model are determined with the obtained values $\xi_1, \xi_2, \xi_3, \xi_4, \lambda, R_C, b$. Further, the obtained parameters are used as initial approximations for the next iteration and the process is then repeated. The results of the calculation are the parameters that correspond to the minimum sum of the squared differences, that is, the highest index of coincidence of the correspondence between the experimental CVC and the CVC of the model.

The ranges of variation in the coefficients while selecting the model parameters were as follows:

ξ_1 = from -2 to $+2$	1000 points
ξ_2 = from -0.01 to $+0.01$	1000 points
ξ_3 = from -0.01 to $+0.01$	1000 points
ξ_4 = from -0.1 to $+0.1$	1000 points
λ = from $+14$ to $+23$	1000 points
R_C = from 0 to $+0.2$	1000 points
b = from -2 to $+2$	1000 points

The current varied from 0.1 to 38 A, which corresponded to 50 points on the CVC.

As a result of the approximation by the least-squares method, the following values of the coefficients are obtained: $\xi_1 = -0.98$; $\xi_2 = 3.38 \cdot 10^{-3}$; $\xi_3 = 10 \cdot 10^{-5}$; $\xi_4 = -2 \cdot 10^{-4}$; $\lambda = 22.991$; $R_C = 0$ OM; and $b = -0.028$.

The current–voltage characteristic that is implemented in the model with the obtained coefficients is shown in Figure 4 as a solid curve. As can be seen on the graph, the characteristic of the model coincides with the experimental one in a satisfactory way.

The estimation of the modeling error was carried out by calculating the root-mean-square deviation of the dependence obtained by calculation from the experimentally obtained data:

$$R^2 = 1 - \frac{\sum_{i=1}^{n} (a_i - p_i)^2}{\sum_{i=1}^{n} p_i^2} = 0.999424 \tag{9}$$

where i = VAC FC experimental point number, n= number of experimental points, a_i= VAC experimental points, and p_i = points obtained as a result of a numerical experiment.

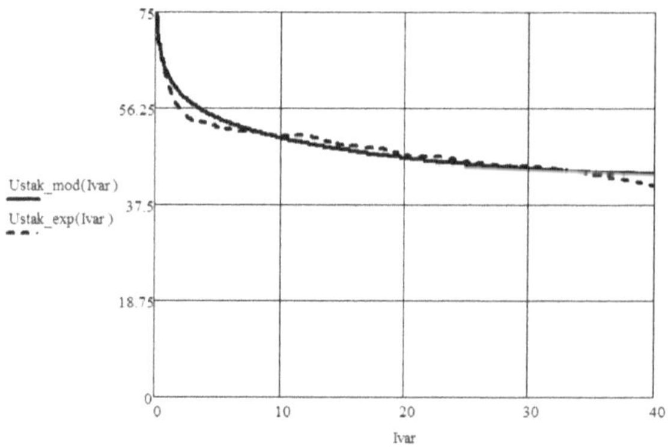

Figure 4. CVC of the stack experimentally obtained (dashed curve) and the CVC of the mathematical model of the stack tuned by the least-squares method (solid curve).

The relative modeling error reaches 5.8% at the beginning of the graph in Figure 4 and then it decreases and does not exceed 3% in the measuring section of the VAC. The obtained modeling accuracy is sufficient for practical use.

Bearing in mind the correlation $U_{FC} = f(I_{FC})$, it is possible to determine the voltage at the input of the converter at the present value of the current consumed. In this case, the current consumed by the converter depends on the input voltage. The problem when the calculation results affect the initial data of the same calculation can be solved by the iteration method. In this case, the algorithm for calculating a converter powered by a fuel cells stack is as follows:

1. Initial data for the calculation are set. Parameters can be both fixed and varied: $U_{OUT}, I_{OUT}, T, L, U_{VT}, U_{VD}$.
2. The initial approximation of the current consumed by the converter is set: $I_{IN} = 0$.
3. The initial approximation of the input voltage of the converter (output voltage of the stack) is determined according to the abovementioned characteristic: $U_{FC} = f(I_{IN})$.
4. The draw of the current consumed by the converter during operation from the voltage obtained at the first iteration is calculated as follows:

$$I_{IN} = f(U_{FC}, U_{OUT}, I_{OUT}, T, L, U_{VT}, U_{VD}). \tag{10}$$

5. The module of the residual between new, obtained, and previous values of the input current is calculated.
6. If the residual modulus exceeds the set current draw value, then a new value of the input current is applied to the CVC of the stack correlation and a new approximation of the input voltage of the converter is determined. Then step 4 takes place and the cycle repeats itself one more time.
7. To exit the iteration loop, it necessary to reach the specified value of the residual modulus between the previous and the successive value of the desired input current or to reach the number of iterations of the maximum permissible value.
8. In case the calculations are correct, the value of the input voltage of the converter is obtained, which is the value of the stack voltage decreased according to the present input parameters. With this value of this voltage, all other parameters of the circuit are determined.

The algorithm is developed in the MathCAD program. All the initial data (except for the desired parameter that is fixed) on the correlation graphs that relate to the amplitude of the choke current, which depicts the ratio between the output current and the output voltage, are obtained. With the obtained graphs, it becomes possible to determine the most intense choke and semiconductor elements operation mode. At the same time, the obtained graph dependencies allow one to select a choke for the saturation current, the induction swing, and the effective current value, as well as to select semiconductor elements for maximum and average currents and calculate the value of the cycle-by-cycle current limiting pickup setting for the control system. Cycle-by-cycle current limitation refers to converter protection that is shaped by the control system and represents the limitation of the PWM pulse width when the choke current amplitude of a given value is reached. The calculation results are shown in Figures 5 and 6.

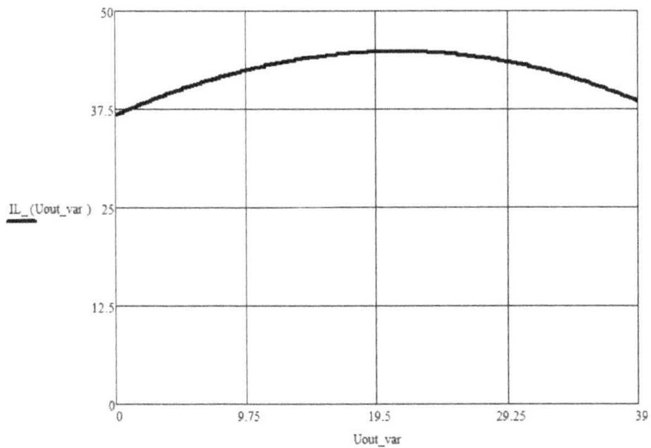

Figure 5. Dependence of the choke current amplitude on the output voltage of the converter at a load current of 36 A.

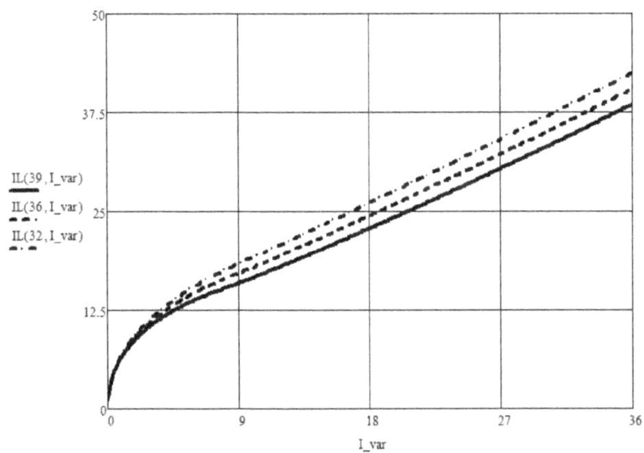

Figure 6. Dependence of the choke current amplitude on the output current of the converter at the output voltages of 32, 36, and 39 V.

As can be seen from the graphs, the maximum value of the choke current corresponds to a voltage of 22 V, which is less than the nominal output voltage of 36 V. At the start of the converter, the output voltage increases and takes its value, so the value of the current protection setting should be higher.

3. Summation of Multiple Power Supplies

In accordance with the proposed structure of the electrical complex, the voltage converter is to be supplemented with a storage battery, which is expected to compensate for peak power overloads. The battery is directly connected at the output of the inverter in parallel with the load. The equivalent circuit of such a converter is shown in Figure 7. Here, the storage battery is presented with a simple IR model [18–20] as an EMF source with Ubat voltage and its internal resistance R2. The reactive elements of the equivalent circuit of the storage battery are not used, since the calculation is carried out for the steady-state operation mode of the converter.

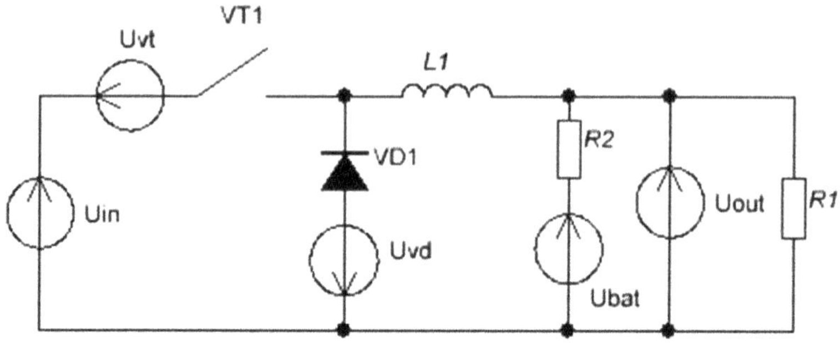

Figure 7. Equivalent circuit of the converter with a storage battery.

We shall consider the integration mode of such a power unit. In the nominal power mode, the battery is fully charged and does not draw any current. As the load current increases, the power drawn from the fuel cells stack accordingly increases to a predetermined value. Typically, the maximum power of the stack is taken to be such an operating mode when the voltage across one interelectrode unit (IEU) decreases down to 0.5–0.6 V. The power value corresponds to certain values of the current I_{INmax} and voltage U_{INmax} on the CVC of the stack. With a further increase in current, the control system begins to reduce the output voltage relative to nominal so that the power consumption from the stack remains constant and equal to the maximum determined value. The correlation for the output voltage of the converter in the input power limitation mode is as follows:

$$U_{OUT} = \frac{I_{IN\max}}{I_{CONV}} \cdot (U_{IN\max} + U_{VD} - U_{VT}) - U_{VD} \ if \ P_{IN} > P_{MAX} \quad (11)$$

$$U_{OUT} = U_{OUT\_nom} \ if \ P_{IN} \leq P_{MAX} \quad (12)$$

where I_{CONV} is the converter output current and U_{OUT_nom} is the specified nominal output voltage of the converter.

The dependence is the same for the modes of discontinuous (DCM) and continuous (CCM) currents of the choke.

If a storage battery is connected to the output, then the integration mode is carried out as follows: In the initial state, the voltages at the output of the converter and the battery are equal. When the current of fuel cell stack exceeds the maximum value, the control system reduces the output voltage and the battery begins to compensate for part of the power consumption. As the discharge proceeds, its EMF decreases. When the overload mode is off, the voltage at the output of the converter rises and becomes higher than the battery voltage, so the process of charging the battery begins. The charge/discharge current is determined by the internal battery resistance and the difference in the values of the battery EMF and the output voltage of the converter.

The converter switches to the power limitation mode when the load current exceeds the following value:

$$I_{OUT\max} = \frac{I_{IN\max} \cdot (U_{IN\max} + U_{VD} - U_{VT})}{U_{OUT\_nom} + U_{VD}} + \frac{U_{BAT} - U_{OUT\_nom}}{R_{BAT}} \quad (13)$$

In accordance with Kirchhoff's laws, a formula is derived to determine the output current of the converter in this mode:

$$I_{CONV} = \frac{-b + \sqrt{b^2 - 4 \cdot a \cdot c}}{2 \cdot a} \tag{14}$$

where $a = R_{BAT}$, $b = -R_{BAT} \cdot I_{OUT} + U_{BAT} + U_{VD}$, and $c = -I_{IN\max}(U_{IN\max} + U_{VD} - U_{VT})$.

With the converter current draw value, it becomes possible to determine its output voltage according to the abovementioned formula to set the power limitation mode and find the magnitude and direction of the current of the battery:

$$I_{BAT} = \frac{U_{BAT} - U_{OUT}}{R_{BAT}} \tag{15}$$

If the load current is lower than the maximum value and the power limitation mode is off, the output load voltage is nominal. In this case, the charge current is constant and does not depend on the load current, in which case, the converter current can be found as follows:

$$I_{CONV} = I_{OUT} - \frac{(U_{BAT} - U_{OUT})}{R_{BAT}}. \tag{16}$$

Combining the equations above, it is possible to set the dependencies for the output current and voltage of the dependencies taking the integration mode into account:

$$I_{CONV} = \begin{cases} \frac{-b+\sqrt{b^2-4\cdot a\cdot c}}{2\cdot a} & if \;\; I_{OUT} > I_{OUT\max} \\ I_{OUT} - \frac{(U_{BAT}-U_{OUT})}{R_{BAT}} & otherwise \end{cases} \tag{17}$$

$$U_{OUT} = \begin{cases} \frac{I_{IN\max}}{I_{CONV}} \cdot (U_{IN\max} + U_{VD} - U_{VT}) - U_{VD} & if \;\; I_{OUT} > I_{OUT\max} \\ U_{OUT\_nom} \quad otherwise \end{cases} \tag{18}$$

Substituting the obtained values of the current and voltage of the converter into the abovementioned algorithm of iterative calculation, it is possible to obtain the voltage value at its input, that is, the point on the CVC of the stack that corresponds to the present load current of the power unit. Considering the input and output voltages and current of the converter, it is possible to determine the amplitude of the choke current and calculate all the other parameters of the power circuit.

Figure 8 shows the external characteristic of the voltage converter in the power limitation mode when a buffer accumulator is absent (dashed line) and present (solid line). The figure shows that the battery makes it possible to raise the output voltage to the required level at the moment of overload. At the same time, a voltage drop to a value below the nominal level is observed a little earlier due to the fact that part of the output energy is spent on charging the battery. When there is no battery backup power supply, in case of overload (34 ... 55 A), the output voltage drops sharply to stabilize the power consumption from the fuel cell.

The calculation formulas are tested on a computer model in the LTspice IV program (Figure 9) and demonstrate good agreement with the simulation results. The deviation of the calculated data obtained by Equations (11–17) from the data obtained by modeling the circuit in the LTspice IV software was estimated by determining the relative magnitude of the error under various operating modes. The calculation error was determined for the amplitude, mean, and effective values of currents and voltages of the circuit according to the equation

$$\Delta = \frac{X_{calc} - X_{model}}{X_{model}} \cdot 100\% \tag{19}$$

where X_{calc} = calculated parameter and X_{model} = parameter obtained as a result of simulation using LTspice software.

The relative error did not exceed the engineering accuracy of 5%, which indicates the correctness of the proposed calculation and the assumptions made.

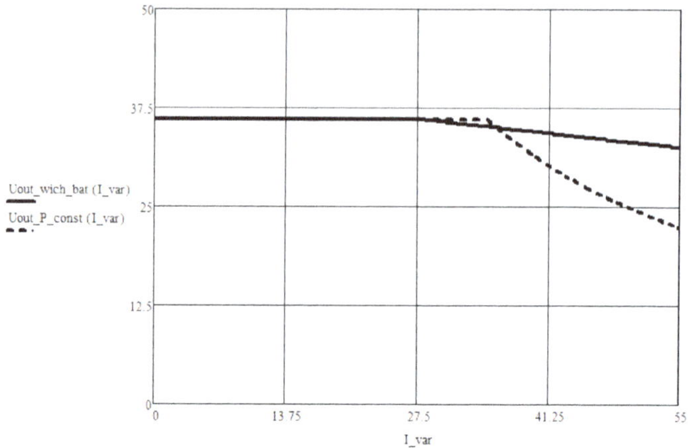

Figure 8. Output voltage of the converter with limited power consumption (dashed line) and considering the influence of the battery (solid line).

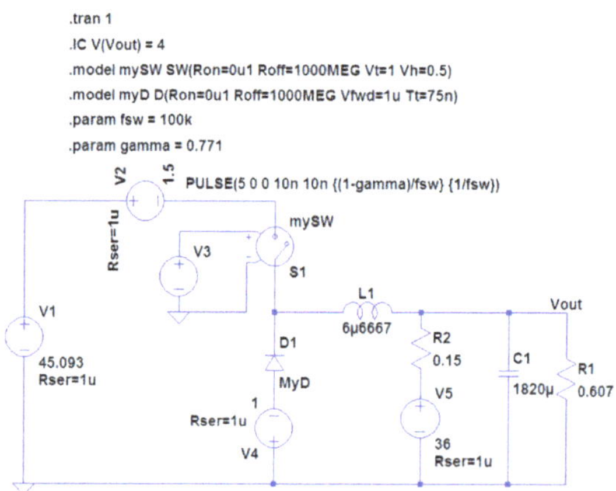

Figure 9. Computer model of the converter with a storage battery.

Thus, the model of a stack of fuel cells was combined with the developed method for calculating the converter, with the mode of integration of power supplies taken into account. A comprehensive mathematical model of the electrotechnical complex of a power unit based on fuel cells was obtained. With the magnitude of the choke current amplitude and the PWM duty cycle in all operating modes, it is possible to calculate the effective and average currents in the circuit elements and the heat losses in them. Thus, due to the fact that all dependencies are obtained in a general form, it becomes possible to figure out the most-loaded modes and optimize power units in terms of efficiency, varying any initial parameters. Thus, determining the current through the battery in the overload and charge modes, it is possible to determine the capacity it requires to ensure the maximum flight duration.

Figures 10–12 show the dependency graphs of some parameters of the converter on the load current when the power consumption is limited to 1300 W. These graphs are given as an example to demonstrate the capabilities of the proposed calculation method. The most loaded operation modes of elements are determined by the maximum or minimum of the corresponding surfaces or two-dimensional graphical dependencies. According to the graph shown in Figure 10, it is possible to determine the highest-amplitude current in the choke for all modes of operation, which corresponds to the maximum output current and the minimum battery voltage.

The first kink, at the beginning of the PWM duty cycle, is explained by the transition from discontinuous mode to continuous mode of the choke current. The duty cycle increases as the load current increases until the maximum power consumption is reached, after which it decreases again to stabilize.

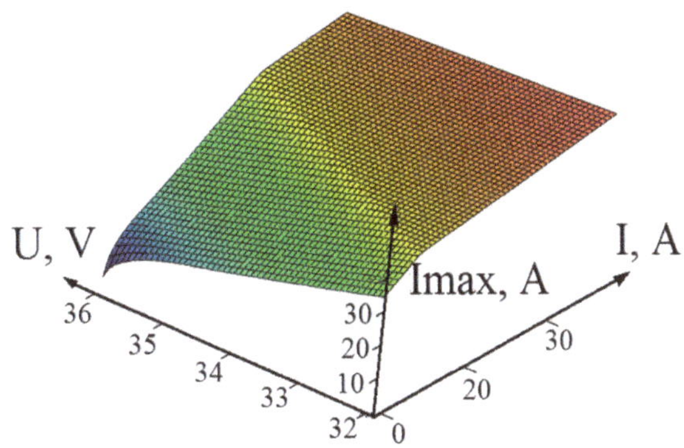

Figure 10. Dependence of the current amplitude of the choke on the load current and voltage of the battery.

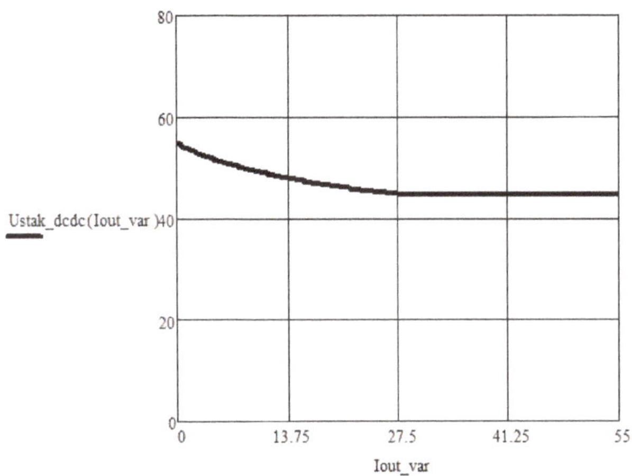

Figure 11. Dependence of the stack voltage on the load current of the power unit.

As is clear from the calculated formulas and graphs, the value of the maximum power setting should be selected so that the output power of the converter is slightly higher than the load required to supply the load in nominal mode in order to ensure the battery charge after it passes the peak overload (Supplementary materials).

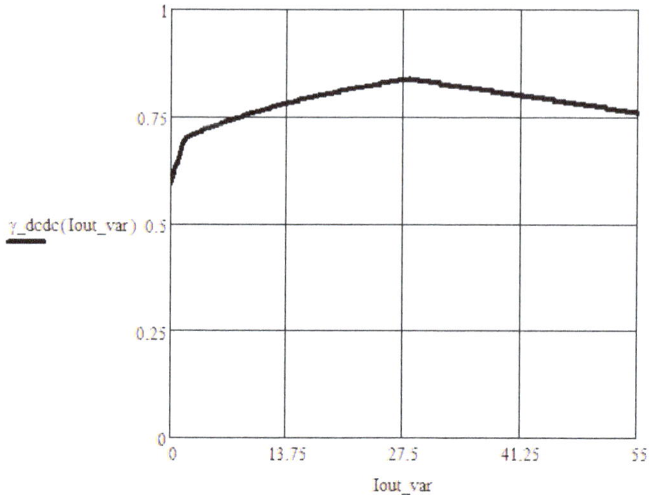

Figure 12. Dependence of the duty cycle of the PWM converter on the load current of the power unit.

4. Design

Based on the proposed mathematical model, an electrotechnical complex of a power unit for an unmanned aerial vehicle is developed, the parameters of which are presented in Table 1.

Table 1. Power unit parameters.

Parameter Name	Value
Nominal output voltage, V	=36 ± 10%
Output voltage range, V	39.6–32.4
Nominal output power, W	1300
Maximum output power, W	2000 (4 times in 2 min)
Operating time in nominal mode, h	2.5
Specific energy consumption, W·h/kg	529.3
Buffer battery capacity (10s1P), A·h	2.7
Fuel tank capacity, l	10 (under 300 bar pressure)
Weight, kg	6398

The voltage converter circuit and its circuit board are shown in Figures 13 and 14, respectively. The electrical complex includes a voltage converter with a control system, a battery unit, and a fuel cell stack. The module of the electrotechnical complex is located on the hull under the UAV fuselage. Figure 15 is a photo of the testing process of the UAV with a power unit built on the proposed electrotechnical complex.

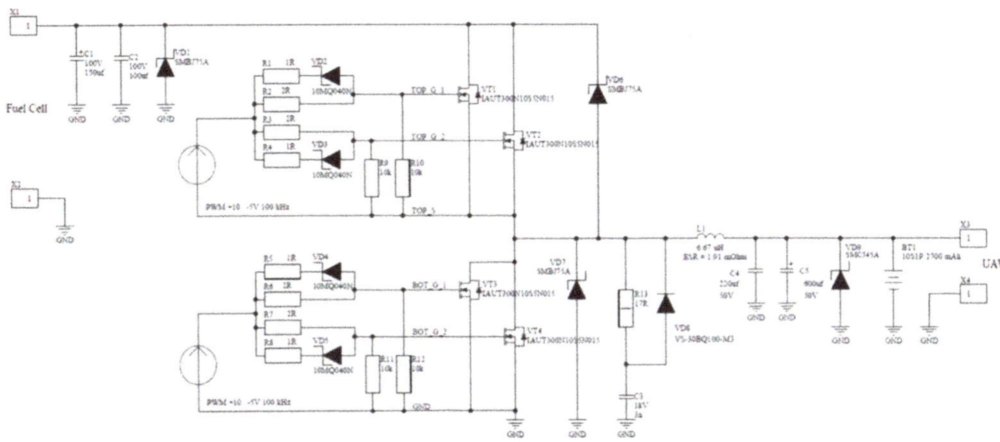

Figure 13. Diagram of the power section of the step-down voltage converter of the power unit.

Figure 14. Step-down voltage converter of the power unit.

Figure 15. UAV with a developed power unit.

5. Conclusions

As a result of the research, a comprehensive mathematical and numerical model of the electrotechnical complex is proposed. It considers the peculiarities of the operation of the step-down converter as part of a power unit based on fuel cells. It also takes into account the parameters of fuel cells and the mode of integration of energy sources from a fuel cell and a storage battery. Calculated ratios are proposed for the output current and voltage of the converter, taking into account the effect of the battery, dependencies for current amplitude of the choke, the duty cycle of the PWM, and the voltage and current of the fuel cell stack when operating on the converter in the integration mode. The correctness of the assumptions made in the process of shaping models has been established.

The obtained dependencies make it possible to determine the most loaded operating mode of the converter elements, select the choke and semiconductor elements according to the maximum, determine the average and effective current, and also select the battery according to the required voltage reduction in the peak overload modes. The proposed mathematical model makes it possible to optimize the entire complex, fuel cell included, due to the fact that all its ratios are presented in general. The correctness of the calculated ratios and the mathematical model is confirmed experimentally. The proposed approach to the design of an electrical complex can be used for other autonomous energy systems based on fuel cells since the proposed model includes all their main elements: a buffer energy storage, a power voltage converter, and a nonlinear power source.

Supplementary Materials: The following are available online at https://www.mdpi.com/article/10.3390/en14216974/s1, Figure S1: Dependence of the power consumed by the UAV propulsion system on the thrust force, Figure S2: Dependence of the mass of accumulator batteries with the 10S1P configuration on their capacity, Figure S3: Experimental current-voltage curve of the stack and its linear approximation.

Author Contributions: Conceptualization, I.V.V. and A.V.P.; methodology, A.V.P.; software, V.S.P.; validation, D.V.B., I.V.V. and I.A.B.; formal analysis, A.V.P.; investigation, I.V.V.; resources, V.S.P.; data curation, A.V.P.; writing—original draft preparation, I.V.V. and I.A.B.; writing—review and editing, A.V.P.; visualization, D.V.B.; project administration, A.V.P. and V.S.P.; funding acquisition, V.S.P. All authors have read and agreed to the published version of the manuscript.

Funding: This research was funded by the Ministry of Science and Higher Education of the Russian Federation, grant number FENN-2020-0022.

Conflicts of Interest: The authors declare no conflict of interest.

References

1. Strele, T. Power Management for Fuel Cell and Battery Hybrid Unmanned Aerial Vehicle Applications. Master's Thesis, Arizona State University, Tempe, AZ, USA, December 2016. Available online: https://repository.asu.edu/attachments/176475/content/Strele_asu_0010N_16473.pdf (accessed on 12 April 2021).
2. Bradley, T.H. Modeling, Design and Energy Management of Fuel Cell Systems for Aircraft. A Dissertation for the Degree Master of Science, The Academic Faculty, Georgia Institute of Technology, Atlanta, GA, USA, December 2008.
3. Guilbert, D.; N'Diaye, A.; Gaillard, A.; Djerdir, A. Fuel cell systems reliability and availability enhancement by developing a fast and efficient power switch open-circuit fault detection algorithm in interleaved DC/DC boost converter topologies. *Int. J. Hydrogen Energy* **2016**, *41*, 15505–15517. [CrossRef]
4. Guilbert, D.; Gaillard, A.; N'Diaye, A.; Djerdir, A. Energy efficiency and fault tolerance comparison of DC/DC converters topologies for fuel cell electric vehicles. In Proceedings of the 2013 IEEE Transportation Electrification Conference and Expo (ITEC), Detroit, MI, USA, 16–19 June 2013. [CrossRef]
5. Kabalo, M.; Paire, D.; Blunier, B.; Bouquain, D.; Simões, M.G.; Miraoui, A. Experimental evaluation of four-phase floating interleaved boost converter design and control for fuel cell applications. *IET Power Electron.* **2013**, *6*, 215–226. [CrossRef]
6. Todorovic, M.H.; Palma, L.; Enjeti, P.N. Design of a wide input range DC–DC converter with a robust power control scheme suitable for fuel cell power conversion. *IEEE Trans. Ind. Electron.* **2008**, *55*, 1247–1255. [CrossRef]
7. Gao, D.; Jin, Z.; Liu, J.; Ouyang, M. An interleaved step-up/step-down converter for fuel cell vehicle applications. *Int. J. Hydrogen Energy* **2016**, *41*, 22422–22432. [CrossRef]
8. Rashid, M.H. *Power Electronics Handbook*; Academic Press: Cambridge, MA, USA, 2001.
9. Benyahia, N.; Benamrouche, N.; Rekioua, T. Modeling, design and simulation of fuel cell modules for small marine applications. In Proceedings of the 2012 XXth International Conference on Electrical Machines, Marseille, France, 2–5 September 2012. [CrossRef]
10. Outeiro, M.; Chibante, R.; Carvalho, A.; Almeida, A. Dynamic modeling and simulation of an optimized proton exchange membrane fuel cell system. In Proceedings of the ASME 2007 International Mechanical Engineering Congress and Exposition, Seattle, WA, USA, 11–15 November 2007; pp. 171–178. [CrossRef]

11. Mann, R.F.; Amphlett, J.C.; Hooper, M.A.; Jensen, H.M.; Peppley, B.A.; Roberge, P.R. Development and application of a generalised steady-state electrochemical model for a PEM fuel cell. *J. Power Sources* **2000**, *86*, 173–180. [CrossRef]
12. Xue, X.; Cheng, K.; Sutanto, D. Unified mathematical modelling of steady-state and dynamic voltage–current characteristics for PEM fuel cells. *Electrochim. Acta* **2006**, *52*, 1135–1144. [CrossRef]
13. Onar, O.; Uzunoglu, M.; Alam, M. Modeling, control and simulation of an autonomous wind turbine/photovoltaic/fuel cell/ultra-capacitor hybrid power system. *J. Power Sources* **2008**, *185*, 1273–1283. [CrossRef]
14. Correa, J.; Farret, F.; Gomes, J.; Simoes, M. Simulation of fuel-cell stacks using a computer-controlled power rectifier with the purposes of actual high-power injection applications. *IEEE Trans. Ind. Appl.* **2003**, *39*, 1136–1142. [CrossRef]
15. Di Dio, V.; La Cascia, D.; Liga, R.; Miceli, R. Integrated mathematical model of proton exchange membrane fuel cell stack (PEMFC) with automotive synchronous electrical power drive. In Proceedings of the 18th International Conference on Electrical Machines, Vilamoura, Portugal, 6–9 September 2008. [CrossRef]
16. Rubio, M.A.; Urquia, A.; González, L.; Guinea, D.; Dormido, S. FuelCellLib—A modelica library for modeling of fuel cells. In Proceedings of the 4th International Modelica Conference, Hamburg, Germany, 7–8 March 2005. Available online: http://modelica.org/events/Conference2005/online_proceedings/Session1/Session1b3.pdf (accessed on 12 March 2021).
17. Xu, L.; Xiao, J. Modeling and simulation of PEM fuel cells based on electrochemical model. In Proceedings of the 2011 International Conference on Remote Sensing, Environment and Transportation Engineering, Nanjing, China, 24–26 June 2011. [CrossRef]
18. Rahmoun, A.; Biechl, H. Modelling of Li-ion batteries using equivalent circuit diagrams. *Przegląd Elektrotechniczny* **2012**, *88*, 152–156.
19. Rahimian, S.K.; Rayman, S.; White, R.E. Comparison of single particle and equivalent circuit analog models for a lithium-ion cell. *J. Power Sources* **2011**, *196*, 8450–8462. [CrossRef]
20. Borisevich, A.V. Modeling Lithium-Ion Batteries for Battery Management Systems: An Overview of the Current State. Modern Technology and Technology. No. 5. 2014. Available online: https://technology.snauka.ru/2014/05/3542 (accessed on 12 April 2021).

energies

MDPI

Article

Optimization of Operating Conditions of a Solid Oxide Fuel Cell System with Anode Off-Gas Recirculation Using the Model-Based Sensitivity Analysis

Eun-Jung Choi [1], Sangseok Yu [2] and Sang-Min Lee [1,*]

1 Department of Clean Fuel and Power Generation, Korea Institute of Machinery & Materials (KIMM), 156 Gajeongbuk-ro, Yuseong-gu, Daejeon 34103, Korea; ejchoi@kimm.re.kr
2 School of Mechanical Engineering, Chungnam University, 99, Daehak-ro, Yuseong-gu, Daejeon 34134, Korea; sangseok@cnu.ac.kr
* Correspondence: victlee@kimm.re.kr; Tel.: +82-42-868-7833

Abstract: Designing a configuration of an efficient solid oxide fuel cell (SOFC) system and operating it under appropriate conditions are important for achieving a highly efficient SOFC system. In our previous research, the system layout of a SOFC system with anode off-gas recirculation was suggested, and the system performance was examined using a numerical model. In the present study, the system operating conditions were optimized based on the system configuration and numerical model developed in the previous paper. First, a parametric sensitivity analysis of the system performance was investigated to demonstrate the main operating parameters. Consequently, the fuel flow rate and recirculation ratio were selected. Then, the available operating conditions, which keep the system below the operating limits and satisfy the desired system performance ($U_{fuel} > 0.7$ and $\eta_{elec} > 45\%$) were discovered. Finally, optimized operating conditions were suggested for three operating modes: optimized electrical efficiency, peak power, and heat generation. Depending on the situation, the demand for electricity and heat can be different, so different proper operating points are suggested for each mode. Additionally, using the developed model and the conducted process of this study, various optimized operating conditions can be derived for diverse cases.

Keywords: SOFC; anode off-gas recirculation (AOGR); hydrogen recirculation; SOFC simulation; sensitivity analysis

Citation: Choi, E.-J.; Yu, S.; Lee, S.-M. Optimization of Operating Conditions of a Solid Oxide Fuel Cell System with Anode Off-Gas Recirculation Using the Model-Based Sensitivity Analysis. *Energies* **2022**, *15*, 644. https://doi.org/10.3390/en15020644

Academic Editor: Orazio Barbera

Received: 20 December 2021
Accepted: 12 January 2022
Published: 17 January 2022

1. Introduction

Carbon-neutral technology issues have become a subject of special interest to cope with global climate change. Solid oxide fuel cells (SOFCs) have received great attention as low-emission power generation devices. They have diverse advantages such as high electrical efficiency, eco-friendliness, and fuel flexibility [1–3]. However, there are still some barriers to commercialization, such as low durability, high cost, and low reliability. To overcome these challenges, various reports have been published on improving system efficiency and finding control points that can ensure system safety [4,5].

An anode off-gas recirculation (AOGR) system can be adopted in the SOFC system as a method to enhance system performance. Generally, anode off-gas (AOG) contains 15–25% residual fuel; therefore, additional benefits of fuel utilization and electrical efficiency can be obtained by recycling AOG [1,6]. Several studies about SOFC systems with AOGR have been conducted. However, most of the studies suggested the effect of the AOGR system compared to the SOFC system without AOGR. System performance can be largely different depending on the system configurations. Therefore, in our previous paper [7], system performances with different system configurations were numerically examined. A survey of literature on SOFC systems with AOGR was conducted in the paper [7].

Additional research papers [1,7–9] have been added to the previous literature survey and are summarized in Table 1.

Selecting the proper operating conditions and system configuration is also important for reliable system performance. The simulation model is widely used to determine the control algorithm and to optimize the operating point because of its cost-effectiveness. As various reactions in the SOFC system are complicated, simulation analysis is more effective in determining the cause-and-effect relationship of the system performance [4,10]. Many studies have conducted sensitivity analysis to determine the key operational and design parameters for a fuel cell system. For sensitivity analysis, a reliable model was first developed for the analysis object's system. Then, a parametric analysis was performed to evaluate the impact of diverse parameters on performance. The dominant parameter can be identified by comparing the sensitivity of each parameter [4,9–12]. Lee et al. [9] numerically examined a combined fuel reforming and SOFC system with AOGR with parametric sensitivity analysis and suggested an optimized point with the highest efficiency. Dhingra et al. [13] developed a model of a 1 kW diesel-fed SOFC system, which was studied using sensitivity analysis. Individual and paired variable sensitivities were examined to study the influence of the key operating variables on the system. The operating conditions of the steam-to-carbon (S/C) ratio, and the fuel and air utilization factor were suggested at the maximum efficiency point. A summary of the related papers bas been suggested in Table 2. Based on the numerical parametric analysis, the effects of system variables were examined in several papers. When the purpose of the study was to select design variables, the number of system variables was higher than others. Among them, two papers suggested the optimum operating condition for the maximum system efficiency.

Table 1. A literature summary of the solid oxide fuel cell (SOFC) system with anode off-gas recirculation (AOGR) (revised based on the Table 1 in ref. [7]).

Authors	SOFC Power	η_{ele} (Max.)	AOGR Device	Comments
Lee et al. [14]	5 kW	64.6%	Ejector	The turbocharger and ejector were used to blow cathode air and AOG. Sensitivity analysis was conducted to determine the optimal operating schemes.
Powell et al. [15]	1.7–2.2 kW	56.6% (LHV)	Blower	The system used an adiabatic external steam reformer and AOGR system. Heat and steam for steam methane reforming were provided by recirculated AOG.
Koo et al. [16]	113.8 kW	66% (LHV)	Blower	A cascade system with double SOFC system and single SOFC system with AOGR was analyzed using the exergy-based analysis method.
Wagner et al. [8,17]	6 kW	66% (LHV)	Fan	A novel micro AOGR fan was introduced and experimentally coupled to SOFC system.
Baba et al. [18]	1 kW	-	Ejector	SOFC system with a variable flow ejector was examined under partial load and full load conditions.
Tanaka et al. [19]	10 kW	58.7% (LHV)	Blower	AOGR blowers were developed and coupled with SOFC system simulator.
Dietrich et al. [20]	0.3 kW	41% (LHV)	Injector	SOFC running on propane with AOGR was experimentally examined and compared to a partial oxidation system.
Choi et al. [7]	1 kW	53.44% (LHV)	Blower	System performances with different system configurations were numerically examined and compared to the SOFC system without AOGR.
Torii et al. [1]	5 kW	69.2%	Blower	Effects of AOGR on system performance were investigated considering carbon deposition on a stack. The possibility of reducing operation cost with a AOGR system was also revealed.
Lee et al. [9]	30 W	-	-	Sensitivity analysis of diverse parameters on a SOFC system's performance with AOGR was conducted to predict key operating parameters and an optimal operating point was suggested.

Table 2. A literature summary of the SOFC analysis for design and optimization.

Authors	System Variables	System Performance	Method	Results
Lee et al. [9]	S/C ratio, stack temperature, AOGR ratio, CO_2 adsorbent	Power density, peak power, efficiency	Numerical individual parametric analysis	Suggested the optimal operating point for maximum efficiency (@ fuel cell temperature = 900 °C with S/C ratio > 3, maximum CO_2 capture, and minimum AOG recirculation)
Kalra et al. [11]	S/C ratio, fuel temperature, pressure, geometry of the SOFC	Power density	Numerical individual parametric analysis	Compared the effects of each parameter
Kupecki [12]	Mass flow rate, pressure ratio and polytrophic efficiency of air compressor, fuel flow rate of combustion chamber, pre-heater heat exchanger UA factor	Electrical efficiency, outlet temperature of SOFC, gas turbine, and combustion chamber	Numerical individual parametric analysis	Compared the effects of each parameter
Dhingra et al. [13]	Air and fuel utilization, oxygen to carbon ratio, S/C ratio, temperature of inlet cathode and pre-heater	System efficiency, stack efficiency, system exhaust temperature	Numerical individual and paired parametric analysis	Suggested the optimal operating point, the maximum system and stack efficiency (@ air utilization = 0.5, fuel utilization = 0.9, S/C ratio = 3)
Lee et al. [14]	Fuel utilization, S/C ratio, external reforming ratio	Electrical and thermal efficiency	Numerical individual parametric analysis	Compared the effects of each parameter

This study is based on the findings of our previous investigation. In the previous paper [7], the performance of an SOFC system with AOGR was numerically analyzed to determine the best system configuration. Based on the proposed system in [7], the objective of this paper is to determine the optimal operating conditions for actual operations. First, to determine the effects of operating parameters on the performance of the SOFC system, a model-based sensitivity analysis was conducted as described in Section 3.1. The control variables of the system were selected through sensitivity analysis. In Section 3.2, the available operating conditions are analyzed to achieve system reliability and target performance. Finally, three operating modes were proposed depending on the situation, and optimal operating points were suggested for each mode.

2. Methodology

Sensitivity analysis was carried out to determine the optimized operating conditions of an SOFC system. The lumped component model of an SOFC with AOGR, which was developed in the SIMULINK environment and validated in our previous paper [7], was used in this study. Detailed equations and explanations of the model were introduced in the previous paper; therefore, they are not repeated in this paper. Only the differences from the previous model are explained.

2.1. A System Configuration

In our previous paper, we evaluated the performance of the reference SOFC system and two different systems with AOGR. It was concluded that the AOGR #2 system, as shown in Figure 1, had the best performance among the three systems [7]. The optimal operating conditions for further analysis in the present study were defined using the AOGR #2 system. A summary of the system components is presented in Table 3. The flows of gases and energy in Figure 1 are described as follows: a mixture of inlet fuel (CH_4), steam, and recirculated AOG gains thermal energy at fuel preheater 1 and 2 from AOG and cathode off-gas (COG). The mixture then flows into an external steam reformer (ESR). After an electrochemical reaction between the reformed fuel and supplied air at an SOFC stack, the heat of COG was used in fuel preheater 2 and a steam generator. A certain amount of AOG, depending on the recirculation ratio (RR), resupplies to the inlet fuel flow by a recirculation blower. The rest of the AOG is used to generate thermal energy at the catalytic combustor (CC) and the heat is transferred to an air pre-heater by the flow of the catalytic combustor off-gas (CCOG). The remaining heat from the CCOG is recovered using a heat recovery heat exchanger (HR-HE). An additional air blower was employed for the catalytic combustor (CC) to prevent an excessive rise in its temperature. In Figure 1, the flows of recirculated AOG and COG are indicated by the orange and green lines, respectively. The blue line depicts the flow of CCOG.

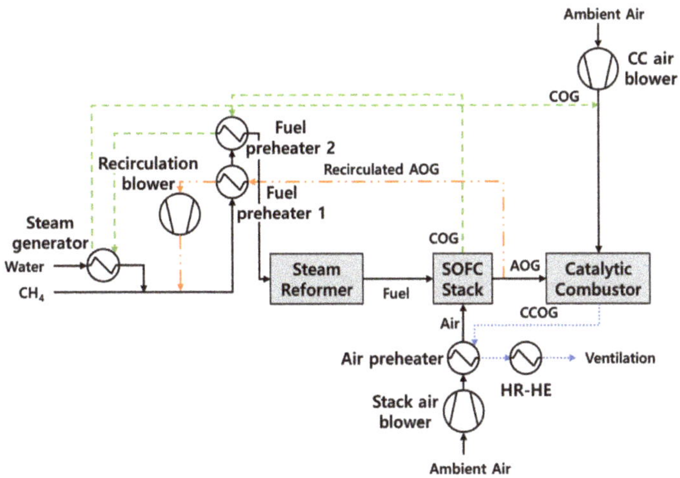

Figure 1. Schematic diagram of a solid oxide fuel cell (SOFC) system with anode off-gas recirculation (AOGR) (AOGR #2 system in ref. [7]).

Table 3. Descriptions of the main components.

Components	Descriptions	Ref. of Detailed Specifications
Stack	1 kW class planar SOFC A direct internal reforming stack Anode-supported cell type	Table 2 in the paper [7]
ESR	An adiabatic reformer A shell and tube type catalytic reformer	Table 4 in the paper [7]
CC	A Pt-catalyzed monolithic combustor	Table 6 in the paper [7]

2.2. Performance Factor

The fuel utilization factor (U_{fuel}) of the system was calculated using Equation (1). A mixture of supplied fuel (CH_4) and recirculated AOG flowed into the ESR. Methane (CH_4) is the only supplied fuel in this model, and the denominator in Equation (1) is calculated by considering recirculated AOG. The air utilization factor (U_{air}) can be defined using Equation (2).

$$U_{fuel} = \frac{J}{8F\dot{n}_{CH_4} + (8F\dot{n}_{CH_4} + 2F\dot{n}_{CO} + 2F\dot{n}_{H_2})_{recycled}} \quad (1)$$

$$U_{air} = \frac{J}{4Fy_{O2}\dot{n}_{air}} \quad (2)$$

The net power of the system (P_{net}) is calculated by subtracting the power consumption of the balance of plant (BOP, $P_{FCBlower}$, $P_{CCBlower}$, $P_{RBlower}$) from the stack-generated power (P_{stack}), as shown in Equation (3). The electrical, thermal, and total efficiencies are described in Equations (4)–(6). The electrical efficiency of the system (η_{ele}) is defined as the ratio of the net power of the system to the chemical energy (LHV_{CH_4}) of the fuel supplied to the system. For the calculation of the thermal efficiency (η_{th}), the temperature of the system exhaust gas from the CC (T_{vent}) is assumed to be 393.15 K at the HR-HE.

$$P_{net} = P_{stack} - P_{FCBlower} - P_{CCBlower} - P_{RBlower} \tag{3}$$

$$\eta_{ele} = \frac{P_{net}}{\dot{m}_{CH_4} LHV_{CH_4}} \times 100 \tag{4}$$

$$\eta_{th} = \frac{c_{p,CCOG}\dot{m}_{CCOG}(T_{airPH,o} - T_{vent})}{\dot{m}_{CH_4} LHV_{CH_4}} \times 100 \tag{5}$$

$$\eta_{tot} = \eta_{ele} + \eta_{th} \tag{6}$$

2.3. Sensitivity Analysis Parameters

It is necessary to analyze the impact of the operating parameters on the system performance characteristics to determine the optimal operating conditions. There are lots of parameters that affect system performance. We selected three variables by considering the parameters that users can control during the actual operation. Consequently, the fuel flow rate (Q_{fuel}), air flow rate (Q_{air}), and RR were considered under two different load conditions. The changes in stack temperature (T_{stack}), recirculation blower temperature (T_{RB}), U_{fuel}, P_{net}, and system efficiencies (η_{elec}, η_{th}, and η_{tot}) were examined for the system performance characteristics.

The sensitivity of each parameter to the system performance characteristics was obtained from Equation (7). First, the system performance under base conditions (PC_{base}) was computed. Then, the system performance was calculated while varying a single operating variable (PC_{test}). When one parameter was changed, the others were kept constant at the base condition. Second, the change in the system performance characteristics from the results under the base conditions was calculated.

$$Sensitivity(\%) = \frac{PC_{test} - PC_{base}}{PC_{base}} \times 100 \tag{7}$$

The base conditions and variations for the sensitivity analysis are listed in Table 4. The sensitivity analysis was conducted with two different load conditions: current density of 0.3 A/cm^2 and 0.5 A/cm^2. The base conditions of the fuel and air flow rate are the values when U_{fuel} and U_{air} of the SOFC system without AOGR become 0.6 at either value of current density.

Table 4. Operating conditions for base condition and variation for sensitivity analysis.

Parameter		Values	Variation
Inlet fuel		CH_4	-
Operating pressure of the system (bar)		1.2	-
Operating current density (A/cm^2)		0.3, 0.5	-
Exhaust gas temperature (K)		393.15	-
S/C ratio at ESR		2.5	-
Ambient temperature (K)		298.15	-
Fuel flow rate (lpm)	j = 0.3 A/cm^2	3.20	2.00–6.00
	j = 0.5 A/cm^2	5.35	3.50–10.50
Air flow rate (lpm)	j = 0.3 A/cm^2	30.56	20.00–60.00
	j = 0.5 A/cm^2	50.94	30.00–100.00
Recirculation ratio		0.4	0–0.8

2.4. Restrictions and Target Performance of the SOFC System

For the thermal stability of system components such as a stack, recirculation blower, and catalytic combustor, there are restrictions on the operating temperature for each component. The desired operating temperature of an SOFC stack in the current study was set between 973.15–1123.15 K. For a recirculation blower and CC, it is generally known that they can withstand temperatures up to 1073.15 K and 1123.15 K, respectively [7,21]. The additional air flow rate to the CC was adjusted by the PI controller of a CC blower to keep the temperature of the CC below 1123.15 K.

The project, which provided funding for this research, is aimed at developing a highly efficient SOFC system with AOGR. Hence, the target fuel utilization factor and electrical efficiency of the SOFC system were above 0.7 and 45%, respectively. The restrictions and target performance are summarized in Table 5.

Table 5. Restrictions and target performances of the SOFC system in this study.

Parameter	Values
Desired operating temperature of an SOFC stack (K)	973.15–1123.15
Maximum temperature of a recirculation blower (K)	1073.15
Maximum temperature of a catalytic combustor (K)	1123.15
Target fuel utilization factor	0.7
Target electrical efficiency of the system (%)	45

3. Results and Discussion

3.1. The Result of Sensitivity Analysis

Figure 2a–c show the results of the sensitivity analysis of all performance characteristics with changes in the fuel flow rate, air flow rate, and recirculation ratio. Figure 2 intuitively explains the effect of each parameter on various performance characteristics. The results are similar regardless of the current density; therefore, Figure 2 only shows the results when the current density is 0.3 A/cm^2. The colors blue and yellow indicate positive and negative changes in each performance characteristic, respectively. The darker the color becomes, the larger the changes in the sensitivity values it indicates. White indicates the base condition. The grey-colored region is outside of the consideration conditions. The exact values of the sensitivity analysis are presented in Table 6.

First, the effect of the fuel flow rate was examined as shown in Figure 2a. The base condition of the fuel flow rate was 3.2 lpm when U_{fuel} becomes 0.6 for the SOFC system without AOGR. The value was varied from 2.0–6.0 lpm, which corresponded to a change of −37.5% and 87.5% from the base condition. U_{fuel} became 0.9 and 0.3 when AOG was not recycled. Overall, the change in the fuel flow rate significantly affected U_{fuel}, η_{elec}, and η_{th}. The fuel flow rate obviously exhibited a negative correlation with U_{fuel}, T_{stack}, T_{RB}, and η_{elec}. The increased fuel flow rate to the stack enhanced the internal reforming reaction, leading to a reduction in T_{stack} and T_{RB}. On the other hand, P_{net} varied both positively and negatively as the fuel flow rate changed. This was because various factors were related to changes in P_{net}. Increased fuel flow to the stack can improve the stack performance by reducing the concentration overpotential loss when the fuel flow rate is low. However, this effect was cancelled out by the negative effect of the decreased stack temperature as the fuel flow rate increased [22]. For the efficiencies, a higher fuel flow rate negatively affected η_{elec}; however, it improved η_{th}. A slight increase in η_{th} at a low fuel flow rate was caused by a higher system temperature.

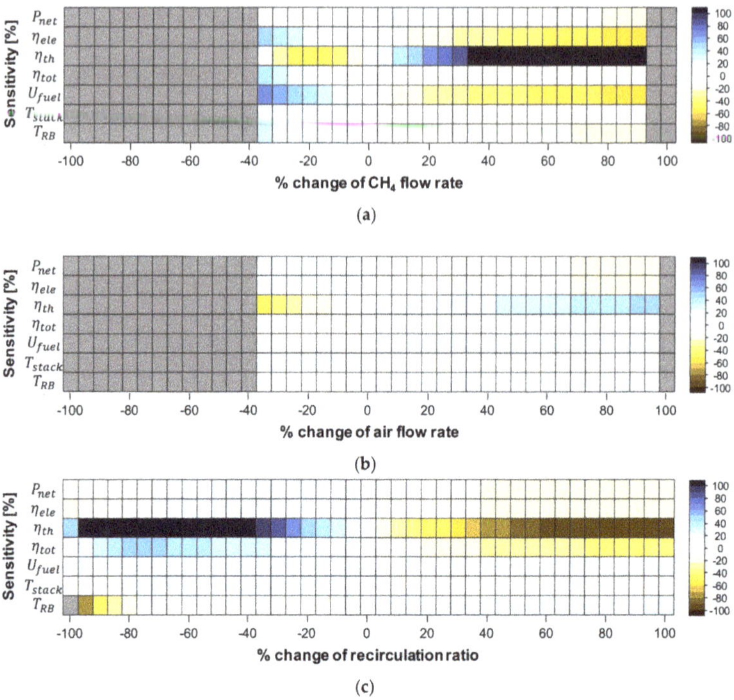

Figure 2. Sensitivity of (**a**) CH_4 flow rate, (**b**) air flow rate, and (**c**) recirculation ratio on performance characteristics.

Table 6. The result of sensitivity analysis of each parameter.

1. Sensitivity of CH_4 Flow Rate to System Performance (%)																						
Load	$j = 0.3$ A/cm^2											$j = 0.5$ A/cm^2										
% Change	−100	−80	−60	−40	−20	0	20	40	60	80	100	−100	−80	−60	−40	−20	0	20	40	60	80	100
P_{net}	-	-	-	-	−8.5	0	2.9	−1.0	−6.1	−10.3	-	-	-	-	-	−12.1	0	2.8	−2.5	−7.6	−12.2	-
η_{elec}	-	-	-	-	14.6	0	−14.5	−28.9	−41.4	−50.6	-	-	-	-	-	9.7	0	−14.4	−30.3	−42.3	−51.2	-
η_{th}	-	-	-	-	−47.6	0	61.8	116	161.8	199	-	-	-	-	-	−47.6	0	57.2	107.5	149.7	182.4	-
η_{tot}	-	-	-	-	4.1	0	−1.8	−4.8	−7.5	−8.9	-	-	-	-	-	-1.4	0	−1.4	−5.4	−7.5	−8.9	-
U_{fuel}	-	-	-	-	33.1	0	−20.4	−33.4	−43.0	−50.5	-	-	-	-	-	33.6	0	−20.3	−33.6	−43.0	−50.1	-
T_{stack}	-	-	-	-	3.2	0	−0.8	−2.7	−4.8	−6.0	-	-	-	-	-	4.4	0	−0.9	−3.8	−5.6	−6.9	-
T_{RB}	-	-	-	-	6.4	0	−3.9	−7.3	−9.6	−10.6	-	-	-	-	-	7.8	0	−4.0	−8.8	−10.4	−11.4	-
2. Sensitivity of Air Flow Rate to System Performance (%)																						
Load	$j = 0.3$ A/cm^2											$j = 0.5$ A/cm^2										
% Change	−100	−80	−60	−40	−20	0	20	40	60	80	100	−100	−80	−60	−40	−20	0	20	40	60	80	100
P_{net}	-	-	-	-	1.7	0	−2.9	−5.9	−8.7	−11.4	-	-	-	-	4.7	3.3	0	−3.2	−6.2	−9.2	−12.0	-
η_{elec}	-	-	-	-	1.7	0	−2.9	−5.9	−8.7	−11.4	-	-	-	-	4.7	3.3	0	−3.2	−6.2	−9.2	−12.0	-
η_{th}	-	-	-	-	−19.7	0	9.4	18.3	27.5	36.6	-	-	-	-	−33.6	−9.2	0	9.3	18.6	27.9	37.1	-
η_{tot}	-	-	-	-	−1.9	0	−0.9	−1.8	−2.6	−3.4	-	-	-	-	−2.3	1	0	−0.9	−1.7	−2.5	−3.1	-
U_{fuel}	-	-	-	-	0	0	0	0	0	0	-	-	-	-	0	0	0	0	0	0	0	-
T_{stack}	-	-	-	-	0.7	0	−1.2	−2.3	−3.2	−4.1	-	-	-	-	2.4	1.7	0	−1.4	−2.6	−3.6	−4.5	-
T_{RB}	-	-	-	-	−1.5	0	0.8	1	1	0.9	-	-	-	-	−1.7	−2.0	0	0.7	0.8	0.8	0.6	-
3. Sensitivity of Recirculation Ratio to System Performance (%)																						
Load	$j = 0.3$ A/cm^2											$j = 0.5$ A/cm^2										
% Change	−100	−80	−60	−40	−20	0	20	40	60	80	100	−100	−80	−60	−40	−20	0	20	40	60	80	100
P_{net}	−9.9	−1.9	2.4	4.8	4.7	0	−4.7	−10.5	−15.2	−17.3	−16.8	−13.3	−2.9	1.6	4.3	4.8	0	−4.8	−11.3	−15.6	−16.9	−16.8
η_{elec}	−10.2	−1.9	2.4	4.8	4.7	0	−4.7	−10.5	−15.2	−17.3	−16.8	−13.3	−2.9	1.6	4.3	4.8	0	−4.8	−11.3	−15.6	−16.9	−16.8
η_{th}	45.3	257.5	200.8	116.8	49.7	0	−40.7	−72.4	−90.1	−97.7	−100.0	368.6	269.2	185.2	110.2	46.8	0	−38.5	−68.8	−84.2	−91.4	−96.3
η_{tot}	−0.9	41.7	35.2	23.5	12.1	0	−10.7	−20.8	−27.7	−30.7	−30.7	55.9	46.2	35	23.6	12.3	0	−10.9	−21.7	−28.1	−30.4	−31.2
U_{fuel}	8.2	5.4	3.1	1.4	0.4	0	0.5	2.3	4.1	4.9	4.7	8.6	5.5	3.1	1.4	0.4	0	0.5	2.4	3.7	4	3.8
T_{stack}	−5.4	0.3	1.5	2.2	2.3	0	−1.5	−2.2	−1.9	−1.2	−0.8	1.8	0.5	1	1.9	2.4	0	−1.5	−2.0	−1.6	−1.0	−0.6
T_{RB}	-	−11.5	−3.6	−1.6	−0.1	0	0.9	2.7	5.3	8.2	11		−12.1	−4.1	−1.8	−0.1	0	0.9	2.9	5.7	8.6	11.5

Second, the effect of the air flow rate was observed as presented in Figure 2b. The air flow rate was varied between −34.6% (20.0 lpm) and 96.3% (60.0 lpm) from the base flow rate (30.6 lpm), and they corresponded to U_{air} values of 0.9 and 0.3 when AOG was not recirculated. The impact of air flow rate variation on the performance characteristics was relatively small compared to the results in Figure 2a,c. Owing to the cooling effect of air, T_{stack} decreased slightly as the air flow rate increased. This yields to small decreases in P_{net} and η_{elec}. Meanwhile, the higher heat capacity of COG and CCOG resulting from the higher air flow rate enhanced the heat transfer amount at the steam generator and HR-HE. This increased T_{RB} and η_{th}.

Finally, the effect of the AOG recirculation ratio was examined. The base condition of the recirculation ratio was 0.4, and it was varied from 0 (no AOG recirculation) to 0.8. The percentage range is −100–100%. As shown in Figure 2c, the recirculation ratio had a strong influence on η_{th}. The mean fuel concentration at the CC caused by the enhanced recirculation ratio led to lower heat recovery at the HR-HE [6]. The increased AOG recirculation ratio also caused lower heat recovery at the fuel and air preheater, which led to a temperature decrease in the fuel and air supplied to the stack. Therefore, the T_{stack} varied within the range of ±5.4% from the temperature under the base condition. P_{net} enhanced gradually as the recirculation ratio increased; however, this effect was overcome with the fuel dilution problem at a higher recirculation ratio [7]. U_{fuel} varied only ±8.2% because the base value of the fuel flow rate was small. The effect of the recirculation ratio on U_{fuel} becomes noticeable at a higher fuel flow rate [6].

3.2. Optimization of Operating Condition

3.2.1. Desired Operating Conditions

Section 3.1 revealed that the sensitivity of the air flow rate to the system performance was weak compared to other operating parameters. Generally, the air flow rate is manipulated to control the stack temperature [4,22]. Therefore, the normal air flow rate was set as the value when U_{air} was 0.6 lpm. The PI controller adjusted the air flow rate when the stack temperature exceeded the available operating temperature, as shown in Table 5.

The purpose of this section is to determine the desired operating conditions that can satisfy the desired system performance. The target performance of the system was suggested in Section 2.4 and Table 5. U_{fuel} and η_{elec} were aimed to be simultaneously higher than 0.7 and 0.45, respectively. The changes in U_{fuel} and η_{elec} with varying fuel flow rates and recirculation ratios are shown in Figure 3a,b when the current density is 0.3 A/cm^2. The colored regions indicate the conditions that satisfy each target performance. When the fuel flow rate is low and the recirculation ratio is high, a significant concentration loss limits fuel cell performance. The results reveal that the system should be operated in a low fuel flow rate region (lower than 2.7 lpm) to meet the targeted U_{fuel} and η_{elec}. As shown in Figure 3, the impact of the recirculation ratio increases as the fuel flow rate increases.

The operating region where the result simultaneously satisfies both requirements can be the desired operating conditions, as shown in Figure 4a. The black region represents the available operating conditions. The desired operating region with different load conditions can be defined using the same process as in Figure 3. Figure 4b displays the desired operating region at 0.5 A/cm^2. At any operating point in the black area in Figure 4, the aimed performance (U_{fuel} > 0.7 and η_{elec} > 45%) can be guaranteed under certain load conditions. The optimal operating point can be determined in the desired operating region depending on the user requirements.

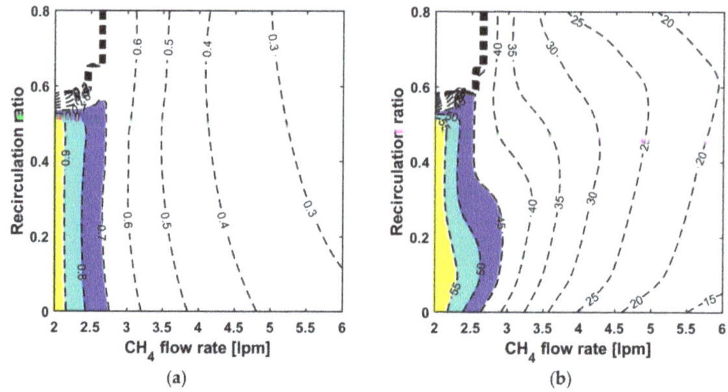

Figure 3. Changes of (**a**) fuel utilization factor (U_{fuel}) and (**b**) electrical efficiency (η_{elec}) with various fuel flow rates and recirculation ratios at current density of 0.3 A/cm^2.

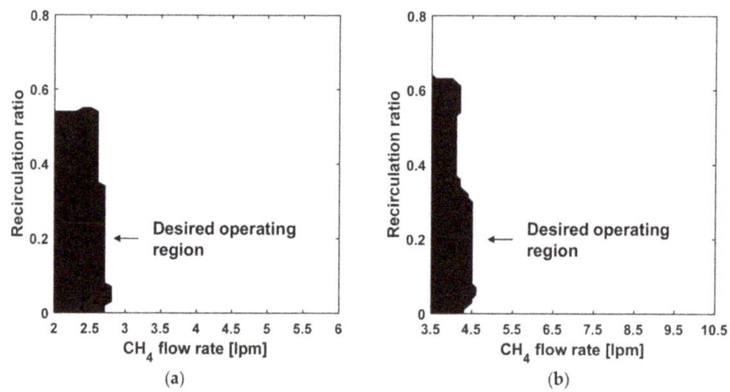

Figure 4. Desired operating conditions of the SOFC system with various fuel flow rates and recirculation ratios at operating current densities of (**a**) 0.3 A/cm^2 and (**b**) 0.5 A/cm^2.

3.2.2. Optimal Operating Point for Different Operating Scenario

Depending on the situation, users need to decide how to operate the SOFC system in the desired region, as suggested in Section 3.2.1. In this section, three scenarios are suggested: optimized electrical efficiency mode, peak power mode, and heat generation mode. The optimized electrical efficiency mode operates at the point when η_{elec} is maximized. Figure 5a,b present the change of η_{elec} when current densities are 0.3 A/cm^2 and 0.5 A/cm^2, respectively. Opaque areas represent the desired operating regions addressed in Section 3.2.1, and the red dot indicates the point with the maximum η_{elec}. In Figure 5a, the maximum η_{elec} was 59.57% when the fuel flow rate and RR were 2 lpm and 0.5, respectively. At that time, P_{net} was 771.5 W. In Figure 5b, as the fuel flow rate and RR reached 3.5 lpm and 0.6, respectively, η_{elec} reached a maximum of 55.25% and P_{net} became 1252.4 W.

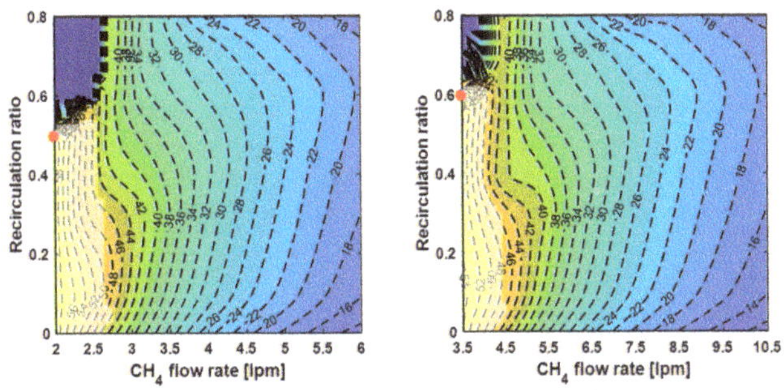

(a) (b)

Figure 5. Change of electrical efficiency with various fuel flow rates and recirculation ratios at operating current densities of (**a**) 0.3 A/cm^2 and (**b**) 0.5 A/cm^2.

The second scenario was the peak power mode. This is the case when maximum power generation is required. The change in net power is shown in Figure 6a,b for each current density, and the red point represents the maximum point in the desired operating region. The maximum net power for each case was 851.2 W at 0.3 A/cm^2 and 1374.8 W at 0.5 A/cm^2. However, higher electrical power is required in some cases. The black points in each figure show the maximum power generated by the system. Maximum power of 863.7 W and 1398.5 W can be generated at 0.3 A/cm^2 and 0.5 A/cm^2, respectively. The system performance and operating conditions at the red and black points in Figure 6 are presented in Table 7.

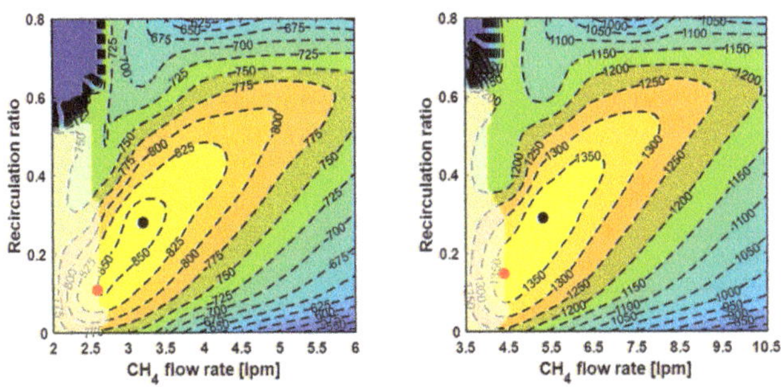

(a) (b)

Figure 6. Change in net power with various fuel flow rates and recirculation ratios at operating current densities of (**a**) 0.3 A/cm^2 and (**b**) 0.5 A/cm^2.

Table 7. Operating conditions and the system performances when net power became maximized in Figure 6a,b.

Operating Conditions and the System Performances				
Load	**j = 0.3 A/cm^2**		**j = 0.5 A/cm^2**	
P_{net} **(W)**	**851.2 (●)**	**863.7 (●)**	**1374.8 (●)**	**1398.5 (●)**
Q_{CH_4} (lpm)	2.6	3.2	4.4	5.3
RR	0.11	0.28	0.15	0.29
η_{elec} (%)	50.4	41.7	48.1	40.7
η_{th} (%)	17.3	14.3	16.1	13.9
η_{tot} (%)	67.9	56.0	64.3	54.7
U_{fuel}	0.72	0.56	0.71	0.56

Finally, there can be a situation in which a large amount of thermal energy is required. For example, a smart farm sometimes needs higher heating energy and lower electrical energy, especially in winter. Figure 7 shows the change in thermal efficiency. As shown in both results, the AOGR system is disadvantageous in terms of thermal performance. Therefore, the maximum thermal efficiency can be obtained when RR is zero, and the fuel flow rate has the highest value in the desired operating region. The maximum η_{th} in Figure 7a was 28.7% and the fuel flow rate was 2.6 lpm. When the current density was 0.5 A/cm^2, the maximum η_{th} was 28.3% at a fuel flow rate of 4.2 lpm in Figure 7b. Additionally, higher thermal energy can be obtained at a higher fuel flow rate, if it is demanded.

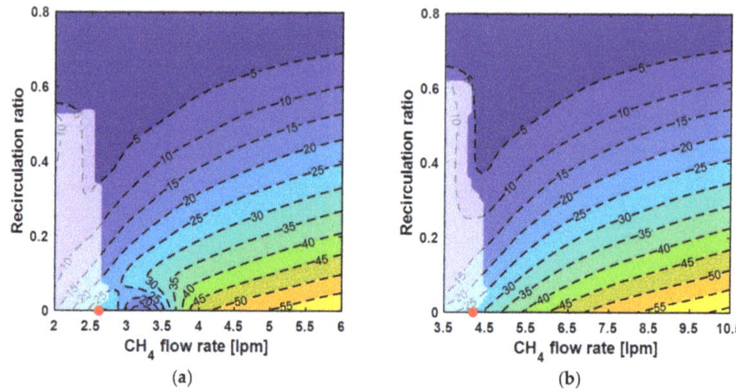

Figure 7. Change in thermal efficiency with various fuel flow rates and recirculation ratios at operating current densities of (**a**) 0.3 A/cm^2 and (**b**) 0.5 A/cm^2.

4. Conclusions

This study is a follow-up to our previous paper [7]. In the previous research, the layout of the SOFC system with AOGR was investigated, and a simulation model was developed for performance analysis. In this paper, the optimized operating conditions for the SOFC system with AOGR were suggested by utilizing the simulation model developed in ref. [7]. A sensitivity analysis of the operating parameters regarding system performance was conducted for the optimization. Through this process, we selected the control parameters for system operation. Second, the desired operating regions were found that satisfied the operating restrictions and target system performance. Finally, three operating scenarios were suggested, and optimized operating points were proposed considering each mode.

As the operating parameters for sensitivity analysis, Q_{fuel}, Q_{air}, and RR were selected. The effects of each parameter on the system performance (T_{stack}, T_{RB}, U_{fuel}, P_{net}, η_{elec}, η_{th}, and η_{tot}) were examined. Consequently, it was revealed that the impact of Q_{air} on the system was relatively small, while those of Q_{fuel} and RR were large. Based on the results, Q_{fuel} and RR were defined as control parameters. Meanwhile, the air flow rate was adjusted by a PI controller to regulate the temperature of the stack. Because of the thermal stability of the components, available temperature ranges exist during the operation. In addition, the project related to this study has a target system performance ($U_{fuel} > 0.7$ and $\eta_{elec} > 45\%$). Using the system performance map depicted in Figure 3 with various Q_{fuel} and RR values, the desired operating conditions, which can guarantee thermal safety of the system and the system performance, were proposed.

Additionally, optimized operating conditions are suggested for three operating scenarios: optimized electrical efficiency, peak power, and heat generation. Depending on the situation, the optimal points of Q_{fuel} and RR differed, and the AOGR system could be effective or not. When the current density was 0.3 A/cm^2, optimal operating conditions were obtained as follows: for the optimized electrical efficiency mode, a maximum η_{elec} of 59.57% was obtained at Q_{fuel} of 2 and RR of 0.5. For peak power mode and heat generation mode, P_{net} and η_{th} reached a maximum of 863.7 W and 27.8%, respectively, at Q_{fuel} of 2, RR of 0.5 and Q_{fuel} of 2.6, RR of 0. While operating SOFC systems, the requirements for electricity and heat can be diverse. Using the developed model and the process of this study, various optimized operating conditions can be derived for various cases, not only for the situation considered in this study.

Author Contributions: Conceptualization, E.-J.C.; simulation, E.-J.C. and S.Y.; investigation, E.-J.C.; writing, E.-J.C.; supervision, S.-M.L. and E.-J.C. All authors have read and agreed to the published version of the manuscript.

Funding: This work was supported by the Technology Innovation Program funded by the Korea Evaluation Institute of Industrial Technology (KEIT) and the Ministry of Trade, Industry & Energy (MOTIE) of the Republic of Korea (No. 20004963). This work was also funded by the Korea Institute of Energy Technology Evaluation and Planning (KETEP) and the Ministry of Trade, Industry & Energy (MOTIE) of the Republic of Korea (No. 20212020800050).

Institutional Review Board Statement: Not applicable.

Informed Consent Statement: Not applicable.

Conflicts of Interest: The authors declare no conflict of interest.

Nomenclature

c_p	specific heat capacity (J kg^{-1} K^{-1})
F	Faraday constant (C mole^{-1})
J	current (A)
j	current density (A cm^{-2})
$\dot{m}$	mass flow rate (kg s^{-1})
$\dot{n}$	molar flow rate (mole s^{-1})
P	power (W)
Q	volume flow rate (lpm)
T	temperature (K)
U	utilization factor
y	molar fraction
η	efficiency
Subscripts	
elec	electrical
th	thermal
tot	total
o	outlet
Abbreviations	
AOG	anode off-gas
AOGR	anode off-gas recirculation
BOP	balance of plant
CC	catalytic combustor
COG	cathode off-gas
CCOG	catalytic combustor off-gas
ESR	external steam reformer
HR-HE	heat recovery heat exchanger
LHV	lower heating value
RB	recirculation blower
RR	recirculation ratio
S/C	steam to carbon
SOFC	solid oxide fuel cell

References

1. Torii, R.; Tachikawa, Y.; Sasaki, K.; Ito, K. Anode Gas Recirculation for Improving the Performance and Cost of a 5-KW Solid Oxide Fuel Cell System. *J. Power Sources* **2016**, *325*, 229–237. [CrossRef]
2. Xi, H.; Sun, J.; Tsourapas, V. A Control Oriented Low Order Dynamic Model for Planar SOFC Using Minimum Gibbs Free Energy Method. *J. Power Sources* **2007**, *165*, 253–266. [CrossRef]
3. Haberman, B.A.; Young, J.B. Three-Dimensional Simulation of Chemically Reacting Gas Flows in the Porous Support Structure of an Integrated-Planar Solid Oxide Fuel Cell. *Int. J. Heat Mass Transf.* **2004**, *47*, 3617–3629. [CrossRef]
4. Li, S.; Zhan, C.; Yang, Y. Control System Based on Anode Offgas Recycle for Solid Oxide Fuel Cell System. *Math. Probl. Eng.* **2018**, *2018*, 1–16. [CrossRef]
5. Fardadi, M.; Mueller, F.; Jabbari, F. Feedback Control of Solid Oxide Fuel Cell Spatial Temperature Variation. *J. Power Sources* **2010**, *195*, 4222–4233. [CrossRef]
6. Peters, R.; Deja, R.; Engelbracht, M.; Frank, M.; Nguyen, V.N.; Blum, L.; Stolten, D. Efficiency Analysis of a Hydrogen-Fueled Solid Oxide Fuel Cell System with Anode Off-Gas Recirculation. *J. Power Sources* **2016**, *328*, 105–113. [CrossRef]
7. Choi, E.-J.; Yu, S.; Kim, J.-M.; Lee, S.-M. Model-Based System Performance Analysis of a Solid Oxide Fuel Cell System with Anode Off-Gas Recirculation. *Energies* **2021**, *14*, 3607. [CrossRef]
8. Wagner, P.H.; Van herle, J.; Schiffmann, J. Theoretical and Experimental Investigation of a Small-Scale, High-Speed, and Oil-Free Radial Anode Off-Gas Recirculation Fan for Solid Oxide Fuel Cell Systems. *J. Eng. Gas Turbines Power* **2020**, *142*, 041023. [CrossRef]
9. Lee, T.S.; Chung, J.N.; Chen, Y.-C. Design and Optimization of a Combined Fuel Reforming and Solid Oxide Fuel Cell System with Anode Off-Gas Recycling. *Energy Convers. Manag.* **2011**, *52*, 3214–3226. [CrossRef]
10. Testa, E.; Maggiore, P.; Pace, L. Sensitivity Analysis for a PEM Fuel Cell Model Aimed to Optimization. *WSEAS Trans. Power Syst.* **2015**, *10*, 10.
11. Kalra, P.; Garg, R.K.; Kumar, A. Parametric Sensitivity Analysis for a Natural Gas Fueled High Temperature Tubular Solid Oxide Fuel Cell. *Heliyon* **2020**, *6*, e04450. [CrossRef] [PubMed]
12. Kupecki, J. Sensitivity Analysis of Main Parameters of Pressurized SOFC Hybrid System. *J. Power Technol.* **2019**, *99*, 115.
13. Dhingra, H.; Peppley, B.A. Sensitivity Analysis of a 1 KW Diesel-Fuelled SOFC Generator: A Single and Paired-Variable Study. *J. Power Sources* **2013**, *239*, 527–537. [CrossRef]
14. Lee, K.; Kang, S.; Ahn, K.-Y. Development of a Highly Efficient Solid Oxide Fuel Cell System. *Appl. Energy* **2017**, *205*, 822–833. [CrossRef]
15. Powell, M.; Meinhardt, K.; Sprenkle, V.; Chick, L.; McVay, G. Demonstration of a Highly Efficient Solid Oxide Fuel Cell Power System Using Adiabatic Steam Reforming and Anode Gas Recirculation. *J. Power Sources* **2012**, *205*, 377–384. [CrossRef]
16. Koo, T.; Kim, Y.S.; Lee, D.; Yu, S.; Lee, Y.D. System Simulation and Exergetic Analysis of Solid Oxide Fuel Cell Power Generation System with Cascade Configuration. *Energy* **2021**, *214*, 119087. [CrossRef]
17. Wagner, P.H.; Wuillemin, Z.; Constantin, D.; Diethelm, S.; Van herle, J.; Schiffmann, J. Experimental Characterization of a Solid Oxide Fuel Cell Coupled to a Steam-Driven Micro Anode Off-Gas Recirculation Fan. *Appl. Energy* **2020**, *262*, 114219. [CrossRef]
18. Baba, S.; Takahashi, S.; Kobayashi, N.; Hirano, S. Performance of Anodic Recirculation by a Variable Flow Ejector for a Solid Oxide Fuel Cell System under Partial Loads. *Int. J. Hydrogen Energy* **2020**, *45*, 10039–10049. [CrossRef]
19. Tanaka, Y.; Sato, K.; Yamamoto, A.; Kato, T. Development of Anode Off-Gas Recycle Blowers for High Efficiency SOFC Systems. *ECS Trans.* **2013**, *57*, 443–450. [CrossRef]
20. Dietrich, R.-U.; Lindermeir, A.; Immisch, C.; Spieker, C.; Spitta, C.; Stenger, S.; Leithner, R.; Kuster, T.; Oberland, A. SOFC System Using a Hot Gas Ejector for Offgas Recycling for High Efficient Power Generation from Propane. *ECS Trans.* **2013**, *57*, 171–184. [CrossRef]
21. Kong, Q.; Yin, Y.; Xue, B.; Jin, Y.; Feng, W.; Chen, Z.-G.; Su, S.; Sun, C. Improved Catalytic Combustion of Methane Using CuO Nanobelts with Predominantly (001) Surfaces. *Beilstein J. Nanotechnol.* **2018**, *9*, 2526–2532. [CrossRef] [PubMed]
22. Jiang, W.; Fang, R.; Khan, J.A.; Dougal, R.A. Parameter Setting and Analysis of a Dynamic Tubular SOFC Model. *J. Power Sources* **2006**, *162*, 316–326. [CrossRef]

energies

Article

System Simulation and Analysis of an LNG-Fueled SOFC System Using Additively Manufactured High Temperature Heat Exchangers

Jan Hollmann [1,*], Marco Fuchs [1], Carsten Spieker [2], Ulrich Gardemann [2], Michael Steffen [2], Xing Luo [1] and Stephan Kabelac [1]

1 Institute of Thermodynamics, Leibniz University Hannover, 30167 Hannover, Germany; fuchs@ift.uni-hannover.de (M.F.); luo@ift.uni-hannover.de (X.L.); kabelac@ift.uni-hannover.de (S.K.)
2 Zentrum für BrennstoffzellenTechnik GmbH, 47057 Duisburg, Germany; c.spieker@zbt.de (C.S.); u.gardemann@zbt.de (U.G.); m.steffen@zbt.de (M.S.)
* Correspondence: hollmann@ift.uni-hannover.de; Tel.: +49-511-762-4601

Abstract: A laboratory-scale solid oxide fuel cell (SOFC) system using liquefied natural gas (LNG) as a fuel is designed to be used as an energy converter on seagoing vessels (MultiSchIBZ project). The presented system design phase is supported by thermodynamic system simulation. As heat integration plays a crucial role with regard to fuel recirculation and endothermic pre-reforming, the heat exchanger and pre-reforming component models need to exhibit a high degree of accuracy throughout the entire operating range. Compact additively manufactured tube-bundle and plate-fin heat exchangers are designed to achieve high heat exchange efficiencies at low pressure losses. Their heat transfer correlations are derived from experimental component tests under operating conditions. A simulation study utilizing these heat exchanger characteristics is carried out for four configuration variants of pre-reforming and heat integration. Their system behaviour is analyzed with regard to the degree of pre-reforming and the outlet temperature of the fuel processing module. The combination of allothermal pre-reforming with additively manufactured plate-fin heat exchangers exhibits the best heat integration performance at nominal full load and yields a partial load capability to up to 60% electrical load at net electrical efficiencies of 58 to 60% (LHV).

Keywords: solid oxide fuel cell; fuel cell system design; maritime application; steam reforming; anode off-gas recirculation; additively manufactured heat exchangers

Citation: Hollmann, J.; Fuchs, M.; Spieker, C.; Gardemann, U.; Steffen, M.; Luo, X.; Kabelac, S. System Simulation and Analysis of an LNG-Fueled SOFC System Using Additively Manufactured High Temperature Heat Exchangers. *Energies* **2022**, *15*, 941. https://doi.org/10.3390/en15030941

Academic Editor: Orazio Barbera

Received: 18 December 2021
Accepted: 24 January 2022
Published: 27 January 2022

1. Introduction

The decarbonization of the maritime sector is considered a major step to reduce the anthrophogenic greenhouse gas (GHG) emissions responsible for global warming. According to the most recent International Maritime Organization (IMO) study, the global shipping sector caused up to 1.056 billion tons of CO_2 emissions in the year 2018, contributing to 2.89% of the global anthropogenic CO_2 emissions [1]. Heavy fuel oil and marine diesel oil remain up to now the dominant fuel types on seagoing vessels.

The IMO aims to reduce the maritime CO_2 emissions by half in 2050 compared to a 2008 reference level [2]. Besides technical and operational improvements (e.g., hydrodynamic ship designs, propulsion efficiency, voyage optimisation, etc.), the use of alternative fuels as well as the implementation of alternative energy converters are considered to be measures with the most significant impact [3]. The spectrum of fuel candidates ranges from lower-carbon fossil fuels like liquefied natural gas (LNG), methanol and biofuels to their synthetically produced counterparts from renewable energies like synthetic natural gas (SNG) to carbon free fuels like hydrogen and ammonia [4,5].

Apart from the main propulsion engine, auxiliary engines (AE), nowadays mostly heavy duty diesel generator sets are the second dominant GHG emitters on board, which provide electrical energy for the vessel infrastructure (e.g., ship lighting, navigational instruments, manoeuvring thrusters, air conditioning, etc.). The average share of the AE on the overall fuel consumption highly depends on the vessel type and ranges from roughly 10% on seagoing container ships to up to 50% on cruise ships and refrigerated bulk carriers [1,6]. While the main engine is shut down during cargo time at ports, auxiliary engines are operated continuously unless electrical shorepower is available. This leads to continuous emissions of GHG as well as pollutants like nitrogen oxides (NO_x), sulfur oxides (SO_x) and particulate matter that pose a health risk, especially in ports close to residential areas [7].

The substitution of fossil fueled auxiliary combustion engines with more efficient energy converters like fuel cell systems is considered a major lever for reducing pollutants at sea and especially during cargo periods in ports. Compared to the conventional energy conversion pathway of diesel generator sets involving the intermediate steps of heat and mechanical power, fuel cells directly convert the internal chemical energy of a fuel into electrical energy. Above-mentioned by-products of combustion are almost avoided. Fuel cell systems for power generation are ideally implemented in a way that they run on the same fuel as the main propulsion engine and exhibit a broad fuel flexibility.

In the recent past until today, several research and prototype projects have been carried out that focus on the maritime application of fuel cells with present and potential future maritime fuels [8–10]. The majority of the recent projects regarding seagoing vessels focus on the application of high temperature polymer electrolyte membrane fuel cells (HT-PEMFC) [10,11] and solid oxide fuel cells (SOFC).

Commercial SOFCs typically use an oxygen ion conducting electrolyte that allows operation at temperatures of 500–850 °C. Due to the high temperatures, SOFC operation profits from fuel flexibility. It allows the direct supply of hydrocarbons like methane or methanol which are chemically converted to hydrogen at the anode catalyst surface via direct internal steam reforming (DIR). Additionally, SOFCs are capable of cracking ammonia into hydrogen and nitrogen [12].

Apart from several research projects using methanol [13] and ammonia [14] as a fuel, multiple SOFC research and application projects regarding LNG as a fuel have been reported, as an increasing number of modern seagoing vessels are equipped or planned with LNG fueled propulsion engines. Consequently, global LNG bunkering infrastructures at major ports have experienced a strong growth worldwide [15]. The EU project *NAUTILUS* has started in 2020 and enforces the installation of an LNG SOFC battery hybrid genset demonstrator designed for an implementation as an auxiliary power system on cruise ships [16]. A recent cooperation of Bloom Energy, Chantiers de l'Atlantique and MSC confirmed the installation of a 150 kW LNG-fueled SOFC pilot plant on board of the MSC World Europa [17]. Bloom Energy and Samsung Heavy Industries announced the equipping of a South Korean LNG carrier vessel with an SOFC system to substitute the propulsion and auxiliary engines [18].

The *SchIBZ* projects, funded by the German Federal Ministry of Transport and Digital Infrastructure (BMVI), have been working on the development of diesel and LNG based SOFC systems. Initially, research was focused on a diesel based SOFC demonstrator to prove the feasibility of a low-emission and efficient substitution of conventional auxiliary engines [19]. The demonstrator was realized as a container setup with a rated power of 50 kW_{el}. Parallel to experimental system testing, a comprehensive thermodynamic system analysis was carried out to identify suitable operating conditions [20,21]. The experimental phase demonstrated the general proof of concept of the chosen system configuration. The follow-up project *MultiSchIBZ* has been concerned with process optimizations by means of novel high temperature heat exchanger concepts to increase the system compactness and power density. Additionally, scale-up measures to establish a container module with a rated power of 300 kW_{el} as well as an additional pathway of LNG as a second fuel were investigated. The necessity of battery storage systems was analyzed to cope for the SOFC ramp rates [22] and a techno-economic analysis was carried out to outline the economic efficiency of such SOFC systems in contrast to conventional auxiliary engines [23].

In the first *MultiSchIBZ* project phase, diesel and LNG laboratory systems were designed and built consisting of a modular SOFC module with a rated power of 15 kW_{el} provided by the fuel cell manufacturer Sunfire GmbH and a fuel processing module, respectively. The design objective was focused on a scale-up concept that envisages a parallel setup of the modular fuel cell modules and a central fuel processing module being adapted in its size according to the targeted electrical output.

Experimental results of the prior project revealed modeling inaccuracies regarding the performance of the high temperature heat exchangers required for heat integration in the fuel recirculation and pre-reforming unit. In order to ensure the development of robust SOFC systems, a more detailed system model is required capable of adequately mapping heat integration over a wide range of operation. Compared to published models found in literature, the presented heat exchanger models exhibit a higher level of detail. A common assumption in published SOFC system modeling is the use of either a constant heat exchange effectiveness [24,25] or a constant overall heat transfer coefficient [26,27]. While this assumption may be true for the design operating point, the off-design characteristics occurring at flow conditions in partial load or other recirculation ratios cannot be mapped accurately, and the effectiveness is overestimated. The presented heat exchanger models exhibit heat transfer correlations that consider variations in mass flow and thermodynamical gas properties due to changes in gas composition comparable to work from Kupecki et al. [28,29]. The unique feature of this publication is the investigation of two different additively manufactured heat exchanger concepts (3D-HEX) and its impact on the overall system. The paper describes the entire development path from (1) the heat exchanger design optimization to (2) the experimental testing and deduction of suitable heat transfer correlations to (3) SOFC system modeling utilizing these component characteristics to investigate heat integration in full and partial load operation.

2. System Design

The development of fuel cell system designs has been focused on achieving high electrical system efficiencies. Regarding the use of methane as fuel for SOFCs, previous publications presented below already simulated different fuel processing system designs differing in their type of reforming (direct internal reforming (DIR), catalytic partial oxidation (CPOX) and steam pre-reforming (SR)). The latter variant can further be divided regarding the type of steam supply for the pre-reforming process by either recirculating condensed exhaust water or a share of the anode off-gas (AOG). Additionally, SR is either performed in an adiabatic or allothermal reactor by means of heat integration. The efficiencies in this paper refer to the net electrical output of the SOFC divided by the lower heating value (LHV) of the supplied fuel (see Equation (21)).

2.1. Overview of Investigated Methane-Fueled SOFC Systems

DIR concepts do not require an elaborate pre-reforming setup as all hydrocarbons react at the anode catalyst but are limited due to the high thermal gradients inside the fuel cell stack [30]. CPOX systems offer a low system complexity as well but yield a significantly smaller system efficiency compared to SR as experimentally shown by Mai et al. [31].

Regarding steam pre-reforming concepts, Peters et al. performed a comprehensive simulation parameter study for different water supply strategies including a water recirculation and AOG recirculation using a blower or an ejector [32]. The sub-variants differ in system complexity with respect to the number of components, the type of heat integration and the adiabatic or allothermal pre-reforming. The variants were investigated at nominal full electrical load with respect to the net LHV electrical efficiency as a function of the stack fuel utilization (FU_{stack}) and the AOG recirculation ratio (RR, for definitions, see Equations (15) and (16)). Results show that the proposed system configurations with AOG recirculation by means of a blower generally achieve higher electrical efficiencies than the variant with a closed water loop. High FU_{stack} and RR parameter settings are necessary to avoid carbon formation in the components, but should not be increased arbitrarily as the net system efficiency drops with increasing peripheral power demand of the blowers due to higher flow rates. Peters et al. determined a FU_{stack} of 60 to 80% at a RR of about 70% as the optimum operating range. The key parameters affecting the net efficiency are reported to be FU_{stack} and RR, followed by minor factors like the type of pre-reforming. A follow-up publication reports on an AOG recirculation system design with a recirculation temperature of 160 °C yielding net efficiencies of about 60% [33]. Both publications did not consider partial load behaviour and did not assume any heat losses.

Four different system configurations consisting of either allothermal or adiabatic SR and either water or AOG recirculation were investigated by van Biert et al. [34]. Results show that the systems with allothermal pre-reforming achieve higher electrical efficiencies and lower thermal gradients inside the stack than the adiabatic variants. Deviating from [32], the system with allothermal pre-reforming and water recirculation achieves slightly higher efficiencies than those with AOG recirculation. For a reliable system design, the authors consider it necessary to simulate both stack and system components in detail due to large interdependencies.

2.2. System Design Criteria

Apart from electrical system efficiency, the following criteria were chosen during the concept design phase of the laboratory system:

- **product safety** is influenced by the number of critical components which should be reduced to a minimum. Critical components are, for example, heat exchangers with a fuel gas and air or an oxygen-carrying medium, which can pose a safety risk on a ship and result in increased safety requirements.
- **development risk and availability of components** is minimized by selecting only commercially available components. In particular, this limits the AOG recirculation design as hydrogen compatible high-temperature blowers for temperatures above 300 °C are not available. Ejectors are available in principle or can be designed for the particular application but require a high pressure primary fluid (e.g., water or fuel) that ideally has a much higher density than the medium to be recirculated. Another disadvantage is the lack of modulation, which can cause problems especially at partial load operation.
- **robust heat integration** defines the amount of heat provided to temperature sensitive components like the pre-reformer. This influences the degree of pre-reforming and the system partial load capability. Optimal heat integration is done in a way that hot fluids with larger heat capacity rates supply heat to cold fluids with smaller heat capacity rates.
- **complexity of control** can be minimized by selecting a system design that can be operated stably via largely independent control loops.

2.3. Selected System Design

Figure 1 shows the proposed system configuration consisting of the Fuel Processing Module (FPM) and the Fuel Cell Module (FCM). The latter is a 15 kW_{el} module from Sunfire GmbH with 24 stacks of 30 3YSZ electrolyte supported cells each with an LSCF cathode and a Ni/GDC anode. The anode and cathode gas flows are supplied in parallel-flow configuration and are preheated to sufficient stack entry temperatures by counter-flow heat exchangers.

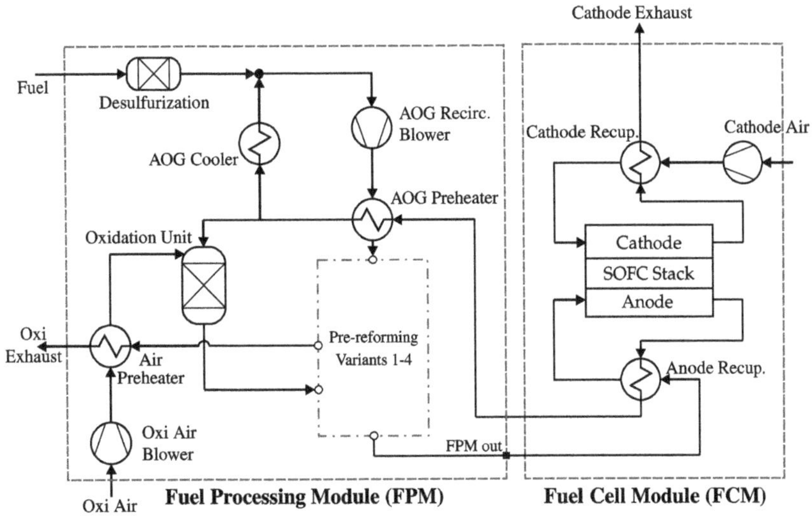

Figure 1. System configuration of the laboratory system including the fuel cell module (**right**) and fuel processing module (**left**) with a placeholder for the studied pre-reforming concept variants.

Following the results from the literature review in Section 2.1, the FPM is designed as a steam pre-reforming layout in combination with AOG recirculation at intermediate temperatures in order to achieve high efficiencies and to avoid the complexity of a closed water loop. In the selected variant, AOG enters the FPM with a temperature of about 600 °C and is cooled in the AOG Preheater. Part of the colder AOG is then passed through the AOG Cooler and mixed with the desulfurized fuel gas. The outlet temperature in the AOG Cooler is adjusted by means of cooling water or air so that the downstream mixture temperature matches the desired recirculation temperature. The AOG recirculation blower then conveys the mixture through the AOG Preheater, where it is heated up again for the following pre-reforming process. The remaining part of the AOG is fed to a catalytic burner (oxidation unit) where burnable components react with preheated air. The burner exhaust gas is used to first provide heat for pre-reforming and then to heat the combustion air in the Air Preheater.

Compared to system designs found in the above-mentioned publications, the FCM and FPM are only coupled in terms of the anode recirculation, whereas the cathode off-gas is not supplied to the oxidation unit. This is due to the fact that the required cathode volume flow to cool the stacks is much larger than the required air flow in the oxidation unit to yield the desired high outlet temperatures necessary for pre-reforming. Especially for a scaled-up configuration with multiple fuel cell modules, a direct cathode off-gas coupling would furthermore lead to larger volumes of the pipes and components downstream of the oxidation unit unnecessarily increasing installation space. Instead, the cathode off-gas flows may be used for cogeneration of useful heat for thermal processes on the vessel by means of additional heat exchangers outside the depicted system boundaries in Figure 1. For the laboratory system scale, these measures are not further investigated and are not part of this study.

2.4. Pre-Reforming and Heat Integration Variants

As literature does not conclude to one optimal system design, different pre-reforming and heat integration concepts were investigated. Pre-reforming in Figure 1 is depicted by a placeholder that represents the four proposed pre-reforming variants shown in Figure 2.

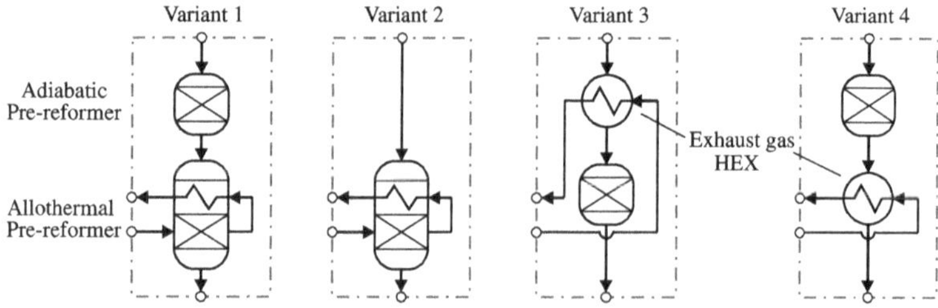

Figure 2. The four proposed pre-reforming variants.

The variants differ with regard to the sequence of heat integration and chemical reactions:

- Variant 1: a combination of adiabatic and allothermal pre-reforming, in which the heat for allothermal pre-reforming is provided by the oxidation unit exhaust gas.
- Variant 2: an allothermal pre-reforming in which the heat is provided by the oxidation unit exhaust gas.
- Variant 3: an upstream oxidation unit exhaust gas heat exchanger, followed by adiabatic pre-reforming.
- Variant 4: an adiabatic pre-reforming with a subsequent oxidation unit exhaust gas heat exchanger.

Each of the variants offers at least one specific advantage. A combination of two pre-reformers as in variant 1 is unusual, but results in an initial temperature drop in the adiabatic stage so that more heat can be transferred in the allothermal stage allowing a higher degree of pre-reforming. This is at the expense of higher and thus more cost-intensive use of catalyst material, higher pressure losses, and assembly space. Variant 2 thus has the advantage of less catalyst material, but at the cost of lower heat transfer in the allothermal pre-reformer and presumably lower outlet temperatures and degrees of pre-reforming. The least amount of catalyst material is used in the adiabatic variants 3 and 4, where the comparatively large allothermal pre-reformer can be replaced by smaller exhaust gas heat exchangers that do not require any catalyst bed. Variant 4 offers the advantage to modulate the FPM outlet temperature over a wider range via the burner temperature than in the other variants. However, this is at the expense of comparatively higher methane concentrations in the product gas. In variant 3, on the other hand, the degree of pre-reforming at the reformer outlet can be further reduced by the upstream exhaust gas heat exchanger compared with variant 4, but the FPM outlet temperature cannot be modulated as high as in variant 4. The impact of the four variants on the system performance will be discussed in Section 5.

Independent from the aforementioned pre-reforming variants, heat exchanger and reactor design were faced with the challenge of manufacturability and a low pressure loss demand in all stages, since the fuel cell stacks only allow pressure levels of 50 mbar (relative). As the installation on board of a vessel already leads to an exhaust gas back pressure of 10 to 15 mbar, this results in an average permissible maximum pressure loss of 5 mbar per FPM component. At the same time, high heat exchanger efficiencies have to be ensured to achieve high pre-reformer temperatures. Due to its flow gas compositions, the allothermal pre-reformer is considered as a critical component, which has to withstand high thermomechanical stress while maintaining gas tightness between the flows.

To meet the pressure loss criteria, the adiabatic pre-reformer is designed as a cylindrical reactor containing a precious-metal packed bed catalyst. The allothermal pre-reformer is designed as a conventional shell-and-tube heat exchanger with two bundles in cross-counter-flow configuration as depicted in Figure 3. The exhaust gas flows through the shell while the reformate gas first flows through one preheating tube bundle with empty tubes and subsequently through the reactor tube bundle containing a second packed bed catalyst. The use of a precious-metal catalyst is favored over Ni as it does not need to be reduced at the initial start-up. Furthermore, the disadvantage and safety risk of a typical Ni catalyst regarding the pyrophoric behavior during an unanticipated exposition to oxygen is avoided. The oxidation unit is designed as a cylindrical reactor containing a monolithic honeycomb structured metal catalyst. Regarding the AOG Preheater and exhaust gas heat exchanger, compact and efficient heat exchangers with low pressure losses are required. Their design by means of additive manufacturing is presented in the following section.

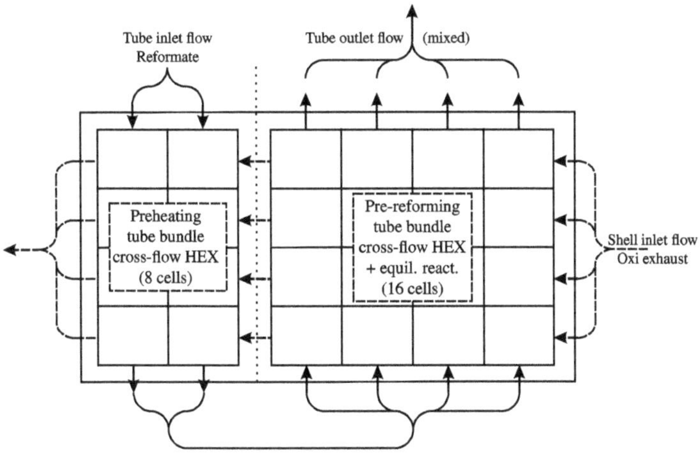

Figure 3. Visualization of the allothermal pre-reformer showing the flow distribution assumptions in the tube flow (solid lines) and shell flow (dotted lines).

3. Design and Characterization of Additively Manufactured High Temperature Heat Exchangers

Due to the limited installation space and increased safety requirements in the fuel processing module, the compact and non-gas leakage design of the heat exchangers is of decisive importance. Therefore, different types of heat exchangers are developed depending on the specific requirements. These include both tube bundle-based and plate-based heat exchangers. Due to the high degree of compactness and the increased safety requirements, these heat exchangers are manufactured using additive manufacturing, or more precisely, selective laser melting (SLM). Production using conventional methods, such as welding, might pose a safety concern as cross leakage might arise in the course of operating time and would not provide the required heat flux density of 0.5 kW/L. The heat exchangers developed are presented in detail below.

3.1. Tube-Bundle Heat Exchanger (3D-TB-HEX)

The tube-bundle heat exchangers used in this project are designed in cross-counter-flow construction as illustrated in Figure 3. In this type of heat exchanger, the hot stream flows around the tube bundles and cools down, while the cold stream heats up accordingly inside the tubes. The number of deflections and thus the number of bundles determines the effectiveness of the heat exchanger. An increasing number of bundles can approximate the ideal counter-flow and the effectiveness increases. Depending on the design, this will happen at the expense of the pressure loss due to a longer "tube section".

In order to meet the required power density, pressure loss and safety requirements, optimisations were made to the basic circular tube shape and measures are taken to increase heat transfer on the outside as well as on the inside surface. Additive manufacturing allows geometries to be produced that would not be feasible using conventional methods.

The general tube geometry is designed in a droplet shape. This reduces the wake area behind the tube for the shell-flow, leading to reduced pressure loss and increased heat transfer due to thinner thermal boundary layers. Additionally, the outside heat transfer area is increased with small fins. The geometry of the droplet shape is developed on the basis of the numerical simulated flow (Ansys Fluent) around a circular tube bundle of five tubes and adapted according to the wake area behind the tubes. At first, numerical simulations of the heat transfer and pressure drop of a circular tube bundle with five tube rows were carried out. These results were then compared with literature data. The heat transfer coefficient differs less than 15% and the pressure drop less than 12% from the literature data, which is in the range of uncertainty for the literature data regarding circular tube banks [35,36]. Next, the geometry of the tube was modified to reduce the wake area behind the tubes. During this process, the mesh quality and independence as well as the flow field of the numerical calculation were permanently checked to ensure a valid calculation. After finding the best fitting tube geometry, additional fins were applied to the outside surface, increasing the heat transfer area by 81% with a fin efficiency of 97%. The pressure drop was increased by only 28% due to higher velocities in the narrowest sections between two tubes compared to the circular tubes.

On the inside of the droplet tube, a twisted tape is implemented that promotes swirl generation and turbulence for an increased heat transfer rate. By means of additive manufacturing, the twisted-tape is also connected to the tube wall leading to an increased heat transfer area in contrast to common tube inserts with a large contact resistance between insert and tube wall. An important parameter for the heat transfer and pressure drop of the twisted tape is the number of twists per unit length. Therefore, numerical simulations and also several experiments were carried out, testing different twist ratios and their influence on the heat transfer and pressure drop, as no valid Nusselt and pressure-drop correlations exist for this design and temperature range. As for the outside optimisation, the numerical and the experimental results are in good agreement with literature data for the basic circular tube, thus ensuring correct experimental and numerical data for the twisted-tape version. From these tests, the version with 10 twists per meter exhibits the best trade-off between the heat transfer and pressure drop increase. The optimized droplet shape and the small fins on the outside as well as the inserted twisted tape are shown in Figure 4. For the so-called 3D-TB-HEX, two bundles are used in a cross-counter-flow arrangement.

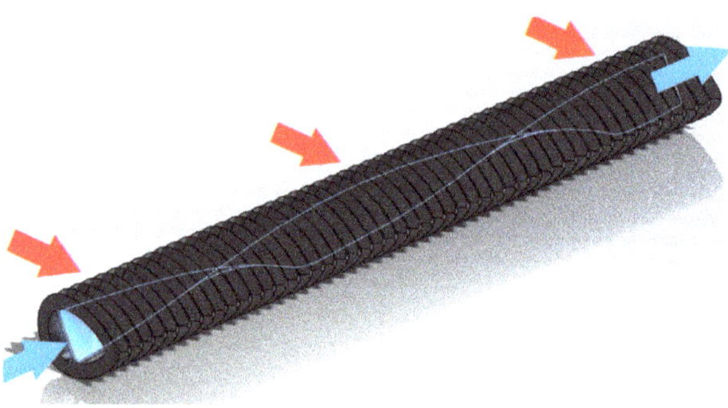

Figure 4. CAD model of a single tube of the 3D-TB-HEX. The helix structure of the internal twisted tape is depicted by means of blue contour lines.

3.2. Plate-Fin Heat Exchanger (3D-PF-HEX)

Another version of heat exchangers are plate-fin heat exchangers as they are characterised by a very high effectiveness due to the counter-flow characteristic [37]. "Wavy-fins" are chosen as the internal structure, as they are easy to manufacture due to their shape and contribute to the stability of the heat exchanger. The design of the heat exchanger, i.e., the determination of the structural parameters, was carried out using equations from literature and accompanied flow simulations with *OpenFoam* to extend the validity of the literature data. The results are also presented in Luo et al. [38]. In summary, the numerical simulation of the heat transfer agrees well with the experimental data from literature, while the numerical obtained pressure drop is lower as the experimental results in literature.

Based on the literature data, the geometric parameters of fin height, fin spacing, wavelength and amplitude are determined by optimisation with the constraints of maximum tolerable pressure loss and the most compact design possible. Further boundary conditions are the minimum wall thickness that can be printed and the maximum tolerable angle between a printed surface and the vertical printing axis. The global Matlab particle swarm algorithm is used for the optimisation. Due to the higher level of compactness of the plate-fin heat exchanger compared to the before presented tube-bundle heat exchanger, the effect of axial heat conduction in the solid parts (fins and plates) has been considered during the design process, as it has a significant negative impact on the overall performance. Figure 5 shows a section of the 3D-PF-HEX and the internal geometry. The construction volume of the plate-fin heat exchanger is approx. 1/4 of the 3D-TB-HEX at the same thermal load.

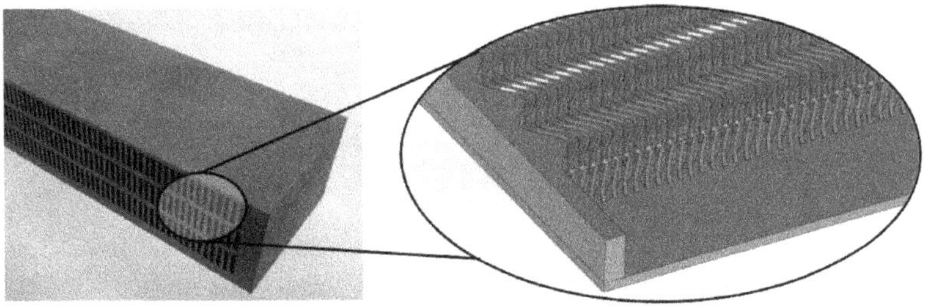

Figure 5. Close-up of the 3D-PF-HEX (left) and a CAD image of the internal fin geometry.

3.3. Experimental Validation

The heat exchangers presented above were manufactured from high-temperature stainless steel (1.4828) and tested experimentally in a separate test rig before being used in the fuel cell system in order to gain precise knowledge of the heat transfer performance as well as the pressure loss. The experimental setup is described in detail in [39]. Furthermore, the gas tightness at high temperatures is investigated by means of special tests. Based on the experimental data, equations for the pressure loss and the heat transfer are derived and then implemented in the later described system model in order to increase the model quality.

The 3D-TB-HEX and 3D-PF-HEX were each investigated at two different temperature levels. The inlet temperature of the hot side was varied between 560 °C and 750 °C by means of an electric heater, and the temperature of the cold side was varied between 200 °C and 250 °C. The mass flow was varied on both sides by two mass flow controllers between 0.2 and 0.7 kg/min, compressed air was used as the test fluid. The pressure loss was measured by two U-tube manometers.

In order to increase the accuracy of the air temperature measurement, all connections (fluid inlets and outlets) of the heat exchanger were additionally equipped with external heating to bring the temperature of the pipe wall closer to the measured gas temperature and thus reduce the influence of thermal radiation. These measures reduce the temperature difference between the respective pipe wall and the gas temperature measurement to less than 2K and significantly reduce negative thermal radiation influences.

To reduce heat losses, both heat exchangers were equipped with insulation made of a microporous insulating material. The insulation material also contained an opacifier to reduce losses due to thermal radiation, as the base material of the insulation becomes partially transparent at high temperatures. Figure 6 shows the insulated 3D-PF-HEX during installation.

Figure 6. Close-up of the insulated 3D-PF-HEX during installation at the described test rig.

3.4. Data Evaluation

From the temperatures obtained, the overall heat transfer coefficient UA_{exp} was obtained for each measuring point using the logarithmic temperature difference ΔT_{log} and a correction factor F_{Corr} for counter-flow or cross-counter-flow characteristic (1):

$$UA_{\text{exp}} = \frac{\dot{Q}_{\text{c}}}{F_{\text{Corr}}\Delta T_{\text{log}}} = \left(\frac{1}{h_{\text{c}}A_{\text{c}}} + R_{\text{wall}} + \frac{1}{h_{\text{h}}A_{\text{h}}}\right)^{-1}, \quad (1)$$

with R_{wall} as the thermal wall conduction resistance and h_{h} and h_{c} as the heat transfer coefficients of the hot and cold fluid side, respectively. The correction factor F_{Corr} for the cross-counter-flow (and others) depends on the characteristic of the heat exchanger and is defined as the ratio of the effectiveness of any heat exchanger (in this paper the cross-counter-flow) compared to the ideal counter-flow heat exchanger:

$$F_{\text{Corr}} = \frac{\epsilon_{\text{CCF}}}{\epsilon_{\text{CF}}} \ (F_{\text{Corr}} \leq 1). \quad (2)$$

Equation (1) cannot be solved directly for the unknown heat transfer coefficients h_{c} and h_{h}. Therefore, an optimisation algorithm is applied, using an error function given in Equation (3), containing the experimental overall heat transfer coefficient and a theoretical overall heat transfer coefficient of all measuring points (multi variable Wilson plot method):

$$\min_{(\alpha_{\text{c}},\alpha_{\text{h}},\beta_{\text{c}},\beta_{\text{h}})\in\vec{x}} f(\vec{x}) = \sqrt{\frac{1}{N-1}\sum_{j=1}^{N}\left[\frac{UA_{\text{th},j}(\vec{x})}{UA_{\text{exp},j}} - 1\right]^2}. \quad (3)$$

The theoretical overall heat transfer coefficient $UA_{\text{th,j}}$ is calculated using the right side of Equation (1). For the two unknown heat transfer coefficients, typical basic Nusselt (Nu) correlations for the heat transfer coefficients are used:

$$Nu_{\text{c/h, th},j} = \frac{h_{\text{c/h, th},j}\cdot d_{\text{h, c/h}}}{\lambda_{\text{m, c/h},j}} = \alpha_{\text{c/h}}\cdot Re_{\text{c/h},j}^{\beta_{\text{c/h}}}\cdot\sqrt[3]{Pr_{\text{c/h},j}}, \quad (4)$$

with d_{h} as the hydraulic diameter, λ_{m} as the mean thermal conductivity, Re as the Reynolds number and Pr as the Prandtl number of the fluid flow. By varying the coefficients α_{c}, α_{h}, β_{c} and β_{h}, the Nusselt correlations for the heat transfer of both fluids are adjusted in such a way that the error value is minimised, i.e., the theoretical overall heat transfer coefficient matches the experimentally determined overall heat transfer as closely as possible. Figure 7 depicts the parity plots comparing the experimentally obtained and the deducted theoretical values of both heat exchangers along with reference lines and percentage deviations. As a result, equations for the heat transfer coefficient as a function of the Reynolds number are determined. The definitions for the Reynolds number, the hydraulic diameter as well as the Nusselt number for the tube-bundle and plate-fin heat exchanger can be found in the Appendices A.1 and A.2, respectively.

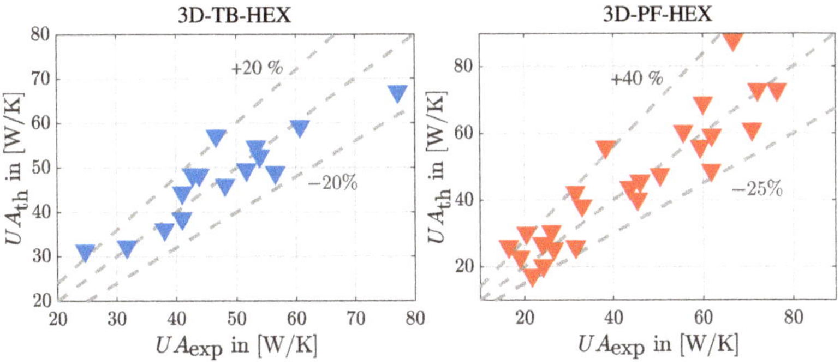

Figure 7. Comparison between the experimentally obtained overall heat transfer coefficient UA_{exp} and the theoretically calculated overall heat transfer coefficient UA_{th} using the parameterized Nusselt correlations (see Equation (4)) obtained from the experimental data.

3.5. Experimental Results

Figure 8 shows the results of the determined Nusselt number for the tube and shell side of the 3D-TB-HEX as a function of the respective Reynolds number. From the data, it can be clearly seen that a lower Nusselt number is achieved with tube side Reynolds numbers below 160 and shell side Reynolds numbers below 75 compared to a conventional circular tube. The reason for this behaviour can be explained by the design of the heat exchanger. Due to the very low velocity, parasitic influences such as axial heat conduction within the tube wall as well as heat radiation gain influence, which have a negative effect on the performance and thus on the heat transfer coefficient. As a result of the integral evaluation procedure, this leads to a lowering of the Nusselt number. If the flow velocity is increased, the Nusselt number increases both on the tube side and on the shell side. At its peak, the Nusselt number is increased by 48% or 38%, compared to a smooth circular tube of the same hydraulic diameter.

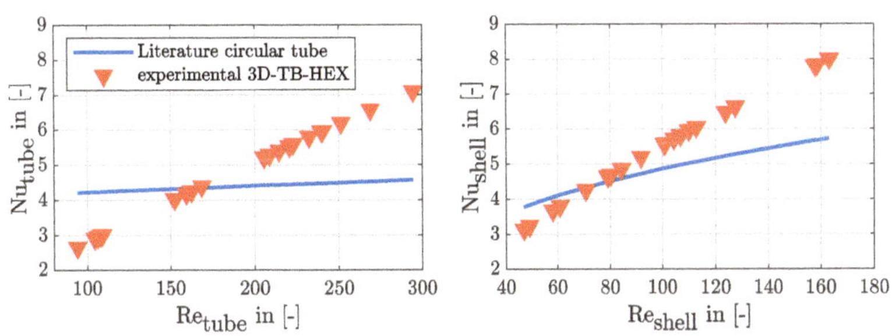

Figure 8. Experimentally derived Nusselt numbers of the 3D-TB-HEX tube (**left**) and shell flow (**right**) as a function of the Reynolds number compared to the Nusselt numbers of a basic circular tube configuration calculated with correlations from literature [36]. The range of Reynolds numbers for the intended system application spans between 150 and 280 for the tube side and between 80 an 140 for the shell side.

Figure 9 shows the curve of the Nusselt number versus the Reynolds number for the 3D-PF-HEX. The blue solid line symbolises the Nusselt numbers determined from literature data [38] and the filled triangles represent the Nusselt numbers determined from the experiments. In contrast to the tube bundle, the literature data available here also correspond to those used for the design. Furthermore, the literature data presented here have already been extended to include the influence of axial heat conduction, but not the thermal radiation which also occurs, but on a significantly smaller scale. In the low Reynolds number range, the Nusselt numbers are lower than the corresponding literature data. With increasing Reynolds number, this behaviour changes and the Nusselt number exceeds the literature data starting at a Reynolds number of $Re = 100$.

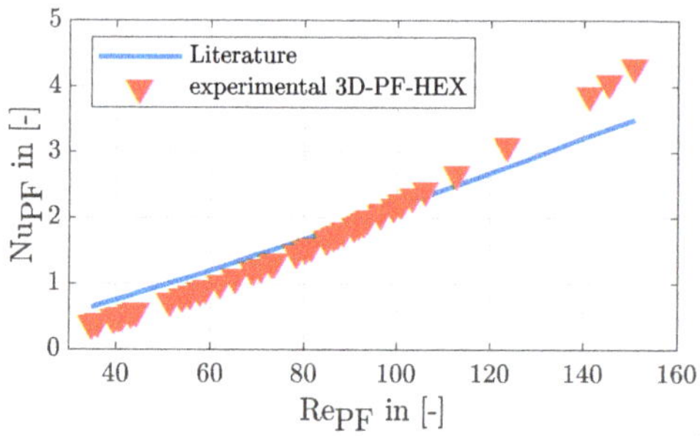

Figure 9. Experimentally determined Nusselt numbers of the hot and cold 3D-PF-HEX flows as a function of the Reynolds number compared to the Nusselt numbers calculated with correlations from literature [38]. The range of Reynolds numbers for the intended system application spans between 80 and 130.

The reason for this slightly different behaviour between literature and measurement is due to an increasing measurement uncertainty in the low and high Reynolds number range. As a result of the low heat capacity currents, very high number of transfer units (NTU; for the definition, see Equation (10)) and thus a very high effectiveness of over 98% is achieved. In combination with the asymptotic behaviour of the effectiveness as a function of the NTU value with a counter-flow apparatus [40], even small deviations or uncertainties in the temperature measurement lead to large changes in the NTU value and thus in the UA value and finally in the calculated heat transfer coefficient.

In the opposite case, i.e., with increasing Reynolds numbers ($Re > 100$), the measurement uncertainty is also the main reason for the increasing deviation compared to the literature. The Nusselt numbers determined in this Reynolds number range were determined with unequal heat capacity ratios, the mass flow of the cold side, and thus its Reynolds number, was significantly smaller than that of the hot side. This also leads to a very high efficiency and thus also to an increasing measurement uncertainty of the UA value and finally to a higher uncertainty of the heat transfer coefficient, or the Nusselt number. In the medium Reynolds number range, the measurement results agree very well with the literature data, so that the underlying calculation model is well suited for predicting the effectiveness of the 3D-PF-HEX. The developed correlations of the Nusselt-Number for the 3D-TB-HEX as well as for the 3D-PF-HEX are used in the system modelling to increase its accuracy and to compare the impact of different kinds of heat exchanger on the overall system efficiency, see Section 5 for detailed comparison.

4. Component Modelling and System Model

Component and system modelling was performed in *Matlab / Simulink* using the modelling infrastructure of the *Thermolib* toolbox version 5.3.2 from EUtech Scientific Engineering GmbH. Thermodynamic property data are provided by the NIST Webbook Database. The simulation setup is, in principle, dynamic in time, but, for this paper, it is used only to simulate steady-state operating points. Regarding the input streams of the simulation model, the supplied LNG fuel is treated as pure methane. Depending on the fuel origin and treatment, LNG may also contain other components like higher hydrocarbons and nitrogen that may account for a share of up to 10 mol% [41] which are neglected in this study. It is assumed that a central LNG evaporation infrastructure is already present on the vessel so that the fuel enters the system fully evaporated at a supply pressure of 5 bar and ambient temperature (20 °C). In the maritime environment, the surrounding air is usually humid and contains NaCl which makes the system more susceptible to corrosion and catalyst poisoning. In this design, it is assumed that the cathode and oxidation blowers withdraw filtered and dehumidified air at ambient temperature from the surroundings of the system.

4.1. Fuel Cell Stack

The fundamental stationary 0D-SOFC model has been presented in a previous publication [20], where it was used in a diesel-fueled SOFC simulation model and validated for stationary operating points. For this study, the model was adapted to the employed stack design regarding the number of cells, the area specific resistance (ASR) and expected heat losses.

The essential features of the model include internal methane steam reforming (MSR) and water gas shift (WGS) reactions (see Equations (11) and (12)) at the anode surface. The remaining methane is converted completely. For the WGS reaction, chemical equilibrium at the outlet temperature is assumed. The cell model comprises an energy balance by means of a thermal mass representing the interconnect and cell material. It is assumed that the anode and cathode outlet flows leave the stack at the thermal mass temperature T_{stack}. Deviating from the previous model description in [20], the electrochemical voltage losses are described in a simplified form using an exponentially temperature dependent area specific resistance (ASR) equation that represents the almost linear current voltage characteristic of electrolyte supported cells [33,42]:

$$ASR(T_{\mathrm{stack}}) = ASR_0 \cdot \exp\left[\frac{E_{\mathrm{a}}}{R_{\mathrm{m}}}\left(\frac{1}{T_{\mathrm{stack}}} - \frac{1}{T_0}\right)\right], \tag{5}$$

with ASR_0 as the reference ASR at reference temperature T_0 in Kelvin, E_{a} as the activation energy and R_{m} as the molar gas constant. The ASR of the electrolyte supported cell at beginning of life is located around the value of 650 mΩ cm^2 at reference temperature of 850 °C [43]. Due to its 0D limitation, the model is not capable of calculating the local temperature, current density, ASR and gas composition distribution along the flow path inside the stack. The presented model was fitted to operation data provided by a 1D stationary model of the stack manufacturer comparable to *along the channel* models shown in [42,44] by adjusting the mean ASR value $ASR_0(T_{\mathrm{stack}})$. Similar temperatures and gas compositions of the anode and cathode outlet flows as well as cell voltages could be mapped compared to the more detailed 1D model. Operating points were defined at given electric currents, inlet temperatures, gas compositions as well as fuel and oxygen utilization.

4.2. Heat Exchangers

All heat exchangers except the anode and cathode recuperators as well as the AOG Cooler are modelled using the *NTU* method. The fluid with the smaller heat capacity rate $\dot{C}_{\mathrm{min}} = \min(\dot{C}_{\mathrm{c}}; \dot{C}_{\mathrm{h}})$ determines the maximum possible heat transfer rate $\dot{Q}_{\mathrm{max}}$ [45]:

$$\dot{Q}_{\mathrm{max}} = \dot{C}_{\mathrm{min}} \cdot \left(T_{\mathrm{h,in}} - T_{\mathrm{c,in}}\right) \tag{6}$$

The performance of the heat exchanger can be expressed as its effectiveness ϵ by relating the actual heat transfer rate $\dot{Q}$ to the maximum heat transfer rate $\dot{Q}_{\mathrm{max}}$:

$$\epsilon = \frac{\dot{Q}}{\dot{Q}_{\mathrm{max}}} = \frac{\dot{C}_{\mathrm{c}}(T_{\mathrm{c,out}} - T_{\mathrm{c,in}})}{\dot{C}_{\mathrm{min}}\left(T_{\mathrm{h,in}} - T_{\mathrm{c,in}}\right)} = \frac{\dot{C}_{\mathrm{h}}\left(T_{\mathrm{h,in}} - T_{\mathrm{h,out}}\right)}{\dot{C}_{\mathrm{min}}\left(T_{\mathrm{h,in}} - T_{\mathrm{c,in}}\right)} \quad . \tag{7}$$

The following effectiveness relations can be derived as shown in [45] for a counter-flow configuration:

$$\epsilon_{\mathrm{co}} = \frac{1-\exp(-NTU(1-C_{\mathrm{r}}))}{1-C_{\mathrm{r}}\cdot\exp(-NTU\cdot(1-C_{\mathrm{r}}))} \quad , \tag{8}$$

and a cross-flow configuration with both fluids unmixed:

$$\epsilon_{\mathrm{cr}} = 1-\exp\left(\frac{NTU^{0.22}}{C_{\mathrm{r}}}\left(\exp(-C_{\mathrm{r}}\cdot NTU^{0.78})-1\right)\right) \quad . \tag{9}$$

Apart from the flow configuration, the effectiveness is a function of the number of transfer units (NTU) by setting the product of the overall heat transfer coefficient U and heat exchange area A in relation to the smaller heat capacity flow as well as the ratio between the smaller and larger heat capacity rate C_{r} given by the formulas:

$$NTU = \frac{UA}{\dot{C}_{\mathrm{min}}} \quad \text{and} \quad C_{\mathrm{r}} = \frac{\dot{C}_{\mathrm{min}}}{\dot{C}_{\mathrm{max}}} \quad . \tag{10}$$

Deviating from the previous publication [20], the overall heat transfer coefficients U required in Equation (10) are not held constant, but are calculated as a function of geometry, flow configuration, mass flow and thermal properties of the fluids. This allows the evaluation of the heat exchanger performance within the entire operating range. The product of the overall heat transfer coefficient U and the heat exchange area A is calculated using Equation (1) with the heat transfer coefficients $h_{\mathrm{c/h}}$ for the cold and hot side and the thermal resistance of the wall R_{wall}.

Different Nusselt correlations are used to calculate these heat transfer coefficients depending on the geometry and flow conditions. An overview of the correlations used for each component is provided in Table 1. The correlation equations are either specified in detail in Appendix A or taken from experimental results (Section 3).

For the anode and cathode plate recuperators, the effectiveness is determined by an operating map deducted from datasets provided by the manufacturer using the mass flows and inlet temperatures of the component as input variables. As the AOG Cooler does not contribute to heat integration, AOG cooling is modelled as a change in enthalpy to the desired temperature.

Table 1. Overview of heat exchanger models and their Nusselt correlations.

Component Name	Model Description	Correlations
Anode and Cathode Recup.	Convent. counter-flow HEX	Operating map
Air Preheater	Convent. cross-counter-flow tube bundle HEX	Appendix A.1
Allothermal Pre-reformer	Convent. cross-counter-flow tube bundle HEX (2 bundles, 1 empty, 1 filled with catalyst) Cell model with 24 HEX and equilibrium reaction cells	Appendix A.1
AOG Preheater	3D cross-counter-flow tube bundle HEX or 3D counter-flow plate fin HEX	Experimental Equation (4)
Exhaust Gas Heat exchanger	3D cross-counter-flow tube bundle HEX or 3D counter-flow plate fin HEX	Experimental Equation (4)

4.3. Pre-Reforming

Both inside the SOFC stack and the pre-reforming reactors, MSR and WGS reactions take place simultaneously:

$$CH_4 + H_2O \rightleftharpoons 3\,H_2 + CO \quad \text{(MSR)}, \tag{11}$$

$$CO + H_2O \rightleftharpoons H_2 + CO_2 \quad \text{(WGS)}. \tag{12}$$

In the pre-reformer stages, it is assumed that chemical conversion occurs until chemical equilibrium is reached. In reality, this assumption may not be reached at lower temperatures and leads to significant deviations between simulated and experimental gas compositions. With regard to the studied laboratory system, the assumption is kept as a precious metal catalyst with a high activity compared to typically used Ni catalysts is used. Additionally, the pre-reformers are designed big enough to ensure a high residence time. The gas mixture leaves the reactor at equilibrium reaction temperature and composition. Carbon formation reactions are not considered. In the Oxidation Unit, the remaining AOG burnable components are catalytically converted completely.

In the allothermal pre-reformer, heat exchange and chemical reactions occur simultaneously and inhomogeneously. Due to the cross-flow configuration and temperature dependent chemical equilibrium, temperature distribution and chemical conversion are expected to be distributed significantly between the tube rows. A discrete 2D cell model was chosen as presented by Engelbracht et al. [33] (see Figure 3). Homogeneous flow distribution and an unmixed shell flow are assumed. Each cell consists of a combination of an *NTU* heat exchanger and an additional equilibrium reactor on the cold fluid side. Based on a grid independence study ranging from six to up to 48 cells, a total of 24 cells was chosen to map the component behaviour at a reasonable computation time.

4.4. Blowers and Pressure Losses

Component pressure losses are modelled as a function of the involved mass flow:

$$\Delta p_{\text{comp},i} = k_{\text{comp},i} \cdot \dot{m}^2_{\text{comp},i} \quad . \tag{13}$$

As specific component pressure losses regarding the FPM heat exchangers, pre-reformers and the oxidation unit are not yet experimentally available, the maximum design value of 5 mbar specified during the component design phase is used for each FPM component in the simulation model (compare Section 2.4). The pressure loss coefficients $k_{\text{comp},i}$ are determined by assuming these maximum permissible pressure losses at the component mass flows $\dot{m}_{\text{comp},i}$ during nominal full load operation.

The peripheral electrical power demand is estimated by considering the cathode, recirculation and combustion air blowers depicted in Figure 1 assuming constant isentropic and mechanical efficiencies as done in [34].

4.5. Heat Losses

All components and pipes of the laboratory-scale FPM are separately thermally insulated to ensure accessibility during testing. Heat transfer between the components and ambient are estimated by means of steady-state thermal analyses of the laboratory setup in ANSYS Mechanical using the geometrical design and insulation properties. Overall heat transfer coefficients are deducted and implemented into the component and pipe models to represent temperature dependent heat losses:

$$\dot{Q}_{\text{loss},i} = (U \cdot A)_{\text{amb},i} \cdot \left(T_{\text{comp},i} - T_{\text{amb}}\right) \quad . \tag{14}$$

4.6. System Parameters and Operating Conditions

A set of widely common system parameters is used to define and confine the system operating range. The required excess of fuel and oxygen flow provided at the respective electrodes is characterized by the electrochemical stack fuel and oxygen utilizations:

$$FU_{\text{stack}} = \frac{\dot{n}_{\text{An,in}}\left(x_{\text{H}_2,\text{in}} + x_{\text{CO,in}} + 4x_{\text{CH}_4,\text{in}}\right) - \dot{n}_{\text{An,out}}\left(x_{\text{H}_2,\text{out}} + x_{\text{CO,out}}\right)}{\dot{n}_{\text{An,in}}\left(x_{\text{H}_2,\text{in}} + x_{\text{CO,in}} + 4x_{\text{CH}_4,\text{in}}\right)} \cdot 100\%, \quad (15)$$

$$OU_{\text{stack}} = \frac{0.21\dot{n}_{\text{Ca,in}} - \dot{n}_{\text{Ca,out}}x_{\text{O}_2,\text{out}}}{0.21\dot{n}_{\text{Ca,in}}} \cdot 100\%. \quad (16)$$

The amount of AOG recirculated is defined by the recirculation ratio:

$$RR = \frac{\dot{n}_{\text{AOG,rec}}}{\dot{n}_{\text{AOG,total}}} \cdot 100\% \quad . \quad (17)$$

Both RR and FU_{stack} impact the overall system fuel utilization:

$$FU_{\text{sys}} = \frac{FU_{\text{stack}}}{1 - RR(1 - FU_{\text{stack}})} \quad , \quad (18)$$

as well as the gas composition of the pre-reforming inlet flow as the recirculated AOG is mixed with the fuel flow. The oxygen to carbon ratio O/C is a parameter typically used in literature to depict the pre-reforming gas composition by means of element balancing:

$$O/C = \frac{2x_{\text{CO}_2,\text{pre}} + x_{\text{CO,pre}} + x_{\text{H}_2\text{O,pre}}}{x_{\text{CO}_2,\text{pre}} + x_{\text{CO,pre}} + x_{\text{CH}_4,\text{pre}}} = f(RR, FU) \quad . \quad (19)$$

The effect of carbon formation has not been modeled in this study and depends on the ternary $H/O/C$ gas composition and temperature [34,46]. The O/C value itself does not describe this relation sufficiently. However, O/C values given in the following section serve as comparative values to the above-mentioned publication results.

Analogous to Equation (16), the oxygen utilization of the oxidation unit OU_{oxi} defines the excess oxygen in relation to a complete stoechiometric combustion and is defined as the ratio of non-recirculated molar flows of combustible AOG components to the provided oxygen molar flow:

$$OU_{\text{oxi}} = \frac{(1 - RR) \cdot \dot{n}_{\text{AOG,total}} \cdot \left(0.5x_{\text{H}_2,\text{AOG}} + 0.5x_{\text{CO,AOG}}\right)}{0.21\dot{n}_{\text{oxi,air}}} \cdot 100\% \quad . \quad (20)$$

It is assumed that the SOFC system feeds into an AC electric board infrastructure requiring a frequency inverter which exhibits a conversion efficiency of $\eta_{\text{pe}} = 95\%$ [34]. Overall system performance is evaluated using the net system energetic efficiency:

$$\eta_{\text{sys,net}} = \frac{\left|P_{\text{el,SOFC,DC}}\right| \cdot \eta_{\text{pe}} - \left(P_{\text{el,bl,ca}} + P_{\text{el,bl,rec}} + P_{\text{el,bl,oxi}}\right)}{\dot{n}_{\text{fuel,in}} \cdot LHV_{\text{fuel,in}}} \quad . \quad (21)$$

To specifically investigate the pre-reforming variants in terms of their capability to convert high amounts of methane into hydrogen, the so-called degree of pre-reforming (DOR_{pre}) is defined as used in [47]:

$$DOR_{\text{pre}} = \frac{\dot{n}_{\text{fuel,in}} - \dot{n}_{\text{CH}_4,\text{FPM,out}}}{\dot{n}_{\text{fuel,in}}} \cdot 100\% \quad , \quad (22)$$

relating the amount of converted methane in all pre-reformer stages via endothermic MSR to the supplied fuel flow. The DOR_{pre} mainly correlates with the achievable pre-reforming outlet temperature. The remaining percentage is converted at the SOFC anode inlet via DIR.

The definition of an operating point is specified by the following system input parameters:

- the electrical current I.
- the stack fuel utilization FU_{stack} and the recirculation ratio RR which determine the molar flow of the supplied fuel $\dot{n}_{fuel,in}$.
- the stack oxygen utilization OU_{stack} which defines the cathode molar flow and is manipulated to control the stack outlet temperature T_{stack} to a set point of 815 °C. OU_{stack} is limited upwards to a value of 35% to ensure a sufficient oxygen partial pressure at the cathode.
- the oxygen utilization of the oxidation unit OU_{oxi} which defines the supplied air flow in the FPM and is manipulated to control the oxidation unit outlet temperature $T_{oxi,out}$ to a set point of 750 °C. OU_{oxi} is limited upwards to a value of 80% to ensure a complete combustion of the non-recirculated combustible components.
- the heat flow withdrawn from the anode off-gas in the AOG Cooler which is manipulated to control the recirculation blower inlet temperature $T_{bl,rec,in}$ to a set point of 300 °C. In case of recirculation temperatures below this value, the AOG Cooler is inactive and does only contribute in terms of its heat losses.

Depending on the electrical current, the stack temperature as well as the anode and cathode inlet properties, the SOFC model determines the resulting cell voltage which acts as output variable along with the electrical DC power $P_{el,SOFC,DC}$ and the net system efficiency $\eta_{sys,net}$. The SOFC inlet temperatures are variable and depend on the performance of the anode and cathode heat exchangers as well as the FPM outlet stream conditions. The stack operation is restricted by the minimum stack temperature and the minimum cell voltage which are given in Table 2 as part of a complete list of system parameter constraints.

Table 2. Constraints and selected operating range of the system parameters. Bold printed values depict the nominal full load operating point.

System Parameters		Unit	Constraints	Input Values/ Set Points
Eletrical current	I	[A]		20, **30**
Recirculation ratio	RR	[%]		56 ... **70** ... 80
Oxygen to carbon ratio	O/C	[-]	≥2.0	2.0 ... **2.55** ... 3.0
Stack fuel utilization	FU_{stack}	[%]	≤75	**75**
Stack oxygen utilization	OU_{stack}	[%]	≤35	*
Oxid. unit oxygen utilization	OU_{oxi}	[%]	≤80	*
Cell voltage	U_{cell}	[V]	≥0.65	**
Recirc. blower inlet temperature	$T_{bl,rec,in}$	[°C]	≤300	300
Stack outlet temperature	T_{stack}	[°C]	750 ... 850	815
Oxidation unit outlet temperature	$T_{oxi,out}$	[°C]	≤750	750
Isentropic blower efficiency	$\eta_{bl,is}$	[%]		70
Mechanical blower efficiency	$\eta_{bl,mech}$	[%]		80
Power electronics efficiency	η_{pe}	[%]		95

*: manipulated within the allowed range to control the respective temperature to its set point. **: output variable.

4.7. Parameter Study

To analyze the steady-state operating behavior, a parameter study is conducted covering an operating range depicted in the right column of Table 2. The system is operated at the maximum $FU_{stack} = 75\%$. Higher utilizations, especially at high recirculation ratios, pose the risk of Ni reoxidation at the anode surface [48], whereas lower utilizations would lead to a decrease in the electrical efficiency and lower stack temperatures due to endothermic DIR. The recirculation ratio is set over a range of 56 to 80%. This corresponds to a system fuel utilization of 87.2 to 93.8% and an O/C ratio of 2.0 to 3.0 at the chosen maximum FU_{stack} operation. Full and partial load performance is considered by lowering the electric current from the nominal operating point of 30 A to 20 A. A maximum recirculation temperature set point of 300 °C is selected for the highest possible heat integration.

Each of the pre-reforming variants presented in Section 2.4 is simulated in two configurations which differ in the heat exchanger models used for the AOG Preheater and the exhaust gas heat exchanger (if implemented). The first configuration employs the cross-counter-flow 3D-TB-HEX described in Section 3.1 and the second configuration the more efficient and compact counter-flow 3D-PF-HEX described in Section 3.2. The properties of the system inlet and outlet streams for variant 1 using the 3D-PF-HEX at nominal full load operation as defined in Table 2 are presented in the Appendix A.3.

5. Simulation Results and Discussion

The overall stationary system behaviour is presented for full and partial load operation followed by the analysis of an additional partial load operating strategy. As not all system parameters mentioned above can be displayed, system performance is discussed by means of the DOR_{pre} (see Equation (22)), the system net efficiency $\eta_{\mathrm{sys,net}}$ (see Equation (21)) as well as temperature levels of the AOG recirculation, the pre-reforming outlet and the stack.

5.1. Full Load Operating Points

At all simulated full load operating points, the stack outlet temperature set point is reached as the exothermic electrochemical reaction dominates the SOFC energy balance, and excess air is needed to cool the stacks. The recirculation ratio defines not only the O/C ratio of the pre-reforming gas mixture, but also the heat capacity rates ($\dot{C} = \dot{n} \cdot C_{\mathrm{p,m}}$) of both fluids involved in the heat transfer taking place in the allothermal pre-reformer or exhaust gas heat exchanger. At RR below 65%, the mean heat capacity rate of the hot oxidation unit exhaust gas is larger than the heat capacity rate of the pre-reforming gas ($\dot{C}_{\mathrm{h}} > \dot{C}_{\mathrm{c}}$). In this case, the heat flow is defined by the heat capacity rate $\dot{C}_{\mathrm{c}} = \dot{C}_{\mathrm{min}}$ according to Equation (6). At larger RR, this behaviour reverses as $\dot{C}_{\mathrm{c}} > \dot{C}_{\mathrm{h}} = \dot{C}_{\mathrm{min}}$. This is caused by the increase of the molar flow of the recirculated stream $\dot{n}_{\mathrm{AOG,rec}}$ with simultaneous decrease of the exhaust gas flow as less fuel is supplied to the oxidation unit and less air is required to remain at the constant oxidation unit exhaust temperature. The mean molar heat capacities $C_{\mathrm{p,m}}$ of the fluids change as well due to changing recirculation gas compositions and different oxygen utilizations at the oxidation unit, but have a significantly smaller and counteracting impact on the heat capacity rates, therefore slightly dampening the impact of the molar flow change.

Figure 10a shows the resulting degree of pre-reforming DOR_{pre}, indicating the overall FPM heat integration performance. The four variant configurations using 3D-TB-HEX are displayed as solid lines, whereas the counterpart models with 3D-PF-HEX are depicted as dotted lines. Variant 1 with the 3D-TB-HEX shows a high methane conversion in the range of $RR < 65\%$ with DOR_{pre} values of 32%. This is due to favorable heat capacity ratios both in the AOG Preheater and the allothermal pre-reformer as mentioned above. No heat has to be withdrawn in the recirculation unit as the recirculation temperature remains below 300 °C (see Figure 10b). As the RR increases above 68%, this temperature limit is reached due to worse heat capacity ratios in the AOG Preheater. The higher the recirculation ratio, the more heat has to be withdrawn in the AOG Cooler resulting in lower DOR_{pre} levels of 28%.

Comparing this behaviour to the configuration with 3D-PF-HEX, one can clearly see the positive impact of the higher effectiveness of the AOG Preheater. At low RR, the pre-reforming inlet temperature increases by 35 K leading to higher DOR_{pre} of up to 34%. The more significant difference occurs at higher RR. The DOR_{pre} remains at a constant level of 33% as the AOG recirculation temperature reaches its limit not before an RR of 76% leading to a higher level of heat integration (see Figure 10b). Regarding the nominal full load operating point at $RR = 70\%$, the configuration with a 3D-PF-HEX AOG Preheater does not require AOG cooling in the recirculation unit.

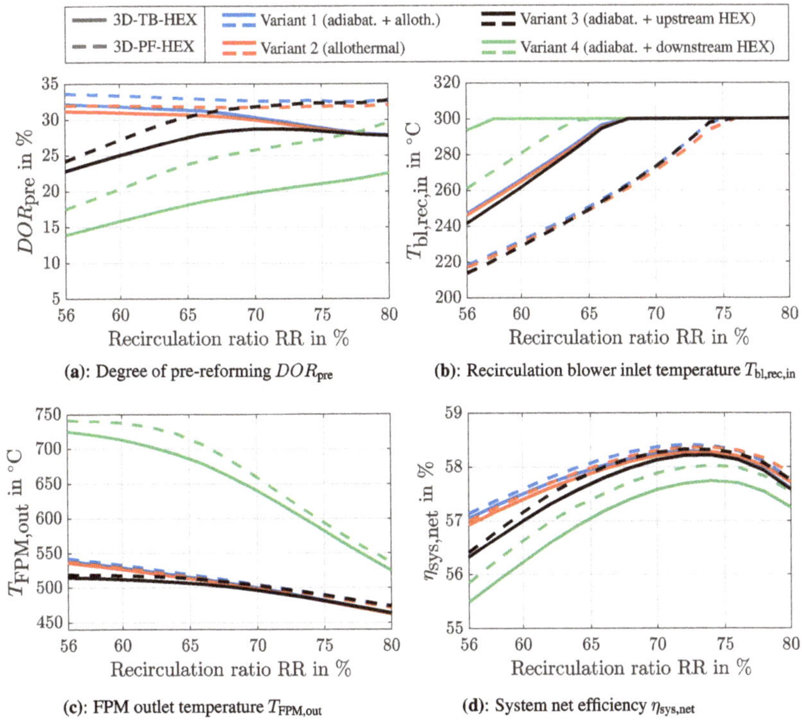

Figure 10. Simulation results for full load operating points ($I = 30$ A) at maximum stack fuel utilization $FU_{stack} = 75\%$ as a function of the AOG recirculation ratio RR.

In general, both heat exchanger configurations of variant 2 show a similar behavior compared to their variant 1 counterparts. Due to the missing adiabatic pre-reforming stage and its favorable temperature drop regarding heat transfer in the following allothermal pre-reformer, both configurations of variant 2 yield lower DOR_{pre} values of 0.5 to 1%-points. The purely adiabatic variants 3 with an upstream exhaust gas heat exchanger behave similarly at high recirculation ratios but lose performance at lower RR. This is due to the fact that less heat can be transferred in the upstream exhaust gas heat exchanger compared to the allothermal heat transfer with a simultaneous endothermic reaction.

As variant 4 features an adiabatic pre-reformer with downstream superheating, the FPM outlet temperatures, displayed in Figure 10c, are located in a much higher range than the previous variants (470 to 550 °C) leading to higher anode inlet temperatures. This is caused by the fact that the required heat of reaction in the adiabatic pre-reformer needs to be supplied indirectly via the anode gas loop and thus heavily depends on heat losses and the effectiveness of the AOG Preheater. As the exhaust gas heat exchanger also lifts the AOG temperature, the AOG recirculation temperature already reaches its limit at much lower recirculation ratios (see Figure 10b), leading to high levels of heat being withdrawn from the system. As a consequence, the DOR_{pre} values of variant 4 with 3D-TB-HEX are significantly lower leading to higher levels of DIR. Out of the four variants, variant 4 features the highest sensitivity towards heat exchanger effectiveness. With both AOG Preheater and exhaust gas heat exchanger designed as 3D-PF-HEX, the DOR_{pre} increases by up to 7%-points but is almost entirely below the levels of the other variants.

The system net efficiencies depicted in Figure 10d generally increase with higher RR due to an increase in the system fuel utilization FU_{sys} but are limited by the electric demand of the recirculation blower leading to the decrease at $RR > 75\%$. Comparing the course of the variants, deviations of up to 0.5%-points at the nominal operating point of RR = 70% are visible.

5.2. Partial Load Operating Points

As the electrical current is decreased, the dissipated heat from electrochemical conversion in the stack decreases accordingly, whereas heat losses remain almost the same. Excess air cannot be decreased arbitrarily due to the OU_{stack} constraint leading to decreasing stack temperatures. For all simulated operating points at 20 A (corresponding to 62 to 68% of electrical net power at full load), the OU limitation already leads to higher cathode air flows than thermally required. Besides artificially heating the cathode inlet flow by means of electric heating, supplying the stack with anode gas at a high temperature and high degree of pre-reforming remains the only option in order to maintain permissible stack temperatures.

Figure 11a shows the DOR_{pre} of the four variants. In general, the depicted values are substantially smaller than for the presented full load operating points, opposing the intended partial load behaviour mentioned above. One main effect is that the heat losses do not decrease proportionately to the component mass flows resulting in higher temperature drops inside the components and pipes. A second impact is the decline of heat transfer coefficients due to smaller mass flows. Variants 1 and 2 no longer remain in ranges above 30% but decrease towards 20% for high RR. Unlike full load operation, both configurations of variant 3 slightly exceed the DOR_{pre} levels of variants 1 and 2. This indicates that heat transfer at partial load in the additively manufactured exhaust gas heat exchangers is more effective than in the allothermal pre-reformer with conventional tube bundles. Variant 4 again yields the lowest DOR_{pre} levels with high FPM outlet temperatures at the same time (see Figure 11c).

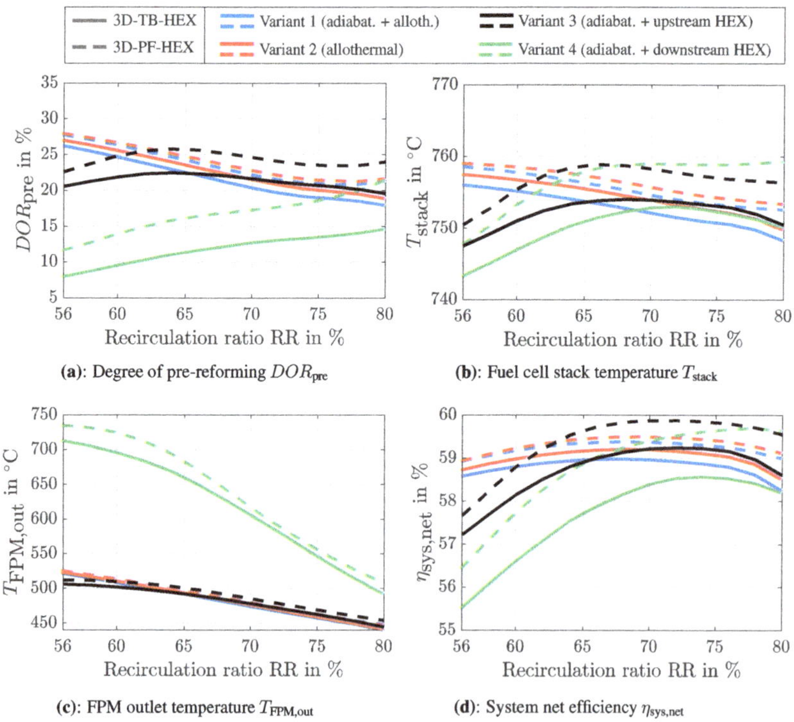

Figure 11. Simulation results for partial load operating points (I = 20 A) at maximum stack fuel utilization $FU_{\text{stack}} = 75\%$ as a function of the AOG recirculation ratio RR.

The resulting stack temperatures are depicted in Figure 11b. All variants show temperatures close to the minimum permissible limit of 750 °C with the adiabatic 3D-PF-HEX variants 3 and 4 yielding slightly higher temperatures at RR around 70%. The partial load efficiencies shown in Figure 11d are generally higher than the full load efficiencies as the cell voltage increases due to lower voltage losses. However, the depicted lower stack temperatures lead to higher ASR values partly counteracting this effect. The highest efficiency of nearly 60% is achieved by variant 3 using 3D-PF-HEX.

As a further load decrease towards 50% of the nominal electrical load would violate the stack temperature constraint, a heat integration countermeasure is investigated to increase the system operating range.

5.3. Measures for Increasing FPM Performance at Partial Load

In order to further increase the partial load capability of the FPM at lower electrical currents, higher degrees of pre-reforming are required. This is realised by increasing the hot mass flow in the allothermal pre-reformer or exhaust gas heat exchanger by means of additional supply of fuel to the oxidation unit being catalytically converted which is already part of the startup heating process. The oxidation unit is further controlled to a set point of 750 °C. Hereafter, only the 3D-PF-HEX variants are considered due to their better impact on FPM heat integration. The amount of additional fuel supplied to the oxidation unit compared to the conventional operation in the previous section is given by the oxidation unit surplus fuel ratio SFR_{oxi}:

$$\dot{n}_{\mathrm{fuel,oxi}} = SFR_{\mathrm{oxi}} \cdot \dot{n}_{\mathrm{fuel,in}} \quad , \tag{23}$$

relating the molar flow of fuel added to the oxidation unit to the molar flow supplied to the AOG recirculation. The system efficiency has to be adapted regarding the increased fuel input:

$$\eta_{\mathrm{sys,net,SFR}} = \frac{|P_{\mathrm{el,SOFC,DC}}| \cdot \eta_{\mathrm{pe}} - (P_{\mathrm{el,bl,ca}} + P_{\mathrm{el,bl,rec}} + P_{\mathrm{el,bl,oxi}})}{\dot{n}_{\mathrm{fuel,in}}(1 + SFR_{\mathrm{oxi}}) \cdot LHV_{\mathrm{fuel,in}}} \quad . \tag{24}$$

Simulations were conducted for the four 3D-PF-HEX variants at 20A partial load, maximum FU_{stack} and a recirculation ratio of 70%. The SFR_{oxi} was varied between 0% (equivalent to operating points depicted in Figure 11 at $RR = 70\%$) and 20%.

Figure 12a depicts the degree of pre-reforming as a function of the surplus fuel in the oxidation unit. An increase of FPM performance is visible for all four variants and stack temperatures rise accordingly (see Figure 12b). Up to an SFR_{oxi} of 5%, variants 1, 2 and 3 behave similarly by increasing to DOR_{pre} values of 33% which is equivalent to the levels achieved in full load operation (see Figure 10a). The behaviour changes as variant 3 converges to a maximum of 39% due to the heat transfer limitation in the exhaust gas heat exchanger. The allothermal variants continue to rise to levels above 50%. Variant 4 remains at rather low pre-reforming levels of 25%.

As displayed in Figure 12b, allothermal pre-reforming of variants 1 and 2 are able to lift the stack temperature to its set point leaving the OU_{stack} control limitation. The adiabatic variants only lead to a maximum increase of 22 K. Regarding the system efficiency, a surplus fuel ratio of 5% decreases the partial load efficiency by 1.5%, whereas a surplus fuel ratio of 15% leads to a decrease of 5% for allothermal variants. The adiabatic configurations decrease further to up to 6.5%-points as the ASR of the SOFC is higher due to lower stack temperatures.

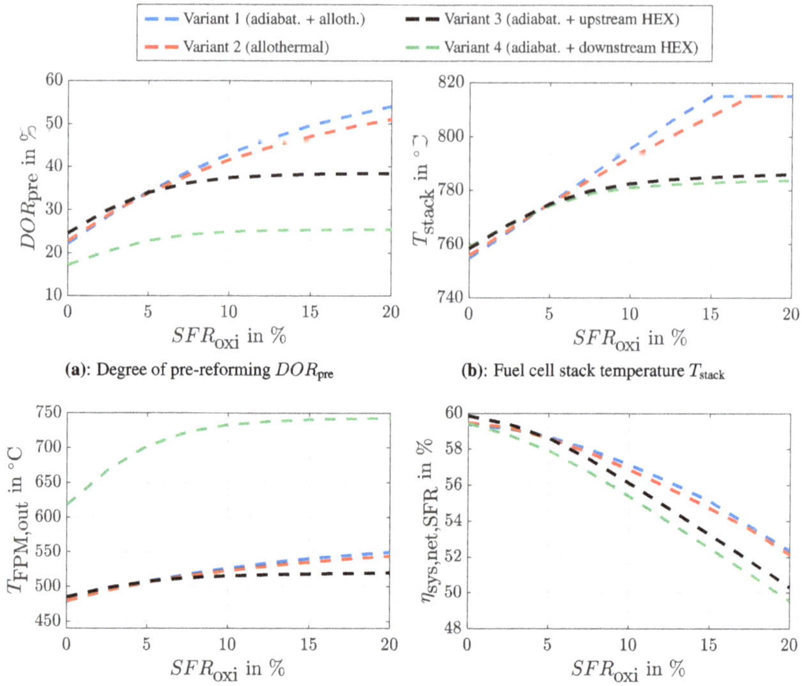

(**a**): Degree of pre-reforming DOR_{pre}

(**b**): Fuel cell stack temperature T_{stack}

(**c**): FPM outlet temperature $T_{FPM,out}$

(**d**): System net efficiency $\eta_{sys,net,SFR}$

Figure 12. Simulation results for partial load operating points with additional fuel supplied to the oxidation unit (I = 20 A) at maximum stack fuel utilization $FU_{stack} = 75\%$ as a function of the surplus fuel ratio SFR_{oxi}.

5.4. Discussion

In accordance with literature, allothermal AOG recirculation variants exhibit the best pre-reforming results regarding full load operation. Operating points with a high fuel utilization and a recirculation ratio of 70% yield the highest full load system efficiencies. A more sophisticated heat integration by means of more efficient heat transfer in the AOG Preheater enables higher degrees of pre-reforming, especially at higher recirculation ratios. All investigated variants are in principle suited for the depicted full load operating range. In terms of heat integration, variant 4 with a downstream exhaust gas heat exchanger is less suitable for the presented FPM configuration as more heat has to be withdrawn in the recirculation loop. Its pre-reforming performance is highly sensitive to heat losses and heat transfer effectiveness.

Regarding the risk of carbon formation in the pre-reformer stages, the operating range might be limited at lower RR. According to more profound equilibrium models found in literature, carbon formation is thermodynamically favourable at O/C values below 2.5 (RR of 66% at maximum FU_{stack}) at a temperature of 500 °C [32,34], which roughly corresponds to the simulated pre-reformer outlet temperatures at nominal full load operation (compare Figure 10c). However, this simplified comparison is limited, as (a) the equilibrium temperatures increase for lower RR to values up to 545 °C and (b) precious-metal catalysts exhibit a higher barrier towards carbon formation compared to Ni catalysts [49]. Eventually, the actual operating range limitation needs to be experimentally validated which will be reported in a future publication.

Regarding the system design criteria described in Section 2.2, the variants 1, 2 and 3 in combination with 3D-PF-HEX offer the advantage of a higher product safety and a less complex control. The AOG Cooler as a potentially critical component (if cooled with air) is not required for a large part of the full load operating range ($RR \leq 76\%$, see Figure 10b) and can therefore be removed. The recirculation temperature does not have to be controlled actively but may be monitored instead. If recirculation temperature reaches high levels, the recirculation ratio may be adapted to slightly lower values by means of the recirculation blower.

In this simulation study, partial load behaviour and its limitation due to thermal restrictions become more apparent than in other publications. This is due to the detailed simulation level of the comparatively high heat losses caused by the laboratory setup as well as the decreasing heat transfer coefficients at lower mass flows. As depicted in Figure 11, partial load is limited to roughly 60% of electrical net power at full load due to low stack temperature levels that can be slightly lifted by means of the more efficient 3D-PF-HEX. Both adiabatic variants show a slight advantage due to a higher heat transfer effectiveness compared to the conventionally built allothermal pre-reformer. A further temperature increase at all studied variants can be realized by means of additional fuel supply in the oxidation unit. As shown in Figure 12b, the allothermal variants 1 and 2 allow a generally higher amount of heat integration, which enables these variants to further decrease the electrical load at the price of decreasing system efficiencies. The thermal robustness and partial load capability are expected to increase substantially when the system is scaled up as described in Section 1.

The system model accuracy will be investigated in detail in a follow-up publication by means of experimental data of stationary full and partial load operating points. Based on the presented simulation results, the laboratory-scale system was planned, designed and built according to variant 1 due to the most efficient full load behavior and the ability to supply the highest amount of heat via additional fuel in the oxidation unit at partial load. Component integration was realized in such a way that a change to the other pre-reforming variants remains possible for further investigation by either removing or replacing the adiabatic or allothermal pre-reformer with an additional heat exchanger. The AOG Cooler is implemented into the recirculation loop so that the entire operating range remains accessible. The component may be removed if experimental results confirm the system behaviour.

Further process improvements regarding heat integration could be performed in future by applying the successful compact additive manufacturing designs to the allothermal pre-reformer. This would merge the higher heat transfer rates of allothermal pre-reforming with a higher heat transfer effectiveness enabling even lower partial load operation without supplying additional fuel to the oxidation unit. In addition, a more detailed 1D-stack model as shown in [44] would provide a more profound knowledge regarding thermal gradients inside the stack along the flow channel.

6. Conclusions

Within the framework of the *MultiSchIBZ* project, an LNG-fueled SOFC laboratory-scale system was simulated, designed and built to investigate the full and partial load behaviour. The objective of the presented work was to choose an efficient and thermally robust fuel processing configuration suitable for a given fuel cell module with a rated electrical power of 15 kW_{el}. The fuel processing module (FPM) was designed considering a scale-up concept for future applications supplying multiple fuel cell modules in parallel in a containerized setup on board of seagoing vessels.

Based on a wide range of fuel processing concepts shown in literature, an anode off gas recirculation concept with methane steam pre-reforming was selected to achieve high system efficiencies. A recirculation temperature of 300 °C was chosen based on the availability of commercial recirculation blowers.

Hydrocarbon-fueled SOFC systems with intermediate temperature anode recirculation and pre-reforming require a high level of heat integration to ensure sufficient pre-reforming temperatures and thus operability without the risk of carbon formation. To ensure this, a system simulation model in *Matlab/Simulink* was equipped with heat exchanger and pre-reformer models exhibiting heat transfer correlation functions to adequately map the mass flow and gas composition dependencies. High temperature heat exchangers were specifically designed and built for the desired application by means of additive manufacturing using selective laser melting. Two types of heat exchangers, a tube bundle cross counter flow (3D-TB-HEX) and a plate fin counter-flow (3D-PF-HEX) configuration, were experimentally tested in a separate test rig at flow conditions similar to the intended application in the FPM. Individual Nusselt correlations were deducted allowing reliable mapping of heat transfer.

Utilizing the obtained heat exchanger characteristics, a system simulation study was performed investigating heat integration of four system designs by means of the degree of pre-reforming, each with a setup of the 3D-TB-HEX or 3D-PF-HEX being used in the anode recirculation. As a key component regarding heat integration, the allothermal pre-reformer was designed as a 2D discretized model assuming chemical equilibrium in each cell, therefore not considering kinetics of the reforming and water-gas shift reaction.

Results show that allothermal variants deliver the highest degree of pre-reforming throughout the investigated range of anode recirculation ratios at nominal electric load. Using the counter-flow 3D-PF-HEX, heat exchange effectiveness and thus the level of overall heat integration is increased both in full and partial load compared to the 3D-TB-HEX. Highest net electrical efficiencies of 58 to 60% are obtained at maximum fuel stack utilization and an anode recirculation ratio of 70%. A suitable strategy to further increase the partial load capability has been presented by supplying a surplus fuel share of up to 5% to the catalytical afterburner which increases the FPM and stack outlet temperatures significantly at the cost of slightly lower net efficiencies.

In summary, system simulation has proven the applicability of the presented system configuration. As a consequence, the allothermal variant with an additional adiabatic upstream stage (called variant 1 in this study) was chosen to be built as a laboratory system for proof of concept and to support the scale-up design by means of model validation. The system was installed at the laboratory of Zentrum für BrennstoffzellenTechnik GmbH and equipped with an appropriate number of temperature and pressure sensors as well as a gas analysis infrastructure. The experimental results will be published in a follow-up publication.

Author Contributions: Conceptualization, J.H., M.F., C.S. and X.L.; methodology, J.H., M.F., C.S. and X.L.; software, J.H. and C.S.; formal analysis, J.H., M.F. and C.S.; investigation, J.H., M.F. and C.S.; writing—original draft preparation, J.H., M.F. and C.S.; visualization, J.H. and M.F.; writing—review and editing, supervision, project administration and funding acquisition, U.G., M.S. and S.K. All authors have read and agreed to the published version of the manuscript.

Funding: The authors gratefully acknowledge the financial support by the Federal Ministry of Transport and Digital Infrastructure, Germany (BMVI, funding code 03B10605H) and the coordination of the *MultiSchIBZ* project by the National Organisation Hydrogen and Fuel Cell Technology (NOW GmbH), Germany.

Acknowledgments: The authors gratefully thank Wolfgang Bender from Hülsenbusch Apparatebau GmbH & Co. KG and Philipp Schwarz from Rosswag GmbH for providing the two heat exchangers for experimental testing.

Conflicts of Interest: The authors declare no conflict of interest.

Abbreviations

The following abbreviations are used in this manuscript:

3D-HEX	additively manufactured heat exchanger
AE	auxiliary engine
AOG	anode off-gas
CPOX	catalytic partial oxidation

DIR	direct internal reforming
FCM	fuel cell module
FPM	fuel processing module
GHG	greenhouse gas
HEX	heat exchanger
IMO	International Maritime Organization
LHV	lower heating value
LNG	liquefied natural gas
MSR	methane steam reforming
MultiSchIBZ	Multiple SchiffsIntegration Brennstoffzellen (German: ship integration of multiple fuel cell (modules))
PF	plate fin
SchIBZ	SchiffsIntegration Brennstoffzellen (German: ship integration of fuel cell (modules))
SLM	selective laser melting
SNG	synthetic natural gas
SOFC	solid oxide fuel cell
TB	tube bundle
WGS	water gas shift

Nomenclature

Roman symbols

A	area [m^2]
ASR	area specific resistance [Ωcm^2]
$\dot{C}$	heat capacity rate [W K^{-1}]
C_r	ratio of heat capacity rates [-]
d_h	hydraulic diameter [m]
DOR_{pre}	degree of pre-reforming [%]
E_a	activation energy [J mol^{-1}]
F_{corr}	correction factor for cross-counter-flow [-]
FU_{stack}	stack fuel utilization [%]
h	heat transfer coefficient [W m^{-2}K^{-1}]
I	electrical current [A]
k	coefficient for component pressure loss [Pa s^2 kg^{-2}]
$\dot{m}$	mass flow [kg s^{-1}]
$\dot{n}$	molar flow [mol
NTU	number of transfer units [-]
Nu	Nusselt number [-] s^{-1}]
O/C	oxygen to carbon ratio [-]
OU	oxygen utilization [%]
p	pressure [Pa]
P_{el}	electrical power [W]
Pr	Prandtl number [-]
$\dot{Q}$	heat transfer rate [W]
Re	Reynolds number [-]
R_m	molar gas constant [J mol^{-1} K^{-1}]
RR	AOG recirculation ratio [%]
R_{wall}	thermal resistance of HEX wall [K W^{-1}]
SFR_{oxi}	Surplus fuel ratio (oxidation unit) [%]
T	temperature [°C]
u	velocity [m s^{-1}]
U	overall heat transfer coefficient [W m^{-2} K^{-1}]
x	molar fraction [-]

Greek symbols

α	proportional Nusselt coefficient [-]
β	exponential Nusselt coefficient [-]
ϵ	heat exchanger effectiveness [-]
η	efficiency [%]
λ	thermal conductivity [W m^{-1}K^{-1}]
ν	kinematic viscosity [m^{2}s^{-1}]
ψ	void fraction [-]

Subscripts

0	reference
amb	ambient
bl	blower
co	counter-flow
comp	component
cr	cross-flow
el	electrical
exp	experimental
fuel	fuel inlet flow
is	isentropic
lam	laminar
mech	mechanical
oxi	oxidation unit
pe	power electronics
pf	plate fin
pre	pre-reforming
rec	recirculation
sys	system
tb	tube bundle
th	theoretical
turb	turbulent

Appendix A. Reynolds- and Nusselt-Number Definitions

Appendix A.1. Definitions for Tube Bundle Heat Exchangers and Reformers

According to [36], the Reynolds number of the tube flow is defined as:

$$Re_{\text{tube}} = \frac{u_{\text{tube}}\, d_{\text{h,tube}}}{\nu_{\text{tube}}\, \psi_{\text{HR}}}, \tag{A1}$$

with the flow velocity u_{tube}, tube-side hydraulic diameter $d_{\text{h,tube}} = d_{\text{i}}$, kinematic viscosity ν_{tube} and void fraction $\psi_{\text{HR}} = 1$ for empty tubes and $\psi_{\text{HR}} = 0.4$ for tubes filled with catalyst bulk (ideal sphere packing).

The shell flow Reynolds-Number is defined analogously [36]:

$$Re_{\text{shell}} = \frac{u_{\text{shell}}\, d_{\text{h,shell}}}{\nu_{\text{shell}}\, \psi_{\text{shell}}}, \tag{A2}$$

with the hydraulic diameter $d_{\text{h,shell}} = \frac{\pi}{2}\, d_{\text{a}}$ and the void fraction ψ_{shell} for the shell side:

$$\psi_{\text{shell}} = 1 - \frac{\pi}{4a}, \tag{A3}$$

with the lateral pitch ratio a of the tube bundle geometry.

On the tube flow side, the Nusselt number Nu_{tube} is deducted from the combined correlations of the two heat transfer boundary conditions (constant wall temperature and constant heat flux) to determine the heat transfer coefficient h_{tube} [36]:

$$Nu_{\text{tube}} = \frac{h_{\text{tube}} \cdot d_{\text{h,tube}}}{\lambda_{\text{tube}}} = 4.01 + 0.00319 \cdot Re_{\text{tube}}^{0.911}. \tag{A4}$$

On the shell flow side, the Nusselt correlation for staggered tube-bundles is used [36]:

$$Nu_{\text{shell}} = \frac{h_{\text{shell}} \cdot d_{\text{h,shell}}}{\lambda_{\text{shell}}} = \left(0.3 + \sqrt{Nu_{\text{lam}}^2 + Nu_{\text{turb}}^2}\right) \cdot f_{\text{A}}, \tag{A5}$$

$$Nu_{\text{lam}} = 0.664 \cdot \sqrt{Re_{\text{shell}}}\,\sqrt[3]{Pr_{\text{shell}}}, \tag{A6}$$

$$Nu_{\text{turb}} = \frac{0.037 Re_{\text{shell}}^{0.8} Pr_{\text{shell}}}{1 + 2.443 Re_{\text{shell}}^{-0.1}\left(Pr_{\text{shell}}^{2/3} - 1\right)}, \tag{A7}$$

$$f_{\text{A}} = 1 + \frac{0.7}{\psi_{\text{shell}}^{1.5}} \frac{b/a - 0.3}{(b/a + 0.7)^2}, \tag{A8}$$

with the longitudinal pitch ratio b and the Prandtl number Pr_{shell}.

The coefficients to describe the Nusselt correlation for the 3D-TB-HEX used in Equation (4) are determined as follows: $\alpha_{c,TB} = 0.402$; $\beta_{c,TB} = 0.86$; $\alpha_{h,TB} = 0.234$; $\beta_{h,TB} = 0.75$.

Appendix A.2. Definitions for Plate-Fin Heat Exchangers

The definition of the Reynolds-Number for the plate-fin type heat exchanger is similar to the tube bundle heat exchangers but is typically defined using the mass velocity (or mass flux) $\dot{G}_{\text{PF}} = \dot{m}_{\text{PF}}/A_c$, with A_c as the free flow area in the finned part:

$$Re_{\text{PF}} = \frac{\dot{G}_{\text{PF}}\, d_{\text{h,PF}}}{\eta_{\text{PF}}}, \tag{A9}$$

with the hydraulic diameter $d_{\text{h,PF}} = \frac{2\, s_{\text{f}}\, h_{\text{f}}}{s_{\text{f}} + h_{\text{f}}}$ with s_{f} as fin spacing and h_{f} as fin height.

The coefficients to describe the Nusselt correlation for the 3D-PF-HEX used in Equation (4) are determined as follows: $\alpha_{c,PF} = \alpha_{h,PF} = 0.00126$; $\beta_{c,PF} = \beta_{h,PF} = 1.64$.

Appendix A.3. Stream Data at Nominal Full Load Operating Point

Stream properties		Cathode Air	Cathode Exh.	Fuel	Oxi Air	Oxi Exhaust
$\dot{n}$	mol/s	1.166	1.109	0.0308	0.113	0.200
T	°C	20.0	175.5	20.0	20.0	253.4
p	bar	1.013	1.036	5.0	1.013	1.028
x_{O_2}	mol%	21	17	0	21	9
x_{N_2}	mol%	79	83	0	79	45
x_{CH_4}	mol%	0	0	100	0	0
x_{H_2O}	mol%	0	0	0	0	31
x_{CO_2}	mol%	0	0	0	0	15

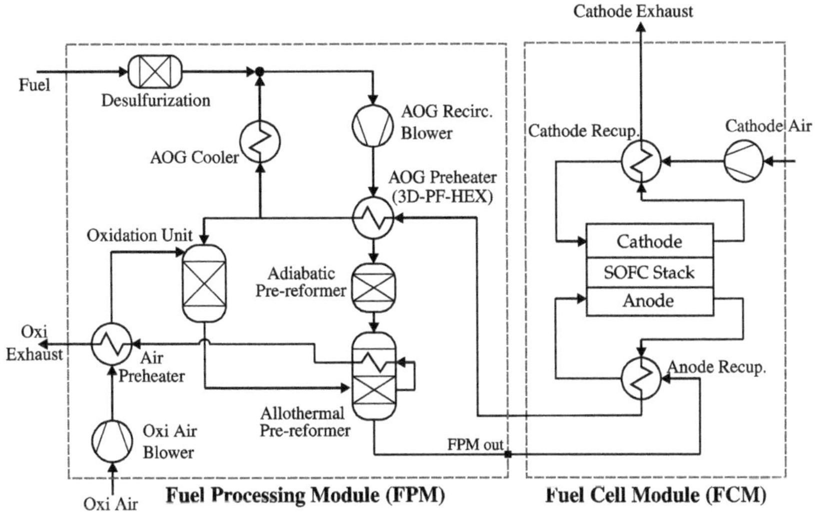

Figure A1. Properties of the system inlet and outlet streams at nominal full load operating point for variant 1 using the counter-flow 3D-PF-HEX.

References

1. International Maritime Organization. *Fourth Greenhouse Gas Study 2020*; International Maritime Organization: London, UK, 2020.
2. International Maritime Organization. *Adoption of the Initial IMO Strategy on Reduction of GHG Emissions from Ships and Existing IMO Activity Related to Reducing GHG Emissions in the Shipping Sector*; International Maritime Organization: London, UK, 2018.
3. Xing, H.; Spence, S.; Chen, H. A comprehensive review on countermeasures for CO_2 emissions from ships. *Renew. Sustain. Energy Rev.* **2020**, *134*, 110222. [CrossRef]
4. DNV GL AS Maritime. *Comparison of Alternative Marine Fuels*. 2019. Available online: https://sea-lng.org/wp-content/uploads/2020/04/Alternative-Marine-Fuels-Study_final_report_25.09.19.pdf (accessed on 18 December 2021).
5. Bicer, Y.; Dincer, I. Clean fuel options with hydrogen for sea transportation: A life cycle approach. *Int. J. Hydrogen Energy* **2018**, *43*, 1179–1193. [CrossRef]
6. European Commission. *2020 Annual Report on CO_2 Emissions from Maritime Transport: Report from the Commission*; European Commission: Brussels, Belgium, 2021. Available online: https://ec.europa.eu/clima/document/download/3eb00f74-c7ba-4dd1-8ea6-18f6f135053e_en (accessed on 18 December 2021).
7. Chu-Van, T.; Ristovski, Z.; Pourkhesalian, A.M.; Rainey, T.; Garaniya, V.; Abbassi, R.; Jahangiri, S.; Enshaei, H.; Kam, U.S.; Kimball, R.; et al. On-board measurements of particle and gaseous emissions from a large cargo vessel at different operating conditions. *Environ. Pollut. (Barking Essex 1987)* **2018**, *237*, 832–841. [CrossRef] [PubMed]
8. van Biert, L.; Godjevac, M.; Visser, K.; Aravind, P.V. A review of fuel cell systems for maritime applications. *J. Power Sources* **2016**, *327*, 345–364. [CrossRef]
9. Xing, H.; Stuart, C.; Spence, S.; Chen, H. Fuel Cell Power Systems for Maritime Applications: Progress and Perspectives. *Sustainability* **2021**, *13*, 1213. [CrossRef]
10. Leites, K. Fuel Cells for Marine Applications. In *Fuel Cells*; Stolten, P.D., Samsun, D.R.C., Garland, D.N., Stolten, D., Garland, N., Samsun, R.C., Eds.; Wiley-VCH: Weinheim, Germany, 2016; pp. 202–207. [CrossRef]
11. Bloom SOFC to power Korean LNG ship. *Fuel Cells Bull.* **2021**, *2021*, 8. [CrossRef]
12. Rathore, S.S.; Biswas, S.; Fini, D.; Kulkarni, A.P.; Giddey, S. Direct ammonia solid-oxide fuel cells: A review of progress and prospects. *Int. J. Hydrogen Energy* **2021**, *46*, 35365–35384. [CrossRef]

13. Barrett, S. (Ed.) METHAPU prototypes methanol SOFC for ships. *Fuel Cells Bull.* **2008**, 5, 4–5. [CrossRef]
14. Barrett, S. (Ed.) ShipFC project on first maritime fuel cell to run on green ammonia. *Fuel Cells Bull.* **2020**, 2, 5–6. [CrossRef]
15. SEA-LNG. LNG as a Marine Fuel—Our Zero Emissions Future Starts Now, 2020. Available online: https://sea-lng.org/wp-content/uploads/2020/02/200214_SEALNG2019reviewDIGITAL_compressed.pdf (accessed on 18 December 2021).
16. Nautical Integrated Hybrid Energy System for Long-Haul Cruise Ships. 2020. Available online: https://cordis.europa.eu/project/id/861647 (accessed on 18 December 2021).
17. Bloom Energy. Bloom Energy, Chantiers de l'Atlantique, and MSC Chart a Course for Cruise Ships Powered by Clean Energy. 2021. Available online: https://www.bloomenergy.com/news/bloom-energy-chantiers-de-latlantique-and-msc-chart-a-course-for-cruise-ships-powered-by-clean-energy (accessed on 18 December 2021).
18. Blue World and Alfa Laval develop marine methanol fuel cell unit. *Fuel Cells Bull.* **2021**, *2021*, 6. [CrossRef]
19. Nehter, P.; Wildrath, B.; Bauschulte, A.; Leites, K. Diesel Based SOFC Demonstrator for Maritime Applications. *ECS Trans.* **2017**, *78*, 171–180. [CrossRef]
20. Valadez Huerta, G.; Álvarez Jordán, J.; Dragon, M.; Leites, K.; Kabelac, S. Exergy analysis of the diesel pre-reforming solid oxide fuel cell system with anode off-gas recycling in the SchIBZ project. Part I: Modeling and validation. *Int. J. Hydrogen Energy* **2018**, *43*, 16684–16693. [CrossRef]
21. Valadez Huerta, G.; Álvarez Jordán, J.; Marquardt, T.; Dragon, M.; Leites, K.; Kabelac, S. Exergy analysis of the diesel pre-reforming SOFC-system with anode off-gas recycling in the SchIBZ project. Part II: System exergetic evaluation. *Int. J. Hydrogen Energy* **2018**, 44, 10916–10924. [CrossRef]
22. Kistner, L.; Bensmann, A.; Hanke-Rauschenbach, R. Optimal Design of Power Gradient Limited Solid Oxide Fuel Cell Systems with Hybrid Storage Support for Ship Applications. *Energy Convers. Manag.* **2021**, *243*, 114396. [CrossRef]
23. Kistner, L.; Schubert, F.L.; Minke, C.; Bensmann, A.; Hanke-Rauschenbach, R. Techno-economic and Environmental Comparison of Internal Combustion Engines and Solid Oxide Fuel Cells for Ship Applications. *J. Power Sources* **2021**, *508*, 230328. [CrossRef]
24. Choi, E.J.; Yu, S.; Kim, J.M.; Lee, S.M. Model-Based System Performance Analysis of a Solid Oxide Fuel Cell System with Anode Off-Gas Recirculation. *Energies* **2021**, *14*, 3607. [CrossRef]
25. Lee, K.; Kang, S.; Ahn, K.Y. Development of a highly efficient solid oxide fuel cell system. *Appl. Energy* **2017**, *205*, 822–833. [CrossRef]
26. Wu, C.C.; Chen, T.L. Design and dynamics simulations of small scale solid oxide fuel cell tri-generation system. *Energy Convers. Manag. X* **2019**, *1*, 100001. [CrossRef]
27. Arteaga-Perez, L.E.; Casas, Y.; Peralta, L.M.; Kafarov, V.; Dewulf, J.; Giunta, P. An auto-sustainable solid oxide fuel cell system fueled by bio-ethanol. Process simulation and heat exchanger network synthesis. *Chem. Eng. J.* **2009**, *150*, 242–251. [CrossRef]
28. Kupecki, J.; Badyda, K. Mathematical model of a plate fin heat exchanger operating under solid oxide fuel cell working conditions. *Arch. Thermodyn.* **2013**, *34*, 3–21. [CrossRef]
29. Kupecki, J. Off-design analysis of a micro-CHP unit with solid oxide fuel cells fed by DME. *Int. J. Hydrogen Energy* **2015**, *40*, 12009–12022. [CrossRef]
30. Powell, M.; Meinhardt, K.; Sprenkle, V.; Chick, L.; McVay, G. Demonstration of a highly efficient solid oxide fuel cell power system using adiabatic steam reforming and anode gas recirculation. *J. Power Sources* **2012**, *205*, 377–384. [CrossRef]
31. Mai, A.; Iwanschitz, B.; Weissen, U.; Denzler, R.; Haberstock, D.; Nerlich, V.; Schuler, A. Status of Hexis' SOFC Stack Development and the Galileo 1000 N Micro-CHP System. *ECS Trans.* **2011**, *35*, 87–95. [CrossRef]
32. Peters, R.; Deja, R.; Blum, L.; Pennanen, J.; Kiviaho, J.; Hakala, T. Analysis of solid oxide fuel cell system concepts with anode recycling. *Int. J. Hydrogen Energy* **2013**, *38*, 6809–6820. [CrossRef]
33. Engelbracht, M.F.A.; Peters, R.; Blum, L.; Stolten, D. Analysis of a Solid Oxide Fuel Cell System with Low Temperature Anode Off-Gas Recirculation. *J. Electrochem. Soc.* **2015**, *162*, F982. [CrossRef]
34. van Biert, L.; Visser, K.; Aravind, P.V. A comparison of steam reforming concepts in solid oxide fuel cell systems. *Appl. Energy* **2020**, *264*, 114748. [CrossRef]
35. Martin, H.; Gnielinski, V. Calculation of Heat Transfer from Pressure Drop in Tube Bundles. In Proceedings of the 3rd European Thermal Sciences Conference, Heidelberg, Germany, 10–13 September 2000; Hahne, E.W.P., Heidemann, W., Spindler, K., Eds.; Edizioni ETS: Pisa, Italy, 2000; pp. 1155–1160.
36. VDI-Gesellschaft Verfahrenstechnik und Chemieingenieurwesen (Ed.) *VDI-Wärmeatlas*, 10th ed.; Springer: Berlin/Heidelberg, Germany, 2013. [CrossRef]
37. Roetzel, W.; Luo, X.; Chen, D. *Design and Operation of Heat Exchangers and Their Networks*; Academic Press an Imprint of Elsevier: London, UK; San Diego, CA, USA; Cambridge, MA, USA; Oxford, UK, 2020.
38. Luo, X.; Zhou, G.Y.; Cong, L.W.; Fuchs, M.; Kabelac, S. Numerical Simulation of Heat Transfer and Fluid Flow in 3D-Printed High-Temperature Plate-Fin Heat Exchangers with OpenFoam. In Proceedings of the 25th National and 3rd International ISHMT-ASTFE Heat and Mass Transfer Conference (IHMTC-2019), Roorkee, India, 28–31 December 2019; Begellhouse: Danbury, CT, USA, 2019; pp. 569–574. [CrossRef]
39. Fuchs, M.; Heinrich, D.; Luo, X.; Kabelac, S. Thermal performance measurement of additive manufactured high-temperature compact heat exchangers. *J. Phys. Conf. Ser.* **2021**, *2116*, 012095. [CrossRef]
40. Kuppan, T. *Heat Exchanger Design Handbook*, 2nd ed.; Mechanical Engineering Series; CRC Press: Hoboken, NJ, USA, 2013; Volume 222.

41. *ISO 16903:2015: Petroleum and Natural Gas Industries–Characteristics of LNG, Influencing the Design, and Material Selection*; International Organization for Standardization: Geneva, Switzerland, 2015. Available online: https://www.iso.org/standard/57891.html (accessed on 18 December 2021).
42. Gubner, A. Non-Isothermal and Dynamic SOFC Voltage-Current Behavior. *ECS Proc. Vol.* **2005**, *2005-07*, 814–826. [CrossRef]
43. Walter, C.; Schwarze, K.; Boltze, M.; Herbrig, K.; Surrey, A. Status of Stack & System Development at Sunfire. In Proceedings of the 14th European SOFC & SOE Forum (EFCF), Lucerne, Switzerland, 20–23 October 2020; pp. 29–38.
44. van Biert, L.; Godjevac, M.; Visser, K.; Aravind, P.V. Dynamic modelling of a direct internal reforming solid oxide fuel cell stack based on single cell experiments. *Appl. Energy* **2019**, *250*, 976–990. [CrossRef]
45. Çengel, Y.A.; Ghajar, A.J. *Heat and Mass Transfer: Fundamentals and Applications*, 6th ed.; McGraw-Hill Education: New York, NY, USA, 2020.
46. Wongchanapai, S.; Iwai, H.; Saito, M.; Yoshida, H. Performance evaluation of a direct-biogas solid oxide fuel cell-micro gas turbine (SOFC-MGT) hybrid combined heat and power (CHP) system. *J. Power Sources* **2013**, *223*, 9–17. [CrossRef]
47. Saarinen, J.; Halinen, M.; Ylijoki, J.; Noponen, M.; Simell, P.; Kiviaho, J. Dynamic Model of 5kW SOFC CHP Test Station. *J. Fuel Cell Sci. Technol.* **2007**, *4*, 397–405. [CrossRef]
48. Sarantaridis, D.; Atkinson, A. Redox Cycling of Ni-Based Solid Oxide Fuel Cell Anodes: A Review. *Fuel Cells* **2007**, *7*, 246–258. [CrossRef]
49. Angeli, S.D.; Monteleone, G.; Giaconia, A.; Lemonidou, A.A. State-of-the-art catalysts for CH4 steam reforming at low temperature. *Int. J. Hydrogen Energy* **2014**, *39*, 1979–1997. [CrossRef]

energies

Article

Modeling of a Grid-Independent Set-Up of a PV/SOFC Micro-CHP System Combined with a Seasonal Energy Storage for Residential Applications

Rahaf S. Ghanem [1], Laura Nousch [2,*] and Maria Richter [2]

[1] Energy Engineering Department, German Jordanian University, Amman 11118, Jordan; rahafghanim0@gmail.com
[2] Fraunhofer IKTS, Fraunhofer Institute for Ceramic Technologies and Systems, 01277 Dresden, Germany; maria.richter@ikts.fraunhofer.de
* Correspondence: laura.nousch@ikts.fraunhofer.de; Tel.: +49-351-2553-7235

Abstract: Renewable energy sources based on solar and wind energy provide clean and efficient energy. The intermittent behaviour of these sources is challenging. At the same time, the needs for efficient, continuous and clean energy sources are increased for serving both electricity and thermal demands for residential buildings. Consequently, complimentary systems are essential in order to ensure a continuous power generation. One of the promising energy sources that helps in reducing CO_2 emissions, in addition to providing electrical and thermal energy efficiently, is a Solid Oxide Fuel Cell (SOFC) system operated in a combined heat and power (CHP) mode, due to high electrical efficiencies (in full and part load) and the fuel flexibility. Currently, most studies tend to focus on fuel cell model details with basic information about the building's energy requirements. Nevertheless, a deep understanding of integrating fuel cell micro-CHP systems with renewable energy systems for the residential sector is required. Moreover, it is important to define an operating strategy for the system with a specific controlling method. This helps in evaluating the performance and the efficiency of the building energy system. In this study, an investigation of different configurations of a hybrid power system (HPS) was carried out. The intended aim of this investigation was to optimize a HPS with minimal CO_2 emissions, serving the energy demands for a single-family house efficiently and continuously. As a result of this study, a photovoltaic (PV)/SOFC micro-CHP system has satisfied the intended goal, where the CO_2 emissions are significantly reduced by 88.6% compared to conventional systems. The SOFC micro-CHP plant operated as a complimentary back-up generator that serves the energy demands during the absence of the solar energy. Integrating the Power to Gas (PtG) technology leads to a similar emission reduction, while the PtG plant provided a seasonal energy storage. The excess energy produced during summer by the PV system is stored in the fuel storage for a later use (during winter). This SOFC micro-CHP configuration is recommended from an energy and environmental perspective. In terms of feasibility, the costs of SOFC based micro-CHP systems are significantly higher than traditional technologies. However, further technology developments and the effect of economy of scale may cause a substantial drop in costs and the micro-CHP shall become economically competitive and available for residential users; thus, enabling a self-sufficient and efficient energy production on site.

Keywords: residential energy systems; hybrid power system; SOFC; power to gas; annual simulation; micro-CHP system; CO_2 emissions

Citation: Ghanem, R.S.; Nousch, L.; Richter, M. Modeling of a Grid-Independent Set-Up of a PV/SOFC Micro-CHP System Combined with a Seasonal Energy Storage for Residential Applications. *Energies* **2022**, *15*, 1388. https://doi.org/10.3390/en15041388

Academic Editor: Orazio Barbera

Received: 20 December 2021
Accepted: 8 February 2022
Published: 14 February 2022

1. Introduction

The worldwide dilemma in today's increase of energy demand is progressing in parallel with the continued increase of fossil fuels demand such as natural gas (NG), crude oil and coal. The combustion of fossil fuels leads to increased emissions of greenhouse gases and noxious pollutants.

Renewable energy technologies, such as solar energy systems in its two types, the photovoltaic (PV) and the solar thermal system, helps to overcome the dilemma as they are assumed as efficient, sustainable and eco-friendly technologies. However, several constraints impede the use of renewable technologies for decentralized generation, mainly the fluctuating and intermittent behavior due to the high dependency on weather patterns. Accordingly, significant research efforts were made in the last decade in combining PV systems with backup power generation, energy storage systems, or both.

However, solar-based power generation has seasonal constraints depending on the installation site. Especially in colder countries, the solar power is available during summer times of low energy demand in the household. In winter times, the power and most importantly the heat consumption is much higher.

The above-mentioned constraints can be efficiently overcome by combining fuel cell (FC) systems to renewable energy technologies. In fuel cells, energy is produced by the direct conversion of the chemical energy into both electricity and useful thermal energy (depending on the type of fuel cell). Therefore, fuel cells offer a more efficient and clean way to produce energy compared to conventional power generation technologies. Fuel cell efficiencies are not limited by the Carnot efficiency as for engine-based conversion technologies, which is due to the avoided energy conversion stages towards heat.

Among fuel cells technologies, the Solid Oxide Fuel Cell (SOFC) is a promising technology compared to others, especially for small and medium power applications (up to 1 MW). The high operating temperature (600–900 °C) in the SOFC allows for the operation using different types of hydrocarbon fuels (e.g., natural gas, biogas, methanol, ammonia) [1–3]. Fueling SOFC systems using natural gas leads to around 54% reduction in CO_2 emissions and the removal of NOx and SOx emissions compared to the traditional fossil fueled energy systems [2]. SOFC systems provide high-temperature exhaust gas that can be utilized for space heating and domestic hot water (DHW) generation in buildings [3]. In addition, for that reason, the SOFC is promising for the residential sector and ideally suited for combined heat and power (CHP) production. By fueling the SOFC with natural gas, an electrical efficiency of 50–60% is reported, compared to 25–35% in Polymer Electrolyte Membrane (PEM) fuel cells [4]. In CHP systems, the full potential of the fuel, including the waste heat, is used, providing a high overall efficiency, which is due to the simultaneously generation of heat and electricity [5]. The electricity and heat demand for dwellings can be primarily covered in an efficient way, reducing fuel consumption and CO_2 emissions [5,6]. Shimoda and Taniguchi [7] compared the use of the cogeneration systems PEM, SOFC and gas engines for different household types in the Japanese residential sector, concluding that SOFC technology can reduce the CO_2 emissions significantly due to the possibility for the electricity dominated mode with a high part load capability and efficiency [8].

However, SOFC systems have limited part load capabilities. Due to the high temperatures, the system's dynamics are slow due to the high thermal inertia. Furthermore, thermal cycles cause power degradation in the cells. Therefore, research activities are ongoing that focus on improved operational strategies like the hot standby [9–12] and improved cycle stabilities.

Additionally, the solar energy constraints can be overcome with a new technology, which is power-to-gas (PtG) technology [13]. PtG technology contributes to pass over energy storing challenges of solar energy systems, which occur due to the seasonal differences of the solar radiation. By using the PtG technology, the surplus PV power is converted into a natural gas, which is compatible with the gas grid. The conversion occurs via two steps: H_2 production by water electrolysis and methane production by reacting H_2 with CO or CO_2 from an external source. By this, a link between the power grid and the natural gas grid can be provided. The resulting CH_4 from the methanation process, known as substitute natural gas (SNG), can be injected into the existing gas distribution grid (if available) or into gas storages or be used as fuel cells' fuel. The produced H_2 from electrolysis can be directly injected into the gas grid or stored in a special storage without passing through a methanation process. However, the amount of H_2 in the gas grid is limited by country specific standards and regulations to a maximum of 0–12 vol.% [14]. Moreover, when it comes to storing the fuel, methane is easier and safer to be stored than hydrogen. It is remarkable that the critical point of hydrogen (p_c = 1.3 MPa, T_c = 33.1 K) is much lower than the critical point of methane (p_c = 4.6 MPa, T_c = 190.6 K), and this is the reason behind the linear behavior the hydrogen gas properties with pressure, while it is non-linear for methane gas [14]. Considering the gas compression, compression of hydrogen requires more energy compared to methane since the lower heating value (LHV) is lower (241.85 MJ/K mol for hydrogen and 802.34 MJ/K mol for methane) and the required mass flow is higher [14]. Based on a recent report from UK's Institute of Mechanical Engineers, it was presented that the methanation process helps in decarbonization, while the carbon, usually carbon dioxide, is captured from an external source. Therefore, it helps in reducing air pollution and CO_2 emissions.

Most early studies, as well as current work, focus on fuel cell model details with basic information of the building energy requirements. However, focusing on the integration of fuel cell micro-CHP systems with renewable energy systems and residential is required. According to that comes the importance of this study, where it takes into consideration the plant and the residential as one object, while it identifies the interaction between the residential and the plant, especially the SOFC micro-CHP. This helps in enhancing the whole system and optimizing a suitable controlling strategy that determines the time of operation for each plant involved in the hybrid power system (HPS) to insure an efficient power generation process. This paper is composed of four sections; in this section, a background has been provided, and related work and the purpose of this study were considered. The methodological approach is described in Section 2. After that, the analysis follows in Section 3, before conclusions are drawn and an outlook is given in Section 4.

Literature Review

In the regard of combining PV–SOFC systems, different generation technologies are investigated for the residential sector to find the most suitable option concerning high energy efficiency and low CO_2 emissions. Sadeghi and Ameri [8] examine a PV/Battery system with different power generation systems, such as SOFC, micro gas turbine, gas generator and diesel generator. In this study, a multi-objective evolutionary algorithm was used to obtain the best solution of the generation system with solar panels and a battery system. The results of this study showed that the SOFC is the most promising power generation technology in the combination with solar panels and a battery. They found that the SOFC has the best economic efficiency, ecological compatibility and reliability.

PEM fuel cells play a role when combined with PV system and electrolyzer. Several studies investigate the coupling of PV systems with PEM fuel cells. Haddad, Ramadan and Khaled [15], for example, analyze coupling parameters for PEM fuel cells, which are the available PV area, the cost of investment and the amount of the produced power. The main drawback using PEM systems is the requirement towards the fuel, which is pure hydrogen in this case. Additionally, Ghnai and Bettayeb [16] present an optimized design and performance of an off-grid PV/PEM/diesel generator power system for a university building. The results of the study based on simulation and performance analysis showed a high renewable fraction of 66.1% of the proposed stand-alone hybrid renewable power system. In addition to that, the system is economically feasible (92 $/MWh) and environmentally friendly (24 $kgCO_2$/MWh). In another study [17], the authors proposed a complete self-sufficient hybrid power system comprised of PV, battery storage system and hydrogen fuel cell. They obtained the value of the recommended HPS from an economic perspective. The proposed HPS is accessible for commercial use. However, the initial cost is relatively high because of the high cost of the hydrogen storage tank. Nonetheless, SOFC can also be used in combination with PV system and electrolyzers. The SOFC has a very high electrical efficiency. If operated with pure hydrogen, an electrical efficiency of 52% are expected. Additionally, SOFC systems simultaneously generate electricity and heat, thus achieving very high overall efficiencies of up to 90% [4]. This may be important for applications in cold countries, thus supplying heating during cold seasons, but also for application in hot countries. Hosseini, Dincer and Rosen [18] suggest a SOFC/electrolysis system, where the produced steam of the SOFC system can be used for an absorption chiller to supply the cooling demand of the house during summer time. In the study of Haddad, Ramadan and Khaled [15], a green energy system is proposed based on coupling a fuel cell either with a PV system or a parabolic trough (PT) solar-thermal system for generating electricity through hydrogen storage and an electrolyzer. Mathematical models were used for PV panels, fuel cells, solar-thermal panels and the electrolyzer. The mathematical models are used to calculate the output power of each system and the produced amount of hydrogen. The selection was made based on different criteria which are: the annual total energy, the most secured system that supplies power greater than the minimum limit or the maximum power during a fixed period, where maximum power is needed for a determined period of time.

Solar energy can be harvested directly as in PV systems or indirectly as in solar thermal systems for generating electricity or for heating/cooling issues. Several studies investigate solar energy systems coupling with fuel cells considering PEM and SOFC.

2. Materials and Methods

In this work, the aims of modelling include the estimation of the environmental benefits in terms of CO_2 emissions, primary energy savings, operating costs and energy storage capacity reduction. Generally, two main approaches are followed for evaluating energy systems; either a simulation method [5] or an experimental method. A simulation method for a HPS in a domestic application is used in this work, where four different scenarios are evaluated and examined according to the targeted objectives of this work.

Within this section, the used inputs and assumptions for the HPS, such as load profiles, solar radiation data and CO_2 emissions formation, are presented in detail. The developed models of the various components which composed the HPS are listed. Then, the control strategy that defines the operation for each integrated plant in the HPS is clarified. The implementation of each simulation scenario is explained in detail as well.

2.1. Inputs and Assumptions

2.1.1. Assumptions

This section presented the assumed values considered in this study, as shown in Table 1.

Table 1. Assumptions used in this study for several parameters. Abbreviations: photovoltaic (PV), lower heating value (LHV).

Assumptions	
PV Panel's Optimal Angle (°)	30
Electrolyzer's Electrical to Fuel Efficiency (%)	75
Electrolyzer's Rated Power (kW)	5
Methanation Process Efficiency (%)	82
Methane Heating Value (MJ/kg)	50
Hydrogen LHV Enthalpy (kJ/mol)	−241
Solar-thermal System Efficiency (%)	75

2.1.2. Load Profiles

The interaction between the building and its energy load profiles is considered in the HPS of this study. A good approximation of how the system is operating is essential in order to be able to model a HPS in any simulation software. Accordingly, the performance of the building on a daily basis, which describes the electricity, space heating and DHW needs, should be defined. In this study, the building's performance is evaluated based on scaled heat and electricity load profiles according to VDI-4655 (Association of German Engineers) reference guidelines [19].

VDI-4655 guideline is defined to determine CHP-system's efficiencies that are designed to supply heat and electrical energy for a residential building. The VDI-4655 guideline is developed by forming a reference load profile based on measured data for electrical, space heating and DHW energy consumptions of several single-family houses located in Germany during ten typical day categories, as shown in Table 2. Each reference load profile represents the daily energy consumption in a one-minute average with a cumulative consumption of 1 kWh/day. The mentioned day-type categories are defined according to the air temperature, clouds and whether the considered day is a working day or a holiday [19].

Table 2. Different day-type categories according to VDI-4655 guidelines.

Season	Air Temperature (°C)	Day Category Abbreviations			
		Working Day		Holiday	
		Cloudy	Sunny	Cloudy	Sunny
Winter	$T_{Air} < 8.5$	WWC	WWS	WHC	WHS
Transition	$8.5 \leq T_{Air} \leq 22.8$	TWC	TWS	THC	THS
Summer	$T_{Air} > 22.8$	SWX		SHX	

Even though VDI-4655 reference load profiles are generated at German sites, they can be used for the Jordan case since this study is made for a single-family house with similar electrical appliances in the single-family house as in Germany. The difference in energy consumption between the case in Germany and Jordan is noticed by the daily demand (c.f. Table 3), which is due to the difference in the geographical location. In Jordan, more electrical energy is needed during summer due to the use of the cooling systems, while less heat energy is needed during winter times. Table 3 shows the daily electrical and thermal energy demands in different seasons.

Table 3. The daily electrical and thermal energy demands for the single-family house during summer, winter, and transition seasons. Abbreviations: domestic hot water (DHW).

Season	Number of Days during the Year	Electrical Demand (kWh/Day)	Space Heating Demand (kWh/Day)	DHW Demand (kWh/Day)
Summer	156	24	-	10
Transition	71	22	-	10
Winter	138	21	38	10

Figures 1–3 represent the scaled electricity, space heating and DHW energy load profile for the chosen single-family house of this study. The load profiles are for SHX, THC and WHC day categories, and the presented patterns follow the house activities.

In general, electricity demands are higher during the day times compared to the night times. In summer, the electricity demand is high during the morning, while it shifts to evening hours during winter days. The heat during summer and transition are only required for the DHW, where the peaks are during morning and evening hours, and where the heat demand for space heating during summer and transitions are zero.

The heat and electricity demand data (in hourly time steps) are used as inputs for the simulation models described below.

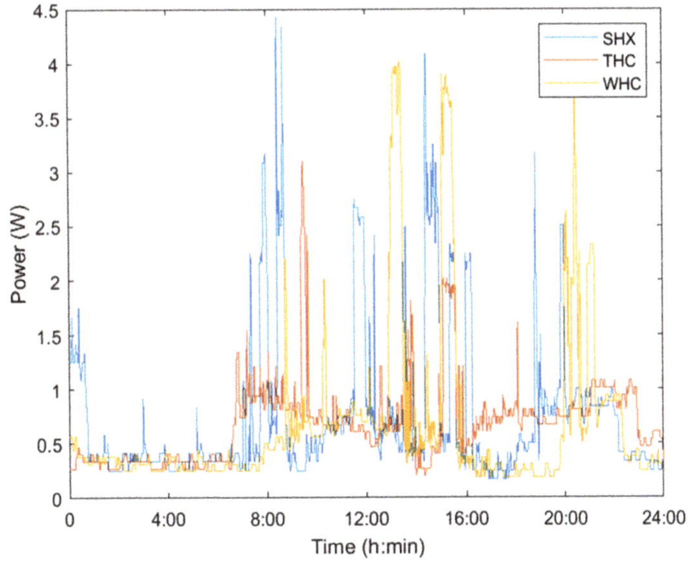

Figure 1. Daily electrical scaled load profile for the single-family house in Jordan during day types SHX, THC and WHC, scaled according to VDI-4655 guidelines.

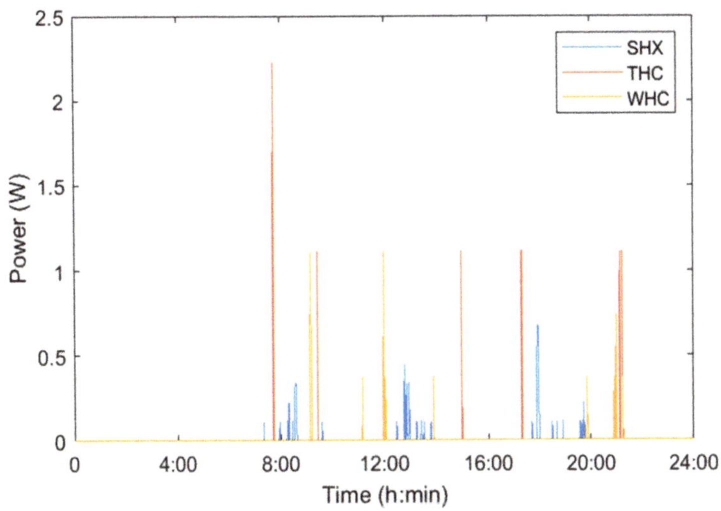

Figure 2. Daily DHW scaled load profile for the single-family house in Jordan during day types SHX, THC and WHC, scaled according to VDI-4655 guidelines.

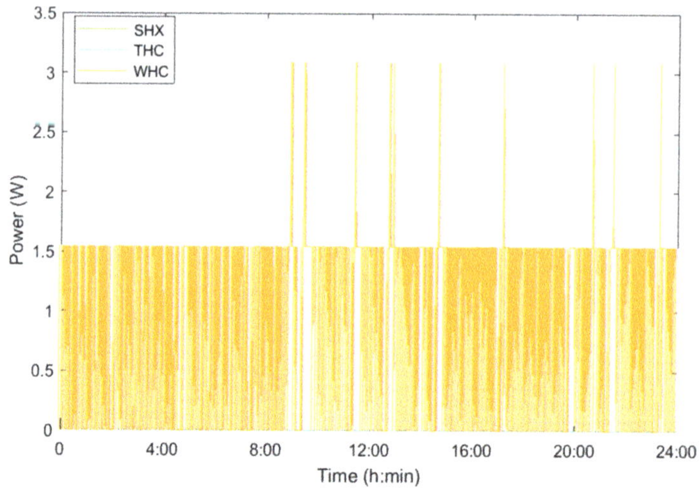

Figure 3. Daily space-heating scaled load profile for the single-family house in Jordan during day types SHX, THC and WHC, scaled according to VDI-4655 guidelines. Space-heating demand during THC and WHC day types is zero.

2.1.3. Solar Radiation Data

The solar radiation data is obtained from Photovoltaic Geographical Information System (PVGIS) tools [20]. The data represents a one-year hourly solar radiation of the location with a latitude and longitude of 31°57′47.3688″ N and 35°55′49.2924″ E (Amman, Jordan), respectively.

Figure 4 shows the solar radiation data for the location of this study. The annual average of solar radiation is estimated to be 5.9 kWh/(m^2, day).

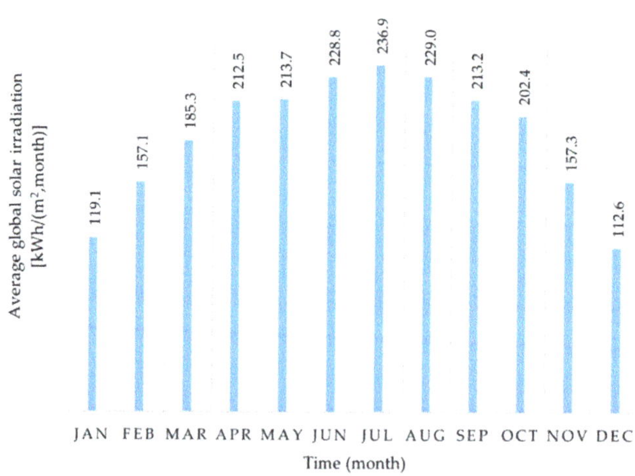

Figure 4. Average monthly global solar irradiation for the selected location with 31°57′47.3688″ N and 35°55′49.2924″ E latitude and longitude, respectively, obtained from PVGIS. Abbreviations: Photovoltaic Geographical Information System (PVGIS).

It is noticed that solar radiation for the selected location is high during the months of May, June, August and September, while it is the highest during July. In contrast, the solar irradiance is low in January, February, November and December.

2.1.4. CO_2 Emissions Calculations

The basic principle is to satisfy heat and electricity demands at all times. Heat demand is covered by the fuel-cell, the solar thermal system, and the diesel heater (DH), while electricity is covered by the fuel cell, the PV system, and the utility-grid. The aim of this study is to minimize the total CO_2 emissions z [5] produced from the operation of the house, where the fuel cell, the diesel heater and the utility-grid have the main impact.

$$z = g_g G_t + g_e E_t \tag{1}$$

where, G_t is calculated as follows [5]:

$$G_t = G_{t,FC} + G_{t,DH} = \frac{Q_{t,FC,\,elec}}{\eta_{FC,\,elec}} + \frac{Q_{t,\,DH,\,ther}}{\eta_{DH}} \tag{2}$$

The amount of produced CO_2 emissions from each source are presented in Table 4.

Table 4. The amount of the produced kgCO_2 from utility-grid, diesel fuel and methane fuel, evaluated for the Jordan case.

CO_2 Emissions Data	
Utility Grid	0.690 kg CO_2/kWh
Diesel Fuel	2.63 kg CO_2/L
Methane Fuel	2.7 kg CO_2/kgCH_4

2.2. *Simulation Models in Modelica*

The developed simulation model that combines the HPS simulation and the mathematical modelling is presented in this section. In this study, several submodels are included in the simulation model, such as the building energy model, the micro-CHP SOFC system model, the PV system model, the PtG system model, the solar thermal system model, and the energy storage systems model, as shown in Figure 5.

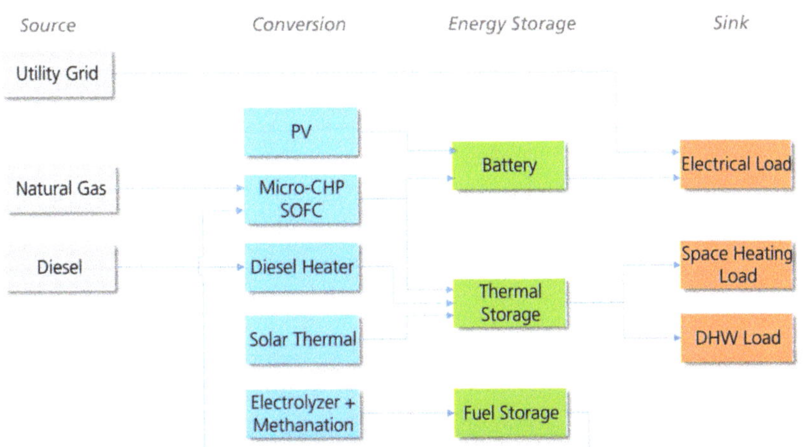

Figure 5. Hybrid power system (HPS) model showing all the integrated components to the system.

2.2.1. Building Energy Model

A single-family house is selected in this study and its energy consumption patterns are referred to VDI-4655 guidelines in order to create electricity, DHW and heat demand data for the building (regardless of the used heating system). The generated data represent the required inputs by the heating and electricity systems that are included in the HPS simulation model.

The selected single-family house located in Amman-Jordan covers a floor area of 180 m^2 with two stories and occupied by five persons. In the assumed reference scenario, the used space heating system is a conventional heater fueled by diesel, while the DHW is covered by a solar thermal system. The electricity source is dual for the reference scenario: 50% is covered by the utility grid and 50% is covered by a PV system.

2.2.2. Micro-CHP SOFC System Model

In this study, the micro-CHP SOFC system is modelled as a power generator fueled with natural gas. The system is modelled as a black box, which converts the fuel energy into electricity and useful heat pursuant to its efficiency. To avoid thermal cycles, a hot standby is assumed for the SOFC system. The operation of the fuel cell is constrained by the minimum and maximum SOC of the battery. The characteristics for the used fuel cell in this study are presented in Table 5.

Table 5. The main characteristics of the modelled fuel cell system.

Fuel Cell Characteristics	
Type	SOFC
Fuel Type	Methane
Rated Power (kW)	2.5
Electrical Efficiency (%)	40
Thermal Efficiency (%)	55
Fuel Utilization (%)	85
Controller Set Value	Battery SOC
SOFC Hot Stand-by Fuel Input (W)	800
SOFC System Internal Power Consumption (W)	30

2.2.3. Photovoltaic System Model

The PV system is sized considering the available area and the tilt angle of the PV-module at the selected location. The output power of the PV system is calculated in terms of the direct solar radiation at the selected location and the number of PV-modules [15]. The optimal tilt angle at the selected location is $\theta_{optimal} = 30°$.

The used PV module is the simplified model from the Modelica BuildingSystems Library [21], which produces the electrical power P_{MPP} for a MPP-controlled PV module, and has the characteristics shown in Table 6.

Table 6. PV-module characteristics according to TSM230PC05 from the Modelica BuildingSystems Library.

PV Module Characteristics	
Rated Power (W)	230
Open Circuit Voltage (V)	37
Short Circuit Current (A)	8.26
Dimensions L × W (m)	1.65 × 0.992

2.2.4. PtG System Model

In the PtG process, the surplus power is converted into gas, providing a link between the power grid and the gas grid. This conversion occurs via two steps: H_2 production by water splitting into H_2 and O_2, and methane production by reacting H_2 with CO or CO_2, which may result from the SOFC flue gas, as assumed in this study.

- Electrolyzer

The electrolyzer is sized based on the maximum DC power generated by the PV system in June (the highest solar radiation in Amman), adding to it 20% as a reserved capacity. The model of the electrolyzer is represented as a black box that converts the surplus electrical energy into fuel energy based on its efficiency. The efficiency of the electrolyzer is calculated as follows [22]:

$$\eta_{EZ} = \frac{\Delta H_{LHV_{H2}}}{U_{cell}} = \frac{1.25V}{U_{cell}} \quad (3)$$

The H_2 mass flow rate produced by electrolysis is calculated as follows [23]:

$$m_{H_2} = \frac{P_{EZ}\eta_{EZ}}{HHV_{H_2}} \quad (4)$$

The operation of the electrolyzer is constrained by the maximum SOC of the battery as the priority is to fully charge the battery. The used type of electrolyzer in this study is the polymer electrolyte membrane electrolyzer (PEME) with an electrical-to-fuel efficiency of 75% [24] and 5 kW rated power. The PEME has a quick start-up time that ranges from seconds to minutes [24] compared to the other types. The quick start up time allows a quick response for converting the surplus energy produced by the PV system without any delays. Using the SOFC in the reversible mode for electrolysis would be another option, which is neglected here due to the lack of technology maturity. As for today, long-term stable reversible Solid Oxide Cells (rSOC) systems are not available yet, but might be considered in future research.

- Methanation Process

The methanation process is modelled based on the given chemical reaction.

$$4H_2 + CO_2 \overset{yields}{\rightarrow} CH_4 + 2H_2O \quad (5)$$

The above chemical reaction indicates that each 1 kg of the produced H_2 by electrolysis needs around 5 kg of CO_2 to form 2 kg of CH_4. The formed CH_4 can be stored in the fuel storage to operate the fuel cell during the absence of the solar radiation. In this study, the considered methanation process is biological methanation (BM) with a hydrogen-to-fuel efficiency of 78% according to the reference [13].

2.2.5. Solar Thermal System Model

In this work, the solar thermal system is modelled as a unit that converts the solar power into thermal energy according to the solar radiation, the efficiency of the system and the total area of the system. As in the PV system model, the solar radiation data is obtained from the PVGIS at the location of the study and is used as an input for the solar thermal system. The efficiency of the solar thermal system is set to 75%.

2.2.6. Energy Storage System Model

Due to the intermittent behavior of the PV system and the solar thermal system, energy storages are essential to meet the load demand at night and during the low solar radiation. Different types of energy storages are integrated to the HPS model in this study: thermal storage, fuel storage and battery.

- Thermal Storage

In this work, the thermal storage model is represented by a hot water storage tank that is characterized based on the volume and energy content. The model is controlled by the maximum energy content and the energy flowing from the plant to the storage and from the storage to the single-family house. For the HPS in this study, the energy content of the thermal storage should serve both the space heating and the DHW demand. The heat energy required to maintain the storage temperature at 60 °C is calculated as follows [25]:

$$q = mc_p\Delta T \quad (6)$$

- Fuel Storage

Similar to the thermal storage, the model of the fuel storage is represented by a special storage tank characterized by the volume and the energy content. This storage is used to store the produced methane from the PtG process. The fuel should be stored under certain conditions in order to ensure a long term and a safe storing process (pressure and temperature). The storage is controlled by the maximum energy content, the heating value of methane fuel (50 MJ/kg), and the flow of energy from the PtG plant to the storage and from the storage to the SOFC plant.

- Battery

For the defined HPS, the battery bank is used as a backup device to overcome the weakness of the SOFC that has a slow dynamic response during sudden peak demands. The battery is also used to store the excess electrical power produced by the PV system. In this work, the battery bank simulation model calculates the actual state of charge (SOC) as follows [26]:

$$\mathrm{SOC} = 100\left(1 + \frac{\int I_D dt}{C}\right) \quad (7)$$

The available battery capacity is controlled by its maximum $P_{C,\ max}$ and minimum $P_{C,\ min}$ charging power and given as follows [26]:

$$P_{C,\ min} \leq \mathrm{SOC} \leq P_{C,\ max} \quad (8)$$

$$\mathrm{SOC} = (1 - \mathrm{DOD}) \quad (9)$$

where DOD is the depth of discharge for the battery bank. The discharging power P_D of the battery is constrained by the maximum hourly discharging power $P_{D,\ max}$ and given as [16]:

$$0 \leq P_D \leq P_{D,\ max} \quad (10)$$

2.3. Control Strategy

Two different proposed algorithms are used in the HPS of this study to indicate which subsystem is operating at each time step. The first algorithm is used to control the electrical part of the HPS and the controller is the battery bank. The second algorithm is used to control the thermal part of the HPS and the controller is the thermal storage. Figures 6 and 7 represent the flow charts of the proposed algorithms for the battery and the thermal storage.

2.4. Simulation Scenarios

In this study, four different simulation scenarios are modelled, as presented in Table 7. In each scenario, a plant is added/removed from the HPS model. All simulation scenarios are analyzed for a full year and for the proposed ten day-type categories mentioned in Section 2.1.2. Later, these scenarios are compared regarding the CO_2 emissions of the HPS, the capacity of the energy storages and the required land area. Exemplarily, Figure 8 represents a simplified schema of Scenario 4.

Scenario 1 refers to the reference system used in the single-family house; 50% of the electricity is served by the PV system and 50% by the utility grid, while the space heating demand is covered by the diesel heater and the DHW is served by the solar thermal system. Scenario 1 is taken as a reference for evaluating the other Scenarios (2, 3 and 4). Scenario 1 is optimized in Scenario 2 in terms of reducing the CO_2 emissions, where both the utility grid and the diesel heater are eliminated. Scenario 3 is then optimized in terms of the total land area by adding the SOFC micro-CHP system. The inclusion of the SOFC system in Scenario 3 causes a reduction in the required capacities of both the PV system and the solar thermal system since now it is covering part of the thermal and electrical demands. Scenario 4 is an optimization of Scenario 3 in terms of offering a seasonal energy storage and a reduction in the CO_2 by integrating the PtG technology to the HPS.

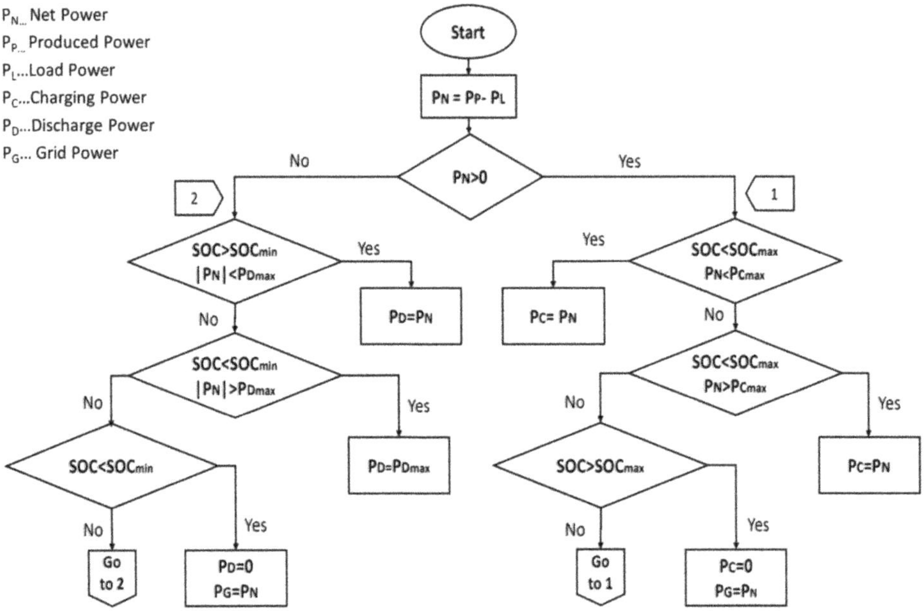

Figure 6. Flow chart of the proposed algorithm for the electrical part of the HPS controlled by the battery bank.

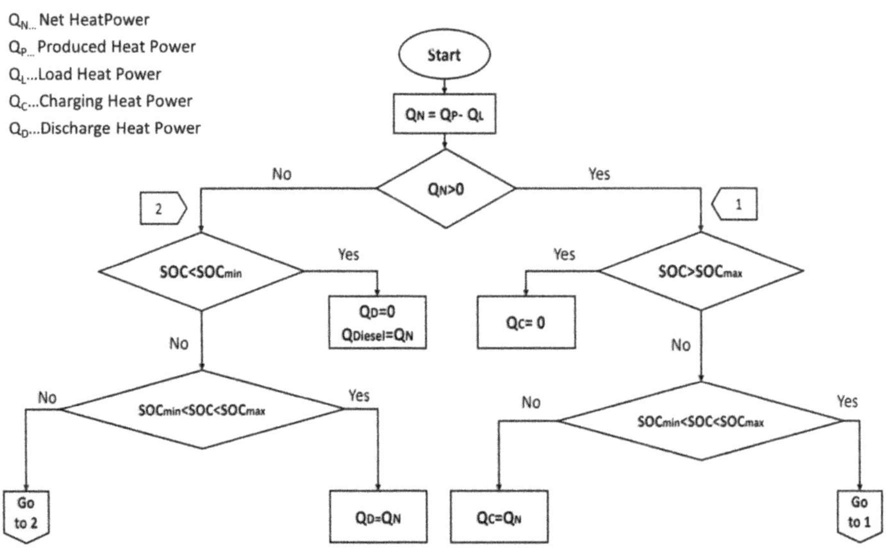

Figure 7. Flow chart of the proposed algorithm for the thermal part of the HPS controlled by the thermal storage.

Table 7. The four different developed simulation scenarios representing the included plants in each scenario.

	Electrical Power Source	Space Heating Source	DHW Source
Scenario 1 (reference)	PV System Utility Grid	Diesel Heater Thermal Storage	Solar Thermal System Thermal Storage
Scenario 2	PV System	Solar Thermal System Thermal Storage	Solar Thermal System Thermal Storage
Scenario 3	PV System Battery System SOFC System	SOFC System Thermal Storage	Solar Thermal System Thermal Storage
Scenario 4	PV System Battery System SOFC with PtG system	SOFC with PtG system Thermal Storage	Solar Thermal System Thermal Storage

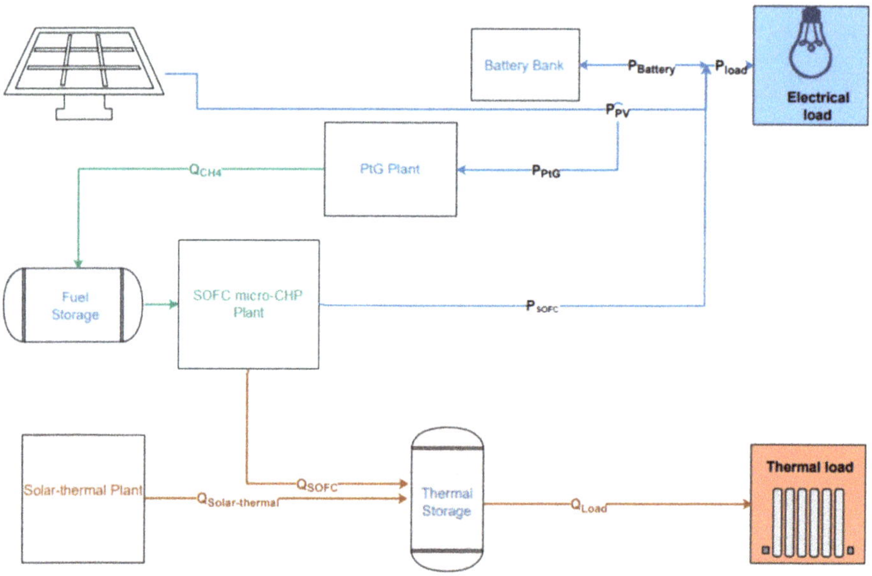

Figure 8. The HPS model with all the integrated components, Scenario 4.

Analyzing the models according to the ten day-type categories enables observation of the results for different time periods in addition to their variation with weather conditions and associated energy demand. Moreover, it helps in identifying how the HPS operates in essentially different day-type categories. The analysis of the winter day represents a high heat demand, while the summer day analysis represents a low heat demand with high electricity demand. In contrast, a full year analysis gives a design proposal that better shows the different requirements of the whole year and provides a general guideline on how the plant should operate.

3. Analysis and Results

In this section, the findings of the study concerning the needed dimension of plants and storages capacities, the total CO_2 emissions, and the total land area of the HPS are listed.

3.1. Energy Production

As shown in Table 8, the primary source for serving electricity is the PV system, while the SOFC micro-CHP operates as a back-up generator. The operation of SOFC micro-CHP system is dominated by the electrical demand, while the thermal energy is a by-product; if needed it will be stored in the thermal storage, otherwise, it dissipates to the surroundings. The produced thermal energy by the SOFC micro-CHP system is higher than the electrical energy due to the higher thermal efficiency. The excess electrical energy produced by the PV system is consumed by the electrolyzer in order to form a useful hydrogen fuel.

Table 8. Annual energy demands and simulated annual energy productions of the PV, the SOFC system, the utility grid, the solar thermal system, the diesel heater and the electrolyzer for the different simulation scenarios.

Annual Demand	**Electricity (kWh)**				**Thermal Load (kWh)**		
	8080				**8926**		
Annual Production	**PV System (kWh/Year)**	**Utility Grid (kWh/Year)**	**SOFC Electrical (kWh/Year)**	**SOFC Thermal (kWh/Year)**	**Solar Thermal System (kWh/Year)**	**Diesel Heater (kWh/Year)**	**Electrolyzer (kWh/Year)**
Scenario 1	3770	4300	-	-	3720	5640	-
Scenario 2	8280	-	-	-	30,650	-	-
Scenario 3	6020	-	2330	3520	21,780	-	-
Scenario 4	7540	-	3520	5130	21,820	-	−2810

Concerning the thermal part, the solar thermal system is the primary source for covering the thermal demand during the whole year. Due to the seasonal behavior of the solar energy, the solar thermal system produces higher thermal energy than the demand in some cases. In times of low solar power (particularly winter), a high capacity of solar thermal system is needed to ensure a full coverage for the thermal demand. This leads to a high amount of excess energy in times of high solar power (summer).

In Scenario 3, the inclusion of the SOFC micro-CHP system cancelled the need for both the diesel heater and the utility grid import.

In Scenario 3 and 4, the generated electrical energy by the SOFC system results in a needed amount of fuel of about 910 kg CH_4/year and 1090 kg CH_4/year, respectively. It is also noticed that the produced hydrogen by the PV system's surplus energy through the electrolyzer in Scenario 4 is about 70 kg H_2/year. Accordingly, in order to produce the above-mentioned amount of hydrogen, 740 L of water is required per year. Additionally, the formed methane fuel through the methanation process is about 120 kg CH_4/year.

3.2. Component Dimensions and CO_2 Emissions

As an intended aim of this work, the area of the HPS should be appropriate compared to the available area of the single-family house and the total CO_2 emission, and the battery capacity should be minimized as well. These criteria are satisfied in Scenario 4, according to the results listed in Table 9 and Figure 9.

Table 9. The total area in m^2 for each simulation scenario according to the number of PV-panels used and the solar-thermal system area.

	No. of PV Panels (Parallel/Series)	**PV Panel Area (m^2)**	**Solar Thermal System Area (m^2)**	**Total Area (m^2)**
Scenario 1	3/5	25	4	29
Scenario 2	3/11	55	30	83
Scenario 3	3/8	40	21	61
Scenario 4	3/10	50	21	71

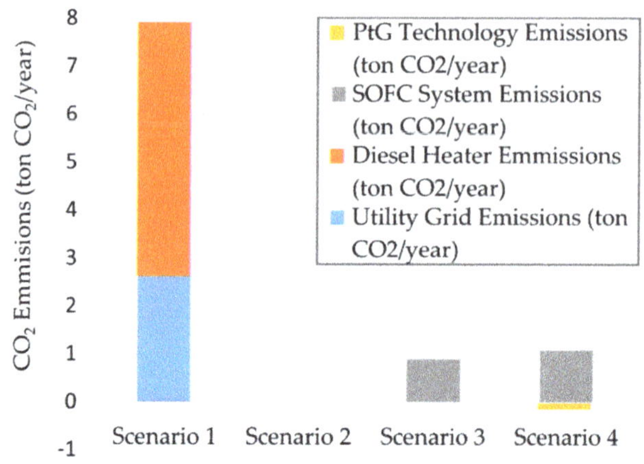

Figure 9. The total amount of the produced CO_2 emissions in each scenario according to the integrated system.

Figure 9 shows the total CO_2 emissions during operation for each scenario divided into the emissions caused by the utility grid, the diesel heater and the fuel cell. The effect of the diesel heater and the utility grid is noticeable when comparing the emissions in Scenario 1 to the other scenarios. The inclusion of the SOFC micro-CHP system in Scenario 3 results in a significant reduction in CO_2 emissions compared to the conventional power generators (utility grid and diesel heater). Furthermore, the integration of the PtG technology in Scenario 4 offers a seasonal energy storage, which is also a CO_2 sink. Summing up the CO_2 source and sink gives a total emission of 0.96 t CO_2/year, which is slightly higher than the CO_2 emission for Scenario 3 of 0.91 t CO_2/year.

The results obtained in Table 10 for the energy storages are based on the assumption that there are no losses to the surroundings. In Scenario 2, a huge capacity of the battery is required in order to compensate the seasonal behavior of the solar energy. As a result of adding the SOFC micro-CHP to the HPS, the needed capacity of the battery bank is reduced significantly; whereas, in Scenario 3, the SOFC compensated the seasonal behavior of the solar energy instead of the battery bank. Integrating the PtG technology with fuel storage to the HPS offers a long term and seasonal energy storage with a proper capacity compared to the battery bank. Figure 10 shows the behavior of both, the battery bank and the fuel storage, throughout the year, starting from the 1st of January.

Table 10. The needed storage capacities for the battery, the thermal storage and the fuel storage for each scenario.

	Battery Capacity (kWh)	Thermal Storage Volume (m^3)	Fuel Storage (kWh)
Scenario 1	-	1.2	-
Scenario 2	640	4.0	-
Scenario 3	25	0.75	-
Scenario 4	14	0.75	6

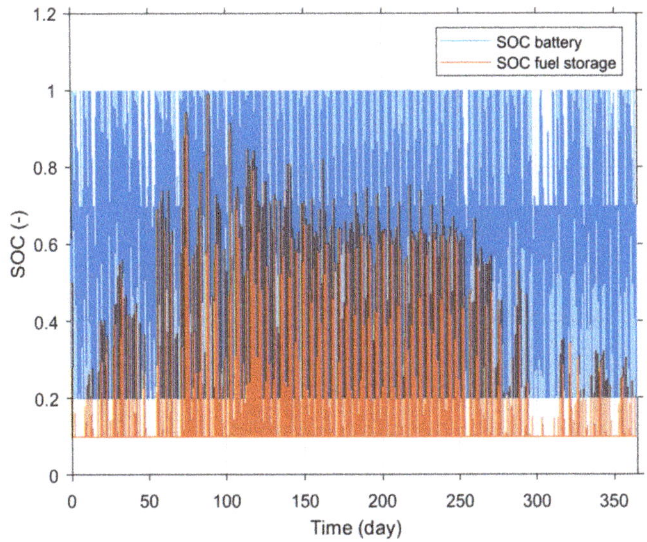

Figure 10. The seasonal behavior of the battery and the fuel storage in an annual basis starting from January in the Scenario 4.

4. Discussion

This study presented a HPS model for residential use analyzed in terms of the environmental benefits. Data sets for different day-type categories and weather scenarios were developed based on different configurations in order to determine the sizing characteristics and operation profiles for all the plants included in the HPS. The heat demand is covered by the solar thermal system, the fuel cell and the thermal storage, while the electricity demand is covered by the PV system, the fuel cell and the battery bank.

The importance of this study is that it takes in consideration the plant and the house as one object, which helps in optimizing the whole system. In addition, it identifies the interaction between the house and the plant, especially the SOFC micro-CHP. This led to the development of a proper controlling strategy that determines the operational time for each plant involved in the HPS for better and more efficient power generation. Using the annual simulations, the plant components can be dimensioned and the feasibility of the different scenarios can be proven.

The study showed that the HPS comprised of PV/SOFC micro-CHP provides CO_2 savings of 88.6% compared to the conventional designed system. The addition of a PtG plant provides CO_2 savings in the same range of 87.8%. The HPS with a PtG plant and methanation process in this study has a seasonal energy storage with an acceptable capacity of fuel storage. It shows that the excess energy produced by the PV-array during summer can be stored efficiently with low losses to be used during the low PV power production in winter. This helps in reducing the needed area for the PV panels compared to Scenario 2 and the battery bank capacity. Compared to Scenario 3, the PV area is slightly increased, while the battery capacity is lower and the SOFC system has a slightly higher operational time during the year. With these design numbers, it was found that the PtG plant helps in forming 12% of the needed fuel to operate the SOFC for the residential user by capturing the excess energy from the PV system.

From the author's perspective, Scenario 3 and 4 are promising options for future residential buildings. This study makes clear that SOFC technology can help to make households grid independent. To handle the limited cycle stability, the hot standby operation is assumed. The self-sufficient energy supply for the building has the chance to unburden the electricity grid in case of unstable grid conditions (e.g., peak demands) and in regions with unstable grids. Residential building can also be supplied in remote areas. Assuming Scenario 3 and 4, a grid-independent power supply can be realized with limited area requirements compared to Scenario 2. The PtG technology helps to balance the strongly seasonal characteristic of renewable energies, such as the solar power in this case, with reduced battery capacities compared to Scenario 2 (−98%), and also compared to Scenario 3 (−44%). However, higher fuel inputs are required in Scenario 4 since the conversion losses of the PtG system need to be compensated. Nevertheless, the PtG technology provides the option for the production of fuel on site with the potential of being self-sufficient and independent of external fuel supply, by increasing the PV area, the battery capacity, the PtG system and the fuel storage. This option might be more and more interesting for different applications in the future.

Comparing the amounts of produced and consumed heat, in Scenario 2, 3 and 4, a high optimization potential is obvious, as more than half of the produced heat is not used. Seasonal thermal storages could be a solution to balance the differences in production and consumption, which is not considered here. Furthermore, the controlling strategy of the SOFC system might be adapted, e.g., considering heat-controlled or a combined electrical and heat-controlled operation depending on the individual scenario. Additionally, the overall interaction between the house and the plant could be optimized by sophisticated algorithms in terms of model predictive controls or others.

However, the more components are integrated in the HPS, the more Capital Expenditures (CAPEX) are added. Based on a rough estimation, the cost for the defined HPS in Scenario 4 is around 66,000$. In comparison to this, equipment costs of 70,000$ are expected for an HPS of Scenario 2. More costs details are presented in Table 11.

Table 11. The expected CAPEX of the HPS at each scenario in this study, in addition to the costs per each equipment. Abbreviations: Capital Expenditures (CAPEX).

	Total CAPEX ($)	PV System CAPEX ($/kW$_P$)	SOFC CAPEX ($/kW)	Solar Thermal CAPEX ($/m^2)	Electrolyzer CAPEX ($/kW)	Battery CAPEX ($/kWh)
Scenario 1	11,000	740 [27]	3000–4000 [28]	2000	1800 [29]	700–1000
Scenario 2	71,000					
Scenario 3	59,000					
Scenario 4	68,000					

As a next step, detailed cost models can be developed based on the CAPEX and the OPEX, which results from the simulation, to estimate the overall cost of the different scenarios. This will also be interesting with assumed future equipment costs and subsidies, as well as future fuel and energy cost, as well costs of emissions.

In this study, the CO_2 emissions during the lifetime of the components is analyzed. Additionally, a life cycle assessment (LCA) should be done for the proposed HPSs, proving the overall environmental impact over the whole life cycle, such as green house warming potential. However, solar power systems, SOFCs and PtG systems are considered as key technologies to reach the goal of a low carbon economy, by reducing CO_2 emissions and limiting the production of other pollutants during operation.

Author Contributions: Conceptualization, L.N.; methodology, L.N. and M.R., software, R.S.G.; validation, R.S.G., L.N. and M.R.; formal analysis, L.N. and M.R.; investigation, R.S.G.; data curation, R.S.G.; writing—original draft preparation, R.S.G.; writing—review and editing, L.N. and M.R.; visualization, R.S.G.; supervision, L.N.; project administration, L.N. All authors have read and agreed to the published version of the manuscript.

Funding: This research received no external funding.

Institutional Review Board Statement: Not applicable.

Informed Consent Statement: Not applicable.

Conflicts of Interest: The authors declare no conflict of interest.

Nomenclature

Δ	Difference	-
C	Battery Capacity	Ah
c_p	Specific Heat	kJ/kg·k
E_t	Electricity Consumption Per Time Step	kW
g_e	Grid Electricity Emissions Factor	Kg CO_2/kWh
g_g	Natural Gas Emission Factor	Kg CO_2/kWh
G_t	Total Gas Consumption Per Time Step	kW
$G_{t,DH}$	Diesel Heater Gas Consumption Per Time Step	kW
$G_{t,FC}$	Fuel Cell Gas Consumption Per Time Step	kW
H	Enthalpy	kJ/mol
I_{SC}	Short Circuit Current	A
k	Energy Density	J/m^3
m	Mass	kg
$\dot{m}$	Mass Flow Rate	kg/s
η	Efficiency	-
$N_{PV,Parallel}$	Number of PV Modules Connected in Parallel	-
$N_{PV,Series}$	Number of PV Modules Connected in Series	-
P	Electrical Power	kW
Q	Heat Power	kW
q	Heat Energy	J
T	Temperature	°C
U_{Cell}	Cell Potential	V
V_{OC}	Open Circuit Voltage	V
z	Total CO_2 Emissions	Kg CO_2

Subscripts

C	Charging
D	Discharging
DC	Direct Current
DH	Diesel Heater
elec	Electrical
EZ	Electrolyzer
FC	Fuel Cell
G	Grid
L	Load
max	Maximum
min	Minimum
N	Net
P	Production
t	Time step
ther	Thermal

Abbreviations

CAPEX	Capital Expenditures
CH_4	Methane
CHP	Combined Heat and Power
CO	Carbon monoxide
CO_2	Carbon dioxide
DC	Direct Current
DHW	Domestic Hot Water
DOD	Depth of Discharge
FC	Fuel Cell
H_2	Hydrogen

H_2O	Water
HHV	Higher Heating Value
HPS	Hybrid Power System
LCA	Life Cycle Assessment
LHV	Lower Heating Value
MPP	Maximum Power Point
NG	Natural Gas
NO_X	Nitrogen oxides
OPEX	Operational Cost
PEM	Polymer Electrolyte Membrane
PEME	Polymer Electrolyte Membrane Electrolyzer
PtG	Power to Gas
PV	Photovoltaic
PVGIS	Photovoltaic Geographical Information System
rSOC	Reversible Solid Oxide Cell
SHX	Summer Holiday day type
SNG	Substitute Natural Gas
SOC	State of Charge
SOFC	Solid Oxide Fuel Cell
SO_X	Sulfur oxides
SWX	Summer Working day type
THC	Transition Holiday Cloudy day type
THS	Transition Holiday Sunny day type
TWC	Transition Working Cloudy day type
TWS	Transition Working Sunny day type
VDI	Verein Deutscher Ingenieure
WHC	Winter Holiday Cloudy day type
WHS	Winter Holiday Sunny day type
WWC	Winter Working Cloudy day type
WWS	Winter Working Sunny day type

References

1. Tuyen, N.D.; Fujita, G.; Duc, N.; Fujit, G. Modelling a SOFC Power Unit Using Natural Gas Fed Directly: 20. In *Advances in Natural Gas Technology*; Al-Megren, H., Ed.; IntechOpen: Rijeka, Croatia, 2012.
2. Naimaster, E.J., IV; Sleiti, A.K. Potential of SOFC CHP systems for energy-efficient commercial buildings. *Energy Build.* **2013**, *61*, 153–160. [CrossRef]
3. Zhang, L.; Xing, Y.; Xu, H.; Wang, H.; Zhong, J.; Xuan, J. Comparative study of solid oxide fuel cell combined heat and power system with Multi-Stage Exhaust Chemical Energy Recycling: Modeling, experiment and optimization. *Energy Convers. Manag.* **2017**, *139*, 79–88. [CrossRef]
4. Ellamla, H.R.; Staffell, I.; Bujlo, P.; Pollet, B.G.; Pasupathi, S. Current status of fuel cell based combined heat and power systems for residential sector. *J. Power Sources* **2015**, *293*, 312–328. [CrossRef]
5. Adam, A.; Fraga, E.S.; Brett, D.J. Modelling and Optimisation in Terms of CO_2 Emissions of a Solid Oxide Fuel Cell based Micro-CHP System in a Four Bedroom House in London. *Energy Procedia* **2013**, *42*, 201–209. [CrossRef]
6. Frenzel, I.; Loukou, A.; Trimis, D.; Schroeter, F.; Mir, L.; Marin, R.; Egilegor, B.; Manzanedo, J.; Raju, G.; De Bruijne, M.; et al. Development of an SOFC based Micro-CHP System in the Framework of the European Project FC-DISTRICT. *Energy Procedia* **2012**, *28*, 170–181. [CrossRef]
7. Shimoda, Y.; Taniguchi-Matsuoka, A.; Inoue, T.; Otsuki, M.; Yamaguchi, Y. Residential energy end-use model as evaluation tool for residential micro-generation. *Appl. Therm. Eng.* **2017**, *114*, 1433–1442. [CrossRef]
8. Sadeghi, S.; Ameri, M. Comparison of different power generators in PV-battery-power generator hybrid system. *J. Mech. Sci. Technol.* **2014**, *28*, 387–398. [CrossRef]
9. Nousch, L.; Hartmann, M.; Michaelis, A. Improvements of Micro-CHP SOFC System Operation by Efficient Dynamic Simulation Methods. *Processes* **2021**, *9*, 1113. [CrossRef]
10. Hanasaki, M.; Uryu, C.; Daio, T.; Kawabata, T.; Tachikawa, Y.; Lyth, S.M.; Shiratori, Y.; Taniguchi, S.; Sasaki, K. SOFC Durability against Standby and Shutdown Cycling. *J. Electrochem. Soc.* **2014**, *161*, F850–F860. [CrossRef]
11. Xiao-Long, W.; Yuanwu, X.; Tao, X.; Dongqi, Z.; Jianhua, J.; Zhonghua, D.; Xiaowei, F.; Xi, L. Standby and Shutdown Cycles Modeling of SOFC Lifetime Prediction. *Energy Procedia* **2019**, *158*, 1573–1578. [CrossRef]

12. Wu, X.-L.; Xu, Y.-W.; Xue, T.; Zhao, D.-Q.; Jiang, J.; Deng, Z.; Fu, X.; Li, X. Health state prediction and analysis of SOFC system based on the data-driven entire stage experiment. *Appl. Energy* **2019**, *248*, 126–140. [CrossRef]
13. Götz, M.; Lefebvre, J.; Mörs, F.; McDaniel Koch, A.; Graf, F.; Bajohr, S.; Reimert, R.; Kolb, T. Renewable Power-to-Gas: A technological and economic review. *Renew. Energy* **2016**, *85*, 1371–1390. [CrossRef]
14. Tietze, V.; Stolten, D. Comparison of hydrogen and methane storage by means of a thermodynamic analysis. *Int. J. Hydrogen Energy* **2015**, *40*, 11530–11537. [CrossRef]
15. Haddad, A.; Ramadan, M.; Khaled, M.; Ramadan, H.; Becherif, M. Study of hybrid energy system coupling fuel cell, solar thermal system and photovoltaic cell. *Int. J. Hydrogen Energy* **2018**, *45*, 13564–13574. [CrossRef]
16. Ghenai, C.; Bettayeb, M. Modelling and performance analysis of a stand-alone hybrid solar PV/Fuel Cell/Diesel Generator power system for university building. *Energy* **2019**, *171*, 180–189. [CrossRef]
17. Lokar, J.; Virtič, P. The potential for integration of hydrogen for complete energy self-sufficiency in residential buildings with photovoltaic and battery storage systems. *Int. J. Hydrogen Energy* **2020**, *45*, 34566–34578. [CrossRef]
18. Hosseini, M.; Dincer, I.; Rosen, M.A. Hybrid solar–fuel cell combined heat and power systems for residential applications: Energy and exergy analyses. *J. Power Sources* **2012**, *221*, 372–380. [CrossRef]
19. VDI. *Reference Load Profiles of Single-Family and Multi-Family Houses for the Use of CHP Systems VDI-4655*; Beuth: Berlin, Germany, 2008.
20. European Commission, Joint Research Centre Photovoltaic Geographical Information System (PVGIS). Available online: https://ec.europa.eu/jrc/en/pvgis (accessed on 4 May 2020).
21. Fritzson, P. *Introduction to Modeling and Simulation of Technical and Physical Systems with Modelica*; John Wiley & Sons: Hoboken, NJ, USA, 2011. [CrossRef]
22. Scheepers, F.; Stähler, M.; Stähler, A.; Rauls, E.; Müller, M.; Carmo, M.; Lehnert, W. Improving the Efficiency of PEM Electrolyzers through Membrane-Specific Pressure Optimization. *Energies* **2020**, *13*, 612. [CrossRef]
23. Ghenai, C.; Bettayeb, M.; Brdjanin, B.; Hamid, A.K. Hybrid solar PV/PEM fuel Cell/Diesel Generator power system for cruise ship: A case study in Stockholm, Sweden. *Case Stud. Therm. Eng.* **2019**, *14*, 100497. [CrossRef]
24. Kosonen, A.; Koponen, J.; Huoman, K.; Ahola, J.; Ruuskanen, V.; Ahonen, T.; Graf, T. Optimization strategies of PEM electrolyser as part of solar PV system. In Proceedings of the 2016 18th European Conference on Power Electronics and Applications (EPE'16 ECCE Europe), Karlsruhe, Germany, 5–9 September 2016; pp. 1–10. [CrossRef]
25. Sarbu, I.; Sebarchievici, C. A Comprehensive Review of Thermal Energy Storage. *Sustainability* **2018**, *10*, 191. [CrossRef]
26. Chettibi, N.; Mellit, A. Intelligent control strategy for a grid connected PV/SOFC/BESS energy generation system. *Energy* **2018**, *147*, 239–262. [CrossRef]
27. Hilpert, S.; Dettner, F.; Al-Salaymeh, A. Analysis of Cost-Optimal Renewable Energy Expansion for the Near-Term Jordanian Electricity System. *Sustainability* **2020**, *12*, 9339. [CrossRef]
28. Cigolotti, V.; Genovese, M.; Fragiacomo, P. Comprehensive Review on Fuel Cell Technology for Stationary Applications as Sustainable and Efficient Poly-Generation Energy Systems. *Energies* **2021**, *14*, 4963. [CrossRef]
29. Kreidelmeyer, S. *Kosten und Transformationspfade für Strombasierte Energieträger: Studie im Auftrag des Bundesministeriums für Wirtschaft und Energie*; BMWI: Berlin, Germany, 2020.

energies

Article

Dynamic Modeling of a PEM Fuel Cell Power Plant for Flexibility Optimization and Grid Support

Elena Crespi [1,*], Giulio Guandalini [1], German Nieto Cantero [2] and Stefano Campanari [1]

[1] Group of Energy Conversion Systems, Department of Energy, Politecnico di Milano, 20156 Milano, Italy; giulio.guandalini@polimi.it (G.G.); stefano.campanari@polimi.it (S.C.)
[2] Abengoa Innovación, 41014 Seville, Spain; german.nieto@abengoa.com
* Correspondence: elena.crespi@polimi.it

Abstract: The transition toward high shares of non-programmable renewable energy sources in the power grid requires an increase in the grid flexibility to guarantee grid reliability and stability. This work, developed within the EU project Grasshopper, identifies hydrogen Fuel Cell (FC) power plants, based on low temperature PEM cells, as a source of flexibility for the power grid. A dynamic numerical model of the flexible FC system is developed and tested against experimental data from a 100-kW pilot plant, built within the Grasshopper project. The model is then applied to assess the flexible performance of a 1 MW system in order to optimize the scale-up of the pilot plant to the MW-size. Simulations of load-following operation show the flexibility of the plant, which can ramp up and down with a ramp rate depending only on an externally imposed limit. Warm-up simulations allow proposing solutions to limit the warm-up time. Of main importance are the minimization of the water inventory in the system and the construction of a compact system, which minimizes the distance between the components.

Keywords: MW-scale PEM FC; partial load; warm-up; Grasshopper project; scale-up; flexible FC

Citation: Crespi, E.; Guandalini, G.; Nieto Cantero, G.; Campanari, S. Dynamic Modeling of a PEM Fuel Cell Power Plant for Flexibility Optimization and Grid Support. *Energies* **2022**, *15*, 4801. https://doi.org/10.3390/en15134801

Academic Editor: Orazio Barbera

Received: 3 June 2022
Accepted: 27 June 2022
Published: 30 June 2022

1. Introduction and Background

To achieve greenhouse gas emission targets that many countries have set in the last years [1], a significant increase in the penetration of Renewable Energy Sources (RES) in the power grid is required. However, the growth in non-programmable resources such as solar photovoltaics and wind [2,3] hinders the reliability and stability of the electric grid [4] because (i) their discontinuous and uncertain generation causes unforeseeable oscillations in the net load and (ii) their modularity favors the diffusion of distributed generators, connected to the distribution grid and close to the demand. Thus, energy flows become unpredictable both in quantity and direction.

To allow the deployment of RES to grow without affecting the grid reliability, it is necessary to rethink the design, operation, and planning of the power system, both on a technical and on an economic level. On the technical side, it is necessary to increase the system flexibility with fast-ramping dispatchable plants, energy storage systems, demand-side management, and interconnections to the adjacent power grids for trading [5–8]. On the economic side, the regulatory framework must be adjusted to allow new services and business models. New rules for the electricity market have been set in the European Union with the Clean Energy for All European package [9], which aims at helping the energy markets to better manage the energy flows across Europe, to empower the consumers, and to include more renewables. With these rules, the prosumers (both producers and consumers) can vary their consumption and production based on price changes, thus participating in the energy market and contributing to grid efficiency and integration of RES.

At the same time, in this scenario characterized by stringent greenhouse gas reduction targets, the development of a hydrogen economy has gained renewed interest. Indeed, the key role that hydrogen can play in fostering the clean energy transition has been recognized at an international level by the Hydrogen Council, gathering a number of industrial entities [10], as well as by the International Energy Agency [11]. On the one hand, the deployment of hydrogen technologies has the potential to increase the penetration of renewables in the power system, allowing demand–supply balancing and avoiding renewable energy curtailment. On the other hand, hydrogen technologies have sector coupling capability, and clean hydrogen can be used to decarbonize either the industrial sector, the transport sector, or the heating and cooling sector. For instance, hydrogen is used in refinery and for the production of clean fuels (e.g., ammonia and methanol).

At the European level, the European Green Deal [12] includes the European hydrogen strategy, aiming at making clean hydrogen a viable solution to decarbonize the different energy sectors over time across Europe, enabling the European Union to play a global pioneering role in this field. The priority is the development of renewable hydrogen, although in the short-term, other forms of low-carbon hydrogen are necessary to support the creation of a profitable market. Another important aspect is that the targets set by the European hydrogen strategy (e.g., 24% of hydrogen in its energy mix by 2050) also require the import of renewable electricity or of green hydrogen from overseas. Thus, the European hydrogen strategy promotes the deployment of renewable energy sources in developing countries, and especially in the countries of the African Union, bridging the energy transition between advanced and developing countries [13].

In this framework, this work identifies MW-scale Fuel Cell (FC) power plants based on low temperature Proton Exchange Membrane (PEM) cells as a source of flexibility for the future high-RES penetration electric grid, providing grid ancillary services thanks to their fast ramp rate and the load-following capability. The potential of this technology is being investigated within the EU H2020 project GRASSHOPPER, whose main features are described in Section 2. This work focuses on the development of a dynamic numerical model of a flexible PEM FC power plant, based on the layout of the 100-kW pilot unit built within the GRASSHOPPER project. The purpose is to develop an instrument to simulate the system operation, aimed at maximizing the system efficiency and flexibility while limiting the degradation rate.

The topic of dynamic modeling of fuel cell systems has been already addressed in the literature, where the case of stationary fuel cell power plants has been discussed, mostly focusing on the case of high temperature fuel cells, in particular solid oxide fuel cells for stand-alone and hybrid configurations, including combined heat and power applications [14–16]. Their off-design and transient behavior is investigated by the point of view of their safe and reliable operation as well as performance improvement, degradation minimization, and control improvement [17–20]. On the other hand, PEM-based systems have been mostly addressed for mobile applications [21–24], and in few cases for stationary use [25].

This work instead focuses on low temperature systems for stationary use, with the particular target of discussing fast-ramping applications for grid support. A detailed description of the model is presented in Section 3, while in Section 4, the model is tested against experimental data and subsequently applied to simulate plant load-following operation and cold start up.

2. GRASSHOPPER Project

The demonstration of the feasibility of stationary large-scale FC power plants, using PEM cells, has been accomplished in many projects. For example, ref. [26] describes a 70-kW$_{el}$ PEM FC power plant that uses by-product hydrogen from the electrolysis of brine. The pilot plant performance is analyzed over 30,000 h of operation, showing cell irreversible and reversible decay. Ref. [27] shows that the operation of a 50-kW$_{el}$ PEM FC plant over 4400 h, with relatively low-quality hydrogen from a sodium chlorate production process with standard industry purification, does not lead to extensive cell degradation. The feasibility of MW-scale PEM FC power plant has been demonstrated in [28,29], which tested a 2-MW_e plant with by-product hydrogen from a chlor-alkali process. The flexibility of PEM FC stacks is also well demonstrated since PEM FCs are used in the automotive sector. However, the flexibility of stationary FC power plant has never been tested.

The GRASSHOPPER project studies how to use FC power plants, adopting low temperature PEM cells to provide balancing services to the power grid. The aim is to demonstrate MW-scale plant capability to operate dynamically, to realize the next-generation cost-effective FC power plant for stationary application and grid stabilization. The target CAPEX for a production of 25 MW_{el} per year is below 1500 €/kW_{el}, reached through joint design of MEA (membrane electrode assembly), stack, and system.

The project has designed a 100 kW_{el} pilot unit, whose construction was finalized in September 2021. The plant successfully completed factory acceptance tests and is to be demonstrated in the field in Delfzij (NL), with 8 months of continuous operation.

Experience from the 100-kW_{el} pilot plant is the basis for the design of a MW-scale unit. To support the scale-up, important outcomes are also derived from numerical simulations. The work presented in this paper falls within this scope.

2.1. *kW Pilot Plant Layout*

The layout of the GRASSHOPPER 100 kW_{el} plant is shown in Figure 1a, while Figure 1b shows the real plant configuration.

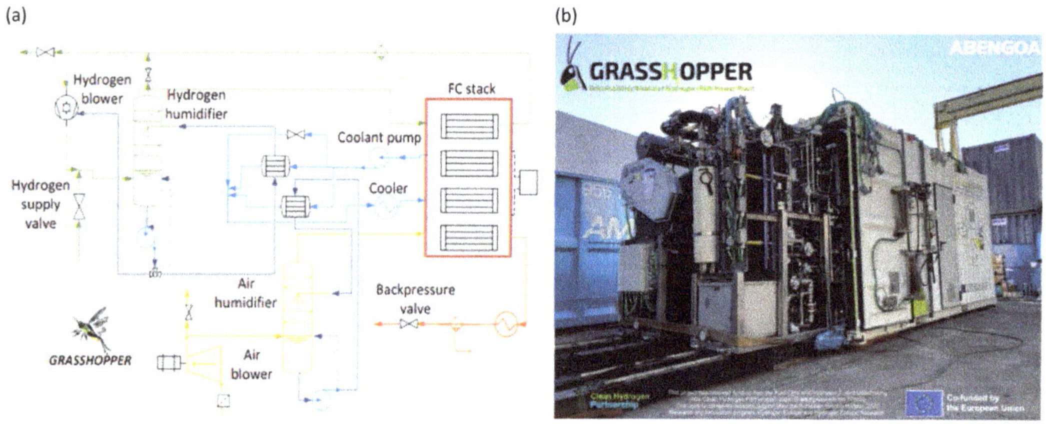

Figure 1. Grasshopper 100 kW_{el} FC power plant (**a**) simplified layout and (**b**) real plant configuration showing the skid mounted and containerized structure of the system.

Filtered air and pure hydrogen are humidified and supplied to the stack. The operating temperature of the stack is maintained at the desired value by the coolant, flowing in a dedicated loop. Heat exchangers allow the recovery of the stack dissipated thermal power for air and hydrogen humidification. The excess heat is dissipated in a dedicated cooler. The cathode backpressure is controlled with a valve, while the regulation of the rotational speed of the air blower and a purge valve allow the control of the 'air stoichiometry' (i.e., the ratio of air flowrate with respect to the stoichiometric flow). Fresh hydrogen is injected into the system through a controlled valve (hydrogen it is available above 4 bar). A liquid ring compressor recirculates the excess hydrogen and the 'hydrogen stoichiometry' (i.e., the ratio of hydrogen flowrate with respect to the stoichiometric flowrate) is controlled with a stack bypass.

The system has been designed to work flexibly between 20% (200 mA/cm^2) and 100% (1000 mA/cm^2) of the nominal current density value. Its performance is optimized and stack degradation is limited thanks to a precise control system.

3. Methods for Dynamic Modeling

This work implements, with the software Simulink, a dynamic model of a PEM FC plant, referring to the Grasshopper 100 kW_{el} plant configuration. Models that solve energy and mass balances during operation with a variable load are built for each plant component, as described in the following paragraphs, reporting also the assumptions for model scale-up. The models of the components are combined to build the model of the entire system. To allow control of the system operation, PI-type and on-off controllers are then implemented.

3.1. Thermodynamic Properties

The model is based on the hypothesis of ideal gas, ideal gas mixture, and ideal liquid. For water and coolant, enthalpy is computed, assuming constant specific heat capacity (Equation (1)), equal to 4.186 kJ/kg/K for water and to 3.52 kJ/kg/K for the water–glycol mixture.

$$h_{liquid} = c_L\,(T - T_{ref}) \tag{1}$$

For the gas, enthalpy is computed with NASA polynomial equations [30] (Equation (2)), where a_1–a_7 are numerical coefficients available on NASA thermodynamic libraries for 1130 solid, liquid, and gaseous pure chemical species.

$$h\left[\frac{J}{mol}\right] = R\,T\cdot\left(a_1 + \frac{a_2 T}{2} + \frac{a_3 T^2}{3} + \frac{a_4 T^3}{4} + \frac{a_5 T^4}{5} + \frac{a_6}{T}\right) \tag{2}$$

When a gaseous mixture of different species is present (e.g., moist air/hydrogen), its enthalpy is computed from the enthalpy (h_J) and the molar fraction (x_J) of each species in the mixture, assuming the ideal gas mixture behavior (Equation (3)). For liquid water, the density is assumed to be constant and enthalpy variation with pressure is taken into account.

$$h_{mix} = \sum_J (h_J\cdot x_J) + \frac{x_{H_2O,liq}}{\rho_{H_2O,liq}}\,(p - p_{atm}) \tag{3}$$

Water in the liquid phase and in the vapor phase is treaded as two different species. When the partial pressure of the water in the mixture is below its saturation pressure, the water is in the vapor phase. Otherwise, it is partially in the condensate liquid phase. The maximum molar fraction of water in the vapor phase coincides with the water molar fraction at saturation. Saturation pressure of water is computed as a function of the temperature, according to a polynomial equation (Equations (4) and (5)) derived in [31] and valid for temperature values ranging from water triple point (T_{tr} = 273.16 K) to the critical point (T_{cr} =647.096 K, p_{cr} = 220.64 bar).

$$p_{H_2Osat,T}[bar] = p_{cr}[bar]\cdot e^{\frac{b_1\cdot\theta + b_2\cdot\theta^{1.5} + b_3\cdot\theta^3 + b_4\cdot\theta^{3.5} + b_5\cdot\theta^4 + b_6\cdot\theta^{7.5}}{1-\theta}} \tag{4}$$

$$\theta = 1 - \frac{T[K]}{T_{cr}[K]} \tag{5}$$

3.2. PEM FC Stack

A single cell is modeled with a lumped-volume approach, considering performance data acquired from detailed cell simulations, thus allowing to simulate large scale effects. Given the stack modularity, the FC stack model is then built, assuming that many identical cells are electrically connected in series.

The model is divided into three subsystems: electrochemical model, fluid dynamic model, and thermodynamic model.

Regarding the electrochemical model, FC voltage and gross electrical power are calculated with semi-empirical current-voltage curves, combining cell theoretical polarization curve equations and experimental datasets, as detailed in [32]. The resulting formulation is shown in Equation (6), where the first term represents an apparent open circuit voltage, the second term the ohmic losses, and the third and the last terms the activation and the concentration overvoltage, respectively. The equation takes into account voltage dependence on backpressure, air stoichiometry, and air relative humidity. It catches the real cells performance with a good approximation, mainly for the range of current density where the cell will generally operate (relative errors <1% for relevant currents values: 200–1000 mA/cm^2). A detailed description of the terms appearing in Equation (6) is presented in [32], together with the coefficients calibration procedure.

$$V_{cell} = V_{OC} + \frac{R_{ohm,1}}{\left(\frac{x_{H_2O}}{x_{H_2O,ref}}\right)^{G_{ohm}}} \cdot i + \left(K_{act,1} + K_{act,2}\left(\frac{p}{p_{ref}}\right)^{\frac{x_{O_2}}{x_{O_2,ref}}}\right) \cdot \ln\left(1 + \frac{i}{i_0}\right) + \left(K_{conc} \cdot \left(\frac{p}{p_{ref}} \cdot \frac{x_{H_2O}}{x_{H_2O,ref}}\right)^{G_{conc}}\right) \cdot \ln\left(1 - \frac{i}{i_L \cdot \frac{p}{p_{ref}}}\right) \quad (6)$$

In order to simulate cold start-up, semi-empirical corrections have been considered to model the effects of the cell temperature on ohmic losses and activation losses. Temperature effects on the concentration losses are not modeled since they impact on the cell performance only with high current density, outside the range of operability of the cells. The ohmic resistance (R_{ohm}) varies exponentially with the temperature [33], according to Equation (7), while the activation losses vary because of the exponential variation with the temperature of the exchange current density (i_0) [33], according to Equation (8).

$$R_{ohm,T} = \left(\frac{R_{ohm,1}}{\left(\frac{x_{H_2O}}{x_{H_2O,ref}}\right)^{G_{ohm}}}\right) \cdot E^{C_{OHM}\left(\frac{1}{T} - \frac{1}{T_{ref}}\right)} \quad (7)$$

$$i_{0,T} = i_0 \cdot e^{\frac{C_{act}}{T}\left(1 - \frac{T}{T_{ref}}\right)} \quad (8)$$

The values of the coefficients C_{ohm} and C_{act} are determined by comparing the regressed polarization curves with an experimental dataset obtained from a similar FC stack, characterized by temperatures ranging between 15 °C and 62 °C, constant stoichiometry, relative humidity, and backpressure. The coefficients are chosen such that the gradient with the temperature for the ohmic resistance and the open circuit overvoltage are equal to the ones measured in the reference database. The value of temperature adopted in the equations is the average temperature of the coolant flowing over the stack, assumed as representative of the stack temperature. The reference temperature is equal to 65 °C. Polarization curves obtained with this model for different values of temperature and air ratio to stoichiometry are shown in Figure 2, where backpressure is 1.35 bar and the average air relative humidity is 100% (i.e., $x_{H_2O}/x_{H_2O,ref} = 1$). The numerical values of the regressed coefficients are summarized in Table 1.

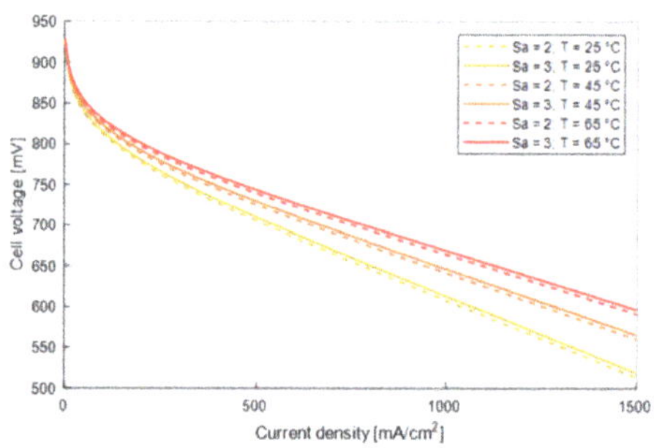

Figure 2. Cell current-voltage polarization curves for different values of operating temperature and air ratio to stoichiometry.

Table 1. Polarization curve coefficients regressed on experimental data.

Coefficient	Regressed Value	
V_{OC}	928.24	mV
R_{ohm}	−0.045	$\Omega \cdot cm^2$
G_{ohm}	0.837	-
C_{ohm}	1700	-
$K_{act,1}$	−41.06	mV
$K_{act,2}$	5.62	mV
i_0	4.86	mA/cm^2
C_{act}	−1047	-
K_{conc}	126.50	mV
i_L	2600	mA/cm^2
G_{conc}	1.183	-

The electrochemical model computes the stack voltage and the gross DC electric power respectively with Equations (9) and (10), where N_{cells} is the number of cells composing the stack and A_{cell} is the active area of each cell.

$$V_{STACK} = V_{CELL} \cdot N_{CELLS} \tag{9}$$

$$P_{gross,stack} = i \cdot A_{cell} \cdot V_{stack} \tag{10}$$

In regard to the cell dynamic behavior, the charge and discharge of the cell double layer is the main phenomenon affecting the cell voltage when the load is varied [34–37]. It can be modeled though a capacitance, whose value varies between 0.01 and 0.05 F/cm^2 according to literature [35]. Since these phenomena are much faster than other dynamic effects, such as the ones associated with the thermal inertia, the double layer capacitance is not included in this model.

Regarding the fluid dynamic model, it solves mass conservation equations over the FC stack for each chemical species (H_2, O_2, N_2, and H_2O) flowing through the anodic and the cathodic channels, to determine flowrate and composition of the anodic and cathodic streams leaving the stacks. Hydrogen and oxygen consumptions and water production are computed as a function of the stack current density, knowing the reaction in the stacks, according to Equations (11)–(13).

$$\dot{m}_{H_2,\,cons} = \frac{i}{2F} \cdot A_{cell} \cdot N_{cells} \cdot 3.6 \cdot MM_{H_2} \left[\frac{kg}{h}\right] \tag{11}$$

$$\dot{m}_{O_2,\,cons} = \frac{\frac{\dot{m}_{H_2,cons}}{MM_{H_2}}}{2} \cdot MM_{O_2} \left[\frac{kg}{h}\right] \tag{12}$$

$$\dot{m}_{H_2O,\,prod} = \frac{\dot{m}_{H_2,cons}}{MM_{H_2}} \cdot MM_{H_2O} \left[\frac{kg}{h}\right] \tag{13}$$

Additionally, according to experimental evidence, the model considers a net transport of H_2O from cathode to anode equal to 0.1 mol/h, a net diffusion of H_2 from anode to cathode corresponding to an internal current density equal to 0.002 A/cm^2, a net transport of N_2 form cathode to anode such that the mass fraction of N_2 in the outlet anodic stream is 10%, and a null net transport of O_2 from cathode to anode or vice versa.

The accumulation of gasses in the cell channels is not included in the model because the channel volume is negligible with respect to the other component volumes.

Pressure drops in the channels are linearly dependent on the volumetric flowrate of coolant and of reactants at the stack inlet, according to prior experimental stack tests and justified by the low velocities in the flow field, which leads to a laminar regime.

Regarding the thermodynamic model, it solves energy balances on the stack (Equation (14)) to determine the average temperature of the cell and the temperature of the flows leaving the FC stack. Temperature dynamic (Equation (15)) is taken into account through the heat capacity of the stack, which is lumped in the Bipolar Plates (BP). The gross DC electric power ($P_{gross,stack}$) is an output of the electrochemical subsystem. Heat transfer from the reactants to the bipolar plants ($Q_{acc\ stack}$) and to the coolant fluid ($\dot{Q}_{to\ coolant}$) is computed, assuming constant heat transfer coefficients. No heat losses to the environment are assumed.

$$\dot{H}_{fuel,out} + \dot{H}_{air,out} = \dot{H}_{fuel,in} + \dot{H}_{air,in} - P_{gross,stackC} - \dot{Q}_{to\ coolant} - Q_{acc\ stack} \tag{14}$$

$$Q_{acc,stack} = C_{BP,anode}\frac{dT_{BPanode}}{dt} + C_{BP,cathode}\frac{dT_{BPcathode}}{dt} \tag{15}$$

The system can be scaled up by increasing the number of cells in each stack or by installing more stacks of the same size. In the first case, more cells are connected in series, resulting in higher voltage and requiring and increase in the flowrate of hydrogen and air supplied to each stack. The resulting wider voltage range is not compatible with commercially available Power Control Systems. In the second case, which is applied in this work, a greater number of stacks is electrically connected in parallel, requiring more converters or a converter able to work with a higher current, although each stack remains identical in geometry and operation.

3.3. Air Blower

The air blower model follows a lumped-parameter approach. It includes dynamic effects related to the mechanical inertia and thermal inertia, while it neglects mass accumulation.

The unit stationary performances are based on the machine performance maps, available on the commercial datasheet and implemented through polynomial functions. The maps give the volumetric flowrate of air processed by the machine, its temperature gain at thermal equilibrium, and the blower mechanical power, as a function of air inlet temperature and pressure (ambient conditions), air outlet pressure (determined by the pressure drops in the downstream components), and the blower rotational speed. Isentropic efficiency and mechanical efficiency are computed.

The blower rotational speed (ω_c [rad/s]) is computed according to Equation (16), representing a balance between the torque generated by the electric motor (τ_m, defined in Equation (17)) and the torque required by the blower (τ_c, defined in Equation (18)), where J_{cm} is compressor-motor block inertia [38], $P_{el,m}$ is the electrical power supplied to the motor (external input), $\eta_{el,m}$ is the motor electrical efficiency, and P_c is the blower mechanical power.

$$J_{cm}\frac{d\omega_c}{dt} = (\tau_m - \tau_c) \tag{16}$$

$$\tau_m = \frac{P_{el,m} \cdot \eta_{el,m}}{\omega_c} \tag{17}$$

$$\tau_c = \frac{P_c}{\omega_c} \tag{18}$$

The blower thermal inertia was initially not included in the model, but experimental data showed the importance of the temperature dynamics: the temperature gain of the air flowing through the blower matches the performance maps value in stationary conditions, but a smaller gain is shown during warm-up.

In general, the thermal power generated in the compression phase ($\dot{Q}_{gen}$) is in part responsible for the air temperature increase ($\dot{Q}_{air}$) and in part transferred to the machine ($\dot{Q}_{trasf}$). This second part is in turn partially absorbed by the machine steel mass ($\dot{Q}_{steel}$), and partially dissipated towards the environment ($\dot{Q}_{diss}$) (Equation (19)).

$$\dot{Q}_{gen} = \dot{Q}_{air} + \dot{Q}_{trasf} = \dot{Q}_{air} + \dot{Q}_{steel} + \dot{Q}_{diss} \tag{19}$$

At thermal equilibrium ($\dot{Q}_{steel} = 0$), the air temperature gain is known from the performance maps. This allows to compute the heat absorbed by the air (Equation (20)), the heat dissipated (Equation (21)), and therefore the generated heat. According to the lumped volume approach, the steel temperature is homogeneous and equal, at thermal equilibrium, to the average air temperature between inlet and outlet (Equation (22)).

$$\dot{Q}_{air,stationary} = \dot{m}_{air}\, c_{p,air}\, \Delta T_{air,\ map} = \dot{m}_{air}\, c_{p,air}\left(T_{air,out,stationary} - T_{air,in}\right) \quad (20)$$

$$\dot{Q}_{diss,stationary} = hA_{ext}\left(T_{blower,stationary} - T_{amb}\right) \quad (21)$$

$$T_{blower,stationary} = \frac{T_{air,in} + T_{air,out,stationary}}{2} \quad (22)$$

During thermal transients, the generated heat ($\dot{Q}_{gen}$) is the same computed for steady state, and Equations (23)–(26) allow computing the air temperature at blower outlet ($T_{air,out}$).

$$\dot{Q}_{air} = \dot{Q}_{gen} - \dot{Q}_{trasf} = \dot{m}_{air} \cdot c_{p,air} \Delta T_{air} \quad (23)$$

$$\dot{Q}_{trasf} - \dot{Q}_{diss} = \dot{Q}_{steel} = hA_{int}\left(T_{air,avg} - T_{blower}\right) = C_{blower} \cdot \frac{dT_{blower}}{dt} \quad (24)$$

$$\dot{Q}_{diss} = hA_{ext}\left(T_{blower} - T_{amb}\right) \quad (25)$$

$$T_{air,avg} = \frac{T_{air,in} + T_{air,out}}{2} \quad (26)$$

The values of the heat transfer coefficients (hA_{int} and hA_{ext}), the blower heat capacity (C_{blower}), and the mechanical inertia (J_{cm}) have been calibrated on experimental data.

The model scale-up procedure assumes:

- Flowrate (V_{air}) , isentropic efficiency (η_{is}), and mechanical efficiency (η_{m}) curves derived from the regressed performance curves, assuming different sizes but the same efficiency profile for the blowers.
- Blower heat capacity proportional to the blower mass (representing the case assembly and excluding the motor, whose mass is expected to increase less than linearly).
- The blower surface in contact with the compressed air (named 'Internal area') and the blower surface in contact with the environment (named 'external area') are functions of the machine diameter. Indeed, by approximating the rotary lobe compressors as two identical cylinders (the lobes, with diameter D_L and height L) contained in a larger cylinder (the case, with diameter D_C and height L), and assuming that the diameter of the single lobe is equal to half the diameter of the case ($D_L = D_C/2$), the relations in Equations (27) and (28) are obtained. Additionally, neglecting the thickness of the case, the volume of the machine can be expressed as the sum of the volume of the lobes and the void volume, which is occupied by air (Equation (29)). Assuming that the void volume is proportional to the nominal volumetric air flowrate and that the ratio between the cylinders diameters remains the same for blowers of different sizes, the ratio between the diameters of the two machines of size A and B is expressed as a function of the ratio between the nominal volumetric air flowrates by Equation (30).

$$A_{ExternalArea} = \pi\, D_C\, L \quad (27)$$

$$A_{InternalArea} = \left(\pi\, N_{lobes}\, D_L + \pi\, D_C\right) L \quad (28)$$

$$\pi \frac{D_C^2}{4} L = \pi \frac{D_L^2}{4} L + V_{void} \quad (29)$$

$$D_{C,B} = D_{C,A} \cdot \sqrt[3]{\frac{\dot{V}_{air,nom,B}}{\dot{V}_{air,nom,A}}} \tag{30}$$

3.4. Humidifier

The humidifier is composed by two sections, the packed-bed column and the water tank. These two sections are modeled separately with a lumped-volume approach.

Regarding the packed-bed column, the gas (hydrogen or air) enters at the bottom and flows upwards, while the water is sprayed at the top and flows downwards. Thus, the humidity of the gas increases while flowing thorough the column because of water evaporation. The column model computes flowrate, composition, pressure, and temperature of the outlet flows by knowing the same properties at inlet. It is assumed, according to industrial experience, that the humidifier effectiveness is 100%: the outlet gas is fully saturated and in thermal equilibrium with the water spray. Additionally, a demister at the top of the column removes the water drops that may be present in the outlet gas. Temperature dynamic in the column is negligible with respect to the water tank and is therefore neglected by the model. For the air humidifier, although the void fraction of the column is the 90% of its total volume, the column model neglects air build-up (the volume of all the air supply line components is lumped in the air manifold). Conversely, for the hydrogen humidifier, the column model includes hydrogen build-up. Pressure drops of the gasses flowing through the humidifier are then taken as constant and set equal to the nominal values reported on the datasheet.

The water tank collects the water that is sprayed in the column. The water tank model includes water build-up and temperature dynamic (the thermal inertia is associated to the heat capacity of the stored water). Perfect mixing in the water tank is assumed. The inputs to the model are the flowrate of the outlet water stream, set by the recirculation pump, and the flowrate and temperature of the water coming from the packed bed column. In addition, a purge or a refill of water can be introduced to regulate the water level in the tank. With the given assumptions, the model computes the amount of water that is accumulated in the tank, its temperature and the temperature of the water that is pumped out.

The model of the humidifier is scaled up, assuming that the size of the water tank is proportional to the system nominal power. The pressure drops in the humidifier column are assumed constant.

3.5. Air Supply Manifold

The air supply manifold model allows to simulate the build-up of the air in the supply line. Indeed, the volume of all components in the supply line is lumped in the manifold, which is located between the air blower and the air humidifier.

The manifold model assumes uniform temperature in the manifold itself (equal to the temperature of the inlet air) and ideal gas behavior. It computes the time variation of the pressure in the air supply line as in Equation (31). The air blower sets the flowrate at manifold inlet ($\dot{m}_{in,manifold}$), while the pressure drops in the components downstream set the air flowrate at the manifold outlet ($\dot{m}_{out,\ manifold}$).

$$\frac{dm_{acc}}{dt} = \frac{V_{manifold}}{R^{*} \cdot T_{manifold}} \cdot \frac{dp_{manifold}}{dt} = \dot{m}_{in,\ manifold} - \dot{m}_{out,\ manifold} \tag{31}$$

To scale up the model, the manifold volume is varied proportionally to the void volume in the air humidifier column, which most contributes to the volume of the air supply line.

3.6. Backpressure Valve

The backpressure valve is modeled as an orifice, through which a one-dimensional isentropic flow of a compressible fluid flows [39,40]. The orifice model reported for subcritical flow is adopted, since the ratio between valve inlet pressure (max 1.6 bar) and outlet pressure (ambient pressure) is always above the critical pressure ratio. According to Equation (32), the flowrate of air through the valve is computed knowing the upstream pressure and temperature, the pressure ratio across the valve, and the valve effective area (A_e).

$$\dot{m}_{out,valve} = \frac{A_e \cdot p_{valve,in}}{\sqrt{R_{air} \cdot T_{valve,in}}} \left(\frac{p_{valve,out}}{p_{valve,in}} \right)^{\frac{1}{\gamma_{air}}} \sqrt{\frac{2\gamma_{air}}{\gamma_{air} - 1} \left(1 - \left(\frac{p_{valve,out}}{p_{valve,in}} \right)^{\frac{\gamma_{air}-1}{\gamma_{air}}} \right)} \quad (32)$$

To scale-up the model, the valve effective area is varied proportionally to the nominal air flowrate.

3.7. Liquid Ring Compressor

The liquid ring compressor is modeled through a stationary model because its rotational speed can vary only in a narrow range, making dynamic effects negligible. Two sequential steps are considered: hydrogen compression and hydrogen-water mixing.

In the first step, the hydrogen stream coming from the FC anode and from the hydrogen humidifier (if hydrogen recirculation over the humidifier is required to guarantee the minimum hydrogen flow to the liquid ring compressor) is compressed. Knowing flowrate, composition, temperature, and pressure of the inlet moist hydrogen, as well as the outlet pressure that the stream has to reach (determined by the hydrogen humidifier located downstream), the model computes the temperature of the hydrogen flow after compression and the electrical power required for hydrogen compression, assuming constant isentropic efficiency (η_{is}) and mechanical efficiency (η_m). The water contained in the inlet hydrogen streams is assumed to be in vapor phase.

In the second step, the compressed hydrogen flow is mixed with the water stream from the hydrogen compressor tank. The mixing process is modeled as an adiabatic process. The flowrate of the water is assumed to be constant since the pump is not controlled and works with a constant rotational speed. The inlet water stream is assumed to be in the liquid phase.

The liquid ring compressor model is scaled up, assuming constant isentropic and mechanical efficiencies and that the water flowrate required by the compressor is proportional to the nominal hydrogen flowrate.

3.8. Heat Exchanger

The heat exchangers are plate-type counter-current heat exchangers, working with liquid on both sides. Given the modular structure of the plates and assuming that the fluids equally distribute among the channels, the model considers a number of identical sub-units. Each of them includes a plate and half of the adjacent hot and cold channels. In each sub-unit, control volumes are then identified through a discretization along the channel direction (1D-model). The model solves mass and energy balances for each control volume, to compute the heat transferred, the plates temperature, and the outlet streams temperature. The model assumes that, in the control volume, the temperature of each plate is uniform, and it considers constant values for the heat transfer coefficients and the heat capacities. The temperature dynamic is associated to the heat capacity of the heat exchanger materials. Conduction along the direction of the flows, thermal resistance of the plate, heat losses toward the environment, and mass accumulation in the channels are neglected. Pressure drops are modeled, assuming the pressure will vary linearly with the volumetric flowrates.

The model is scaled up by increasing the number of plates, so that the total area of heat transfer varies proportionally to the system nominal power. The heat transfer coefficients and the pressure drops are assumed unchanged. The model is representative of a counter-current plate-type heat exchanger or of a more compact plate-and-shell heat exchanger.

3.9. Pumps

A stationary model of the pump is realized, since they are expected to work at constant rotational speed (such as in the air and hydrogen humidifier water circuit) or the dynamic effects are assumed negligible (as for the coolant circuit pump, whose rotational speed is varied to control the coolant temperature gain over the stack). The model computes the pump electrical consumption and the outlet flow pressure and temperature, assuming constant mechanical and isentropic efficiency and knowing flowrate, temperature, and pressure of the inlet flow, as well as the pressure gain that the pump has to provide.

When the model is scaled up, mechanical and isentropic efficiencies are assumed unchanged.

3.10. Pipelines

Pipelines are used to connect the plant components and are responsible of transport delays and pressure drops. These phenomena are included in the pipeline model. The transport delay depends on pipe diameter and length and on the volumetric flowrate, as in Equation (33). Pressure drops depend on the volumetric flowrate, as in Equation (34), where λ is assumed to be constant. The fluid is assumed to be incompressible (as already mentioned, hydrogen build-up is lumped in the hydrogen humidifier volume and air build-up in the air supply manifold volume).

$$t_{delay} = L\,\frac{\pi\,D^2}{4} \cdot \frac{1}{\dot{V}} \tag{33}$$

$$\Delta p = \lambda\,\rho \frac{L}{D}\frac{v^2}{2} = \lambda\,\rho\,\frac{8\,L}{\pi^2 D^5}\,\dot{V}^2 \tag{34}$$

The pipeline model scale-up is based on the assumption that bigger components need to be more distant from each other, therefore the length of the pipes connecting the components increases with the plant nominal power. Additionally, the scale-up assumes that the nominal pressure drops are the same for systems of different sizes. Thus, being the pressure drops calculated with the assumption of turbulent flowrate (Equation (34)), the relationship between the pipes diameters for the two systems of size A and B is given by Equation (35).

$$D_B = \sqrt[5]{D_A^5 \left(\frac{\dot{V}_B}{\dot{V}_A}\right)^2 \frac{L_B}{L_A}} \tag{35}$$

3.11. Power Conversion System (PCS)

The PCS includes the components required to connect the plant to the electric grid, namely a DC/DC converter and an inverter. The PCS is simulated through a constant conversion efficiency, which is set equal to 80%, as experimentally measured on the 100 kW_{el} unit at low current density operation. The choice of these value is conservative since higher values are expected at higher currents, but an experimental performance curve is not available. Additionally, the DC/DC converter is not required for the MW size plant, where a higher stack voltage is expected.

3.12. Control Method

The control strategy is implemented in the model with PID controllers, when a precise control of the variable is required, and with on-off controllers, when it is sufficient to guarantee that the variable remains within a given range of values.

A PID controller is generally characterized by three different control actions: a proportional action, an integral action, a derivative action. In this system, the derivative part in not implemented and a PI configuration is used, characterized by the proportional gain K_P and the integral gain K_I (Equation (36)).

$$K_P + K_I \frac{1}{s} \quad (36)$$

Each controller takes into account the physical characteristic of the system components, allowing the manipulated variable to vary only within a given range of values. To avoid wind-up, clamping of the integral is adopted: the integral in each PI controller is switched off ($K_I = 0$) when the manipulated variable reaches its minimum or maximum limit, allowing a faster response when the error changes its sign.

Table 2 details the controller implemented in the model.

Table 2. Control system: controlled variables, manipulated variables, and controller type.

Controlled Variable	Manipulated Variable	Controller Type
Cathode backpressure	Backpressure valve opening area	PI
Air stoichiometry;	Blower rotational speed + air purge	PI
Air temperature at stack inlet	Coolant flowrate to the air heat exchanger	PI
Hydrogen temperature at stack inlet	Coolant flowrate to the hydrogen heat exchanger	PI
Hydrogen pressure at stack inlet	Fresh hydrogen flowrate	PI
Hydrogen stoichiometry	Hydrogen recirculation	PI
Coolant temperature gain over the stack	Rotational speed of the coolant pump	PI
Coolant temperature at stack inlet	Water flowrate to the cooler heat exchanger	PI
water level in the air humidifier tank	water flow to refill the air humidifier tanks	ON/OFF
water level in the hydrogen humidifier tank	water flow to refill the hydrogen humidifier tanks	ON/OFF

3.13. System Performance Evaluation

The performances of the fuel cell system are evaluated through the stack gross power (defined by Equation (10)), the system net power (Equation (37)), the stack gross efficiency (Equation (38)), and the system net efficiency (Equation (39)). Stack gross efficiency and net efficiency are referred to the Low Heating Value (LHV) of the consumed hydrogen.

$$P_{NET,SYSTEM} = P_{GROSS,STACK} \cdot \eta_{PCS} - P_{AUXILIARIES} \quad (37)$$

$$\eta_{GROSS,\ STACK} = \frac{P_{GROSS,\ STACK}}{\dot{M}_{H_2,CONS} \cdot LHV_{H2}} \quad (38)$$

$$\eta_{NET,SYSTEM} = \frac{P_{NET,\ SYSTEM}}{\dot{M}_{H_2,CONS} \cdot LHV_{H2}} \quad (39)$$

4. Results and Discussion

4.1. Comparison of Simulation Results and Experimental Data

The model is tested against experimental data collected during preliminary testing of the 100 kW_{el} Grasshopper pilot plant. Six datasets are identified, including variable load operation (spanning through the entire range of current density) and plant cold start-up. The data include temperatures, pressures, and flowrates in different points of the system, stack current and voltage, and the power consumptions of the main plant auxiliaries.

To validate and calibrate the model, the following simulation inputs are set equal to the experimental values:

- stack current density,
- coolant flowrate,

- setpoint for the air flowrate entering the humidifier,
- setpoint for the hydrogen flowrate entering the stack.

Additionally, the following parameters are initialized equal to the values measured at test start:

- plant container temperature,
- liquid ring compressor temperature,
- air temperature at the blower outlet,
- water temperature in the humidifier tanks,
- air and hydrogen temperature at the stack inlet,
- coolant temperature at the stack inlet,
- cathode backpressure.

A preliminary calibration of the parameters affecting the warm-up time (e.g., pipes lengths and heat capacities) was performed by minimizing the difference between the simulated and the measured temperature, in different points of the plant, for one of the available datasets. Comparison of experimental data and simulation results for the other five datasets shows the ability of the model to reproduce the real plant behavior, obtaining negligible errors on the prediction of the gross and net power, flowrate, temperature, and pressure in different point of the plant. As an example, for all the six datasets, the RMSE for the gross power is always below 3 kW_{el} and the RMSE for the net power (Figure 3) has a maximum of 4.6 kW_{el} (dataset 6). The plant gross power generation is slightly underestimated by the model, while the net power generation is sometimes slightly underestimated and sometimes slightly overestimated, independent of the power itself. This is related to the impact on the net power of the consumption of the auxiliaries that are not simulated (e.g., fans, coolant circuits auxiliaries).

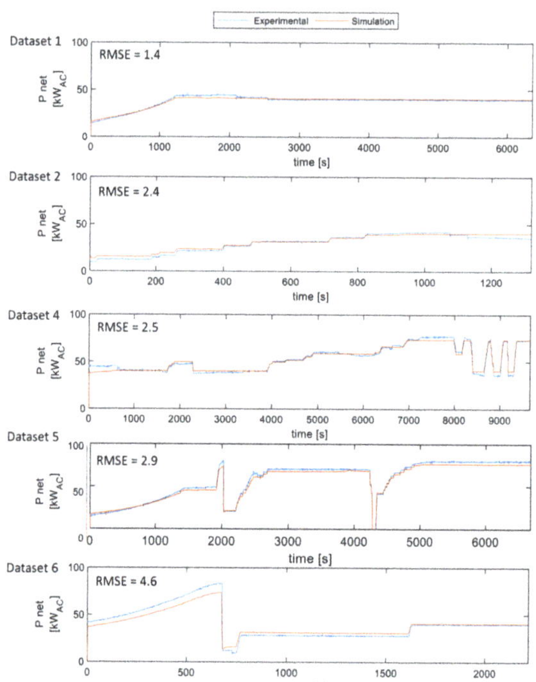

Figure 3. Comparison of experimental and simulated values of net power.

4.2. MW-Size Plant Flexibility Improvementthrough Simulation

The developed model was applied to analyze the performance of a 1-MW PEMFC power plant under a wide range of operating conditions, to point out the main criticalities and to propose solutions to improve the plant design and operation strategy. Simulations include both load-following operation, as required to provide grid ancillary services, and system warm-up.

4.2.1. Load-Following Operation

Load-following operation simulations impose a current setpoint characterized by step changes every 5 min and spanning through the entire operation range. The most extreme case is studied by applying the maximum current step (0.2–1 A/cm^2) in both directions. The actual current variation of the system is limited to 2.5%/s, as decided by the Grasshopper consortium to limit stack degradation.

Figure 4a compares the current density setpoint and the actual value resulting from the simulation. When the setpoint is changed, the current starts changing immediately and reaches the new setpoint in a few seconds (~79 s with the maximum step change). The current density growth results are always limited by the imposed 2.5%/s constraint only, thus reaching the project flexibility targets.

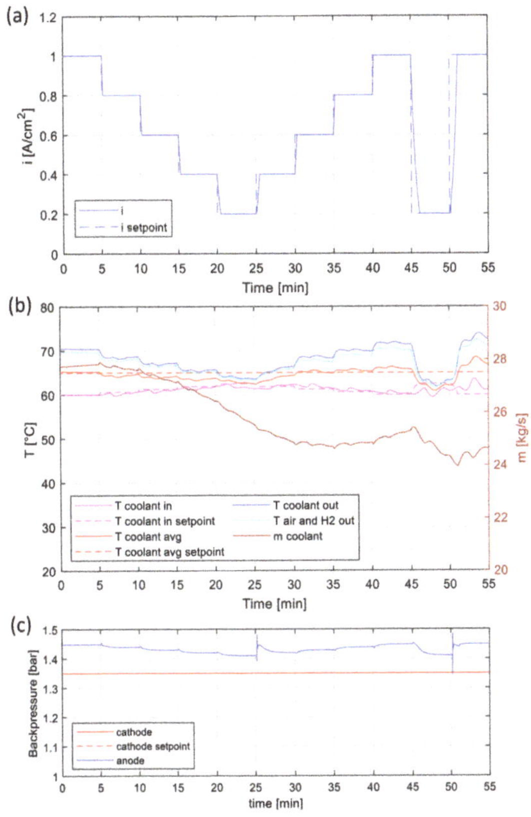

Figure 4. Simulation of load-following operation: time profile of (**a**) simulated current density vs setpoint; (**b**) coolant temperature at stack inlet vs setpoint, coolant average temperature over the stack vs setpoint, coolant temperature at stack outlet, air and hydrogen temperature at stack outlet, coolant flowrate; (**c**) cathode backpressure vs setpoint, and anode backpressure.

No issues emerge from the control of the cathode and anode backpressures (Figure 4c). However, the temperature of the coolant at the stack inlet and the average temperature of the coolant over the stack show a fluctuating behavior (Figure 4b), leading to deviations from the setpoint up to 3 °C. This shows that the PI-type controller is inefficient and suggests the use of more advanced control strategies. Following this conclusion, the Grasshopper pilot plant implements a smith predictor controller, based on the model itself.

A precise control of the air dew-point temperature at the stack inlet is possible (Figure 5a) by regulating the flowrate of the coolant sent to the air humidifier heat exchanger, thus providing the required heat. However, the air average relative humidity over the stack deviates from the setpoint due to delays in the control of the air utilization factor (Figure 5b). Indeed, while the current (and the reactant consumption) changes almost instantaneously, the air flowrate sent to the stack varies more slowly, altering the air utilization factor.

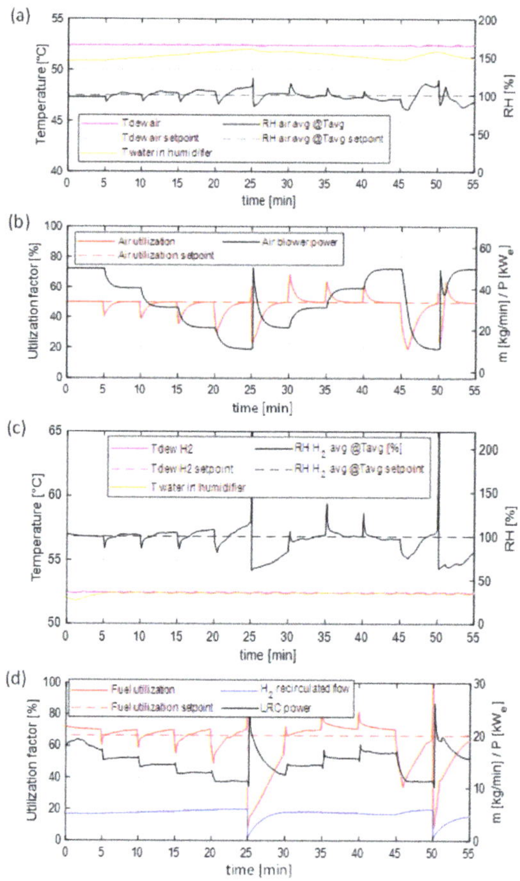

Figure 5. Simulation of load-following operation: time profile of (**a**) air dew point temperature at stack inlet vs setpoint, water temperature in the air humidifier tank, and air average RH vs setpoint; (**b**) air utilization factor vs setpoint, and air blower power consumptions; (**c**) hydrogen dew point temperature at stack inlet vs setpoint, water temperature in the hydrogen humidifier tank and hydrogen average RH vs setpoint; (**d**) fuel utilization factor vs setpoint, hydrogen mass flowrate recirculated over the compressor and LRC power consumptions.

Good control of the hydrogen dewpoint temperature is possible (Figure 5c). Unlike the air side, the water sprayed in the hydrogen humidifier is not heated up (the dewpoint temperature is equal to the temperature of the water in the humidifier tank), and it is necessary to cool the water sent to the Liquid Ring Compressor (LRC) in order to avoid the water in the hydrogen humidifier tank to overcome the desired hydrogen dewpoint temperature. As for the air, a deviation from the RH setpoint is observed when the current is changed, because of the non-instantaneous control of the hydrogen stoichiometry (Figure 5d). Additionally, an unexpected fluctuation of the fuel utilization factor up to 100% happens before the current is increased from 0.2 to 0.4 A/cm^2 and from 0.2 to 1 A/cm^2, because of the too fast control. Otherwise, a slower control was shown to lead to too slow changes in the fuel utilization factor at higher load. This points out the need to use more advanced controllers.

Figure 6 shows the stack gross power, the power consumption of the system auxiliaries, the resulting system net power, and the gross and net efficiency.

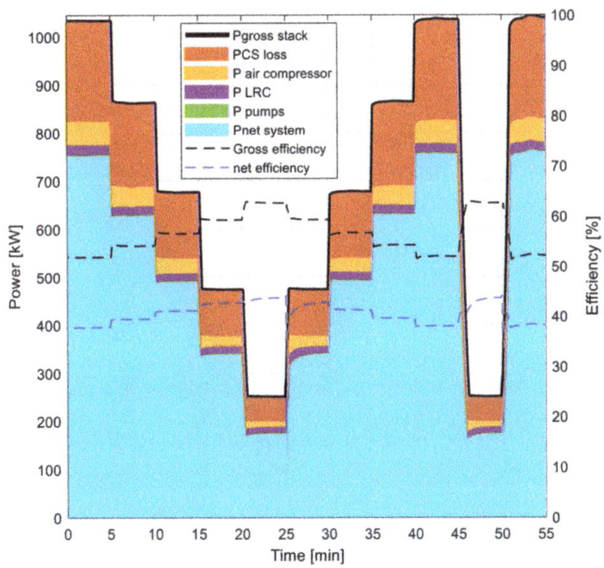

Figure 6. Simulation of load-following operation: time profile of gross power generations, power internal consumptions and net power generation (left axis) and gross stack and net system efficiency (right axis).

Regarding the plant auxiliaries, the consumption of the pumps is negligible, the hydrogen compressor consumption is almost constant at any load (10–20 kW$_{el}$), while the air blower consumption increases proportionally to the load (from 10 kW$_{el}$ to 50 kW$_{el}$). The Power Conversion System (PCS), whose efficiency was assumed equal to 80%, accounts for 2/3 of the total power losses. Indeed, this component represents the major source of losses when its efficiency is lower than 90%.

Regarding the efficiencies, the stack gross efficiency shows a minimum of 52% at the maximum current (1 A/cm^2) and a maximum of 63% at the minimum current (0.2 A/cm^2). The net efficiency reflects the gross efficiency profile: it is 38% at 1 A/cm^2 and 44% at 0.2 A/cm^2. The net efficiency values would increase to 50% at 0.2 A cm^2 and 43% at 1 A/cm^2 with a PCS efficiency of 90%. As a reference, the maximum net efficiency, excluding the PCS losses, would be equal to 56% at 0.2 A/cm^2 and 48% at 1 A/cm^2.

4.2.2. Warm-Up

Cold start-up simulations are performed to understand the time span that the plant requires to reach nominal operating conditions, when it is started from ambient temperature, such as after a long shut-down. This information is important to plan when to start-up the system, according to the services it has to provide.

Warm-up simulations assume a constant ambient temperature equal to 20 °C. In the simulations, the warm-up initial state sets:

- air compressor at nominal speed (giving nominal air stoichiometry at nominal current);
- cathode backpressure 1.35 bar (activated controller);
- LRC at nominal point;
- no hydrogen bypass;
- nominal coolant temperature gain over the stack (activated controller);
- minimum amount of coolant (10%) to heat exchangers to heat up the humidifiers;
- FC stack at its minimum load (0.2 A/cm^2).

Then, the current density is increased up to the nominal value (1 A/cm^2) according to a warm-up procedure proposed by the Grasshopper consortium to limit stack degradation. The rate of increase of the current density is limited to ensure a correct humidification of the membrane: a maximum current density is permitted for each value of the air temperature at the stack inlet and for each value of the average coolant temperature along the stack. Only when current density and stack temperature setpoints are reached, all controllers are activated, thus the choice of the control does not affect the warm-up results.

Simulation results show that the current density is limited by the average coolant temperature over the entire warm-up procedure (Figure 7a). The current remains at its minimum for 71 min, then it starts rising and it reaches the nominal value after 2 h and 28 min. The coolant temperature at stack inlet reaches the setpoint after 5 additional minutes (Figure 7b). Conversely, the pressure reaches the nominal value in about 2 s.

Figure 7c,d show the average air and hydrogen relative humidity (RH) over the stack, computed with respect to the coolant average temperature. The air average RH decreases over time because the average coolant temperature increases faster than the temperature of the air at the humidifier outlet, which determines the air absolute humidity at the stack inlet, but the air average RH is above 100% during the entire warm-up process. The hydrogen average RH is close to 100% during the warm-up process.

The air utilization factor increases with the current (Figure 7e), starting from ~10% and reaching the nominal point (50%) at 1 A/cm^2, as expected. A very low fuel utilization factor (between 6% and 30%, respectively, at 0.2 and 1 A/cm^2) is obtained with the LRC at nominal speed and no hydrogen bypass (Figure 7f). However, when the bypass is activated, the desired setpoint (67%) is reached.

Finally, Figure 8 shows a small increase of both the gross and net efficiency over time at the beginning of the simulation, before the current starts to increase, which is explained by the stack temperature increase. Then, as expected, the gross efficiency decreases when the current increases. Conversely, the net efficiency increases, reflecting the lower impact of the constant internal power consumptions.

The parameters that mostly affect the warm-up time are identified through a sensitivity analysis, performed with the final goal of identifying strategies to accelerate the system warm-up. The analysis considers:

A. A different warm-up strategy, which does not limit the current rise at low temperature. The maximum current rate of increase is set equal to 2.5%/s (same limit set at the nominal point).
B. A reduced amount of water in the humidifiers, assuming a 50% reduction of the water inventory in the humidifier tank when the simulation starts, reducing the humidifier thermal inertia.

C. A more compact system, assuming to connect the system components with pipes three times shorter and with diameters computed to have the same pressure drops as in the base case.

Table 3 characterizes, for each case, the system in terms of water inventory (i.e., water content in the main plant components at the beginning of the simulation). The total amount of liquid in the system is −5% in case B and −38% in case C, with respect to the base case. Additionally, in case C, the amount of coolant in the pipes is −70%. This information is important in order to understand the simulation results since the heat capacity of the water influences the overall thermal inertia of the system.

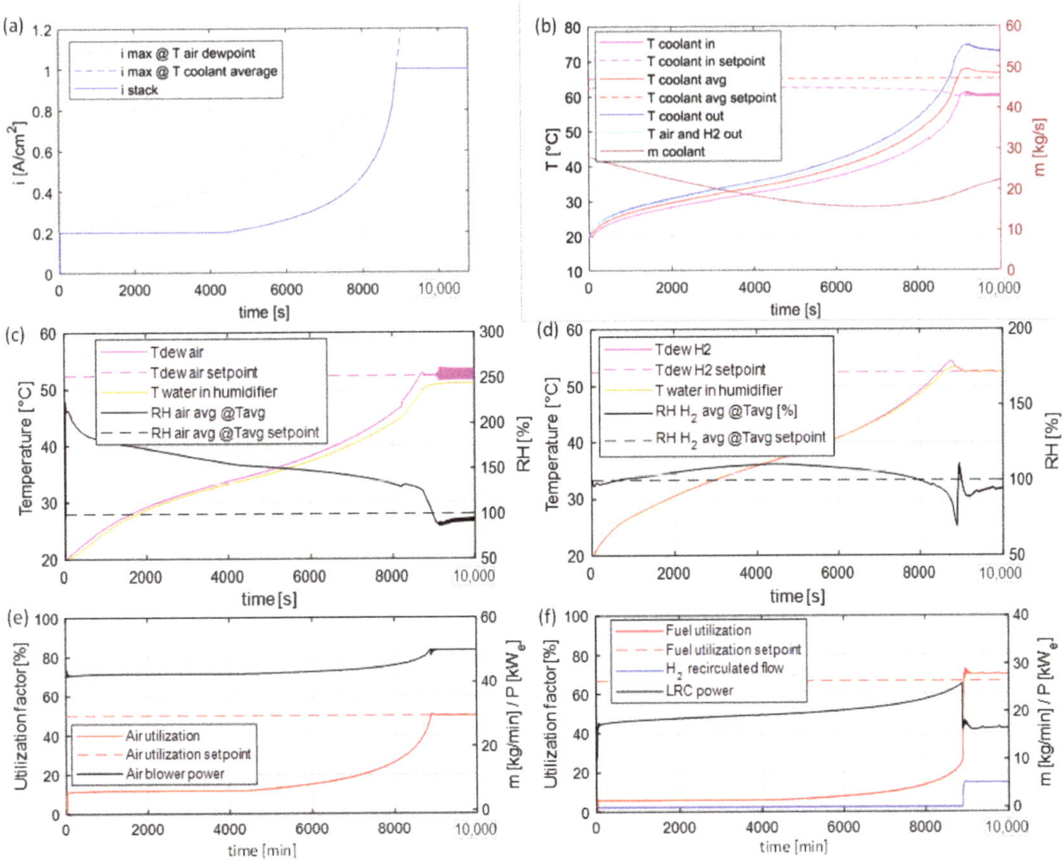

Figure 7. Warm-up simulation: time profile of (**a**) current density vs its maximum values according to the air dewpoint temperature and the average coolant temperature; (**b**) coolant temperature at stack inlet vs setpoint, average coolant temperature over the stack vs setpoint, coolant temperature at stack outlet, air and hydrogen temperature at stack outlet and coolant flowrate; (**c**)/(**d**) air/hydrogen dew point temperature at stack inlet vs setpoint, water temperature in the air/hydrogen humidifier tank, and air/hydrogen average RH vs setpoint; (**e**) air utilization factor and setpoint, and air blower power consumptions; (**f**) fuel utilization factor vs setpoint, mass flowrate of hydrogen recirculated over the LRC, and LRC power consumptions.

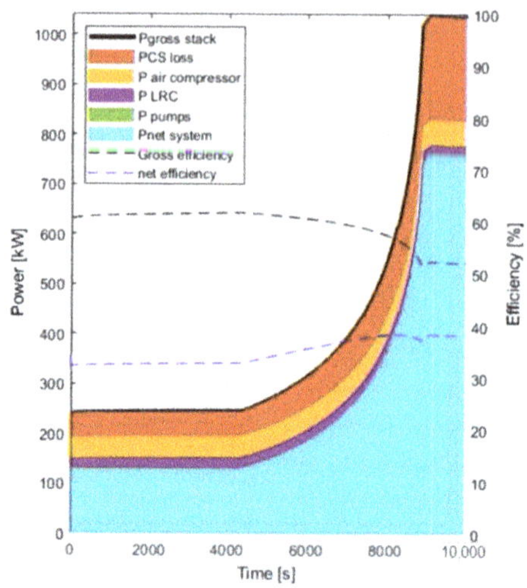

Figure 8. Warm-up simulation: time profile of gross power generations, power internal consumptions and net power generation (left axis) and gross stack and net system efficiency (right axis).

Table 3. Water content (inventory) in the main plant component for the base case and cases A, B, and C of the sensitivity analysis.

	Base—A	**B**	**C**
Coolant in pipes	2416.5 (49.0%$_{tot}$)	2416.5 L (64.0%$_{tot}$)	711.9 L (23.2%$_{tot}$)
Water in air humidifiers pipes	102.1 (2.1%$_{tot}$)	102.1 L (2.7%$_{tot}$)	23.6 L (0.8%$_{tot}$)
Water in H2 humidifiers pipes	102.3 L (2.1%$_{tot}$)	102.3 L (2.7%$_{tot}$)	16.3 L (0.5%$_{tot}$)
Water in air humidifiers tanks	1650 L (33.4%$_{tot}$)	825 L (21.9%$_{tot}$)	1650 L (53.9%$_{tot}$)
Water in H2 humidifiers tanks	660 L (13.4%$_{tot}$)	330 L (8.7%$_{tot}$)	660 L (21.6%$_{tot}$)
Total	4930.9 L	3775.9 L	3061.8 L

Results of the simulations are summarized in Table 4, which characterizes the warm-up, indicating the time required for the system to reach full load and nominal operating conditions ('Warm-up time'), the amount of hydrogen consumed during the warm-up time, the net electric energy that is delivered during the warm-up time (assuming no external limits on the net power output), and the specific hydrogen consumption during the warm-up time.

The warm-up time is shorter in case A with respect to the base case, because of the higher stack thermal losses (higher power and lower efficiency) at higher currents. Indeed, while in the base case, the current rise is proportional to the temperature and the nominal current is reached only at the end of the warm-up process (2 h 38 min), in case A the nominal current is reached in less than 2 min. It can be concluded that the choice of the warm-up strategy strongly affects the warm-up time. The reduction of the amount of water in the humidifier tanks slightly reduces the warm-up time. In case B, the 5% reduction in the total water inventory leads to a 9% reduction of the warm-up time. Conversely, the design of a compact system is very effective. In case C, a 38% reduction in the water inventory leads to a 71% reduction in the warm-up time. In this case, the shorter pipes allow having smaller diameters and higher velocity, keeping constant the total pressure drops. Thus, the reduction of the pipe lengths positively affects the results because of the lower total thermal inertia and because the time needed for the coolant to go from the FC stack to the heat exchangers and vice-versa decreases.

Table 4. Warm-up simulations: warm-up time, amount of hydrogen consumed, delivered electric energy generated and average specific hydrogen consumption during warm-up, for the base case and for cases A, B, and C of the sensitivity analysis.

	Base	A	B	C
Warm-up time	2 h 38 min	34 min	2 h 27 min	46 min
Consumed hydrogen during warm-up [kg_{H2}]	51.2	32.0	49.4	16.1
Delivered energy during warm-up [kWh_{el}]	608.7	372.5	588.3	192.4
Specific consumption during warm-up [kg_{H2}/kWh_{el}]	0.08	0.08	0.08	0.08

5. Conclusions

Future evolution of the electric grid in a low-CO_2 emission and high renewable-based scenario will increasingly require the availability of clean and dispatchable energy resources. In this framework, hydrogen-based fuel cell systems could provide a modular, reliable, and zero-emission solution to the necessity of distributed grid support, providing grid ancillary services thanks to their fast ramp rate and the load-following capability. This work contributes to the analysis of the dynamic behavior of fuel cell power plants based on low temperature Proton Exchange Membrane cells, focusing on the development of a dynamic numerical model of a flexible PEM FC power plant based on the layout of the 100-kW pilot unit built within the EU Horizon 2020 GRASSHOPPER project. The purpose is to develop an instrument to simulate the system operation, aiming at maximizing the system efficiency and flexibility while limiting the degradation rate.

The dynamic model includes FC stack, all main plant components, and controllers. The comparison of simulation results with experimental data from a 100 kW_{el} plant has shown that the model reproduces, with good accuracy, both the components stationary behavior at different load (e.g., stationary values of pressure, temperature, power consumptions, etc.) and the system dynamics (e.g., thermal transients, effects related to delays in mass transfer, etc.)

The validated model has been applied to assess the flexible performance of a 1-MW_{el} unit, providing a suggestion to improve the plant scale-up.

Load-following operation simulations showed that the system is able to follow a load setpoint. When the current setpoint is changed stepwise, a precise backpressure control and a good temperature control are possible. Conversely, deviations of the air and fuel utilization factors from their setpoints affect air and fuel average relative humidity over the stack and suggest the adoption of more advanced controllers (e.g., smith predictors). Both stack gross efficiency and system net efficiency increase while the load decreases. The power conversion system, having an 80% efficiency, results in the main component limiting the system net efficiency, and an improvement of the plant net efficiency up to 10 percentage points is possible by acting on the PCS efficiency only. Indeed, a PCS efficiency higher than 80% is expected for the MW-scale system where, thanks to the higher number of cells and the consequently higher stack voltage, the DC/DC converter is not needed and only the inverter (with efficiency higher than 90%) can be installed.

Simulations of cold start-up show that, starting from 20 °C and limiting the current density at low temperature, the plant takes 2 h 38 min to reach the nominal point. The warm-up time decreases substantially if the stack is operated at maximum current (34 min), thanks to the higher thermal losses of the stack. However, operating at high current and low temperature may reduce the lifetime of the cells. Other solutions to reduce the warm-up time, without affecting the stack, consider a reduction of the water inventory. Halving the water content in the humidifier slightly reduces the warm-up time (−9%), while using pipes three times shorter to connect the plant components appears as the best solution (−70%). The design of a compact system should therefore be a priority in the perspective of building a flexible unit, able to rapidly start-up.

These suggestions have been taken into account in the design of the flexible 1-MW FC power plant, which is the target of the Grasshopper project. Thus, the commercial FC system design will combine the technical improvements deriving from the simulation results with improvements deriving from an economic analysis.

Author Contributions: Conceptualization, E.C. and G.G.; methodology, E.C.; software, E.C.; validation, E.C.; formal analysis, E.C.; investigation, E.C. and G.N.C.; writing—original draft preparation, E.C.; writing—review and editing, G.G. and S.C.; visualization, E.C. and G.G.; supervision, S.C.; funding acquisition, G.G. and S.C. All authors have read and agreed to the published version of the manuscript.

Funding: This project has received funding from the Fuel Cells and Hydrogen 2 Joint Undertaking (now Clean Hydrogen Partnership) under Grant Agreement No 779430. This Joint Undertaking receives support from the European Union's Horizon 2020 Research and Innovation program, Hydrogen Europe and Hydrogen Europe Research.

Institutional Review Board Statement: Not Applicable.

Conflicts of Interest: The authors declare no conflict of interest.

Nomenclature

Acronyms	
FC	Fuel Cell
LHV	Low Heating Value
LRC	Liquid Ring Compressor
PEM	Proton Exchange Membrane
RES	Renewable Energy Sources
RH	Relative Humidity
RMSE	Root Mean Square Error
Symbols	
A	Area, cm^3
C	Heat capacity, kJ/K
c_L	Specific liquid heat capacity, kJ/kg/K
D	Diameter, m
F	Faraday constant, 96485 C/mol
h	Specific enthalpy, J/mol
i	Current density, mA/cm^2
J	Moment of inertia, $kg{\cdot}m^2$
L	Length, m
$\dot{m}$	Mass flowrate, kg/s
MM	Molecular Mass, kg/kmol
p	Pressure, bar
P	Power, kW
R	Universal gas constant, 8.314 J/mol/K
T	Temperature, K
V	Voltage, V
x_J	Molar fraction of species j, mol_j/mol_{tot}
η	efficiency, %
ω	rotational speed, rad/s
ρ	Density, kg/mol

References

1. European Commission. *"Fit for 55": Delivering the EU's 2030 Climate Target on the Way to Climate Neutrality*; European Commission: Brussels, Belgium, 2021.
2. IRENA—International Renewable Energy Agency. Data & Statistics. Available online: https://www.irena.org/Statistics (accessed on 1 June 2022).
3. IEA—International Energy Agency. World Energy Outlook 2021. 2021. Available online: www.iea.org/weo (accessed on 1 June 2022).
4. Impram, S.; Varbak, S.; Oral, B. Challenges of renewable energy penetration on power system flexibility: A survey. *Energy Strateg. Rev.* **2020**, *31*, 100539. [CrossRef]
5. Gils, H.C.; Scholz, Y.; Pregger, T.; De Tena, D.L.; Heide, D. Integrated modelling of variable renewable energy-based power supply in Europe. *Energy* **2017**, *123*, 173–188. [CrossRef]
6. Doyle, J.R.; Johlas, H. Energy Storage Considerations for High Renewable Power Penetration: A Case Study. In *Advances in Smart Grid and Renewable Energy. Lecture Notes in Electrical Engineering*; Springer: Singapore, 2019; Volume 435.
7. Luo, X.; Wang, J.; Dooner, M.; Clarke, J. Overview of current development in electrical energy storage technologies and the application potential in power system operation. *Appl. Energy* **2015**, *137*, 511–536. [CrossRef]
8. Steinke, F.; Wolfrum, P.; Hoffmann, C. Grid vs. storage in a 100% renewable Europe. *Renew. Energy* **2013**, *50*, 826–832. [CrossRef]
9. European Commission. *Clean Energy for All Europeans*; European Commission: Brussels, Belgium, 2019.
10. Hydrogen Council. *How Hydrogen Empowers the Energy Transition*; Hydrogen Council: London, UK, 2017.
11. IEA Report. The Future of Hydrogen. 2019. Available online: https://www.iea.org/reports/the-future-of-hydrogen (accessed on 15 May 2022).
12. European Commission. *The European Green Deal*; European Commission: Brussels, Belgium, 2019.
13. Sadik-Zada, E.R. Political Economy of Green Hydrogen Rollout: A Global Perspective. *Sustainability* **2021**, *13*, 13464. [CrossRef]
14. Nanaeda, K.; Mueller, F.; Brouwer, J.; Samuelsen, S. Dynamic modeling and evaluation of solid oxide fuel cell—Combined heat and power system operating strategies. *J. Power Sources* **2010**, *195*, 3176–3185. [CrossRef]
15. Kang, S.; Ahn, K.-Y. Dynamic modeling of solide oxide fuel cell and engine hybrid system for distributed power generation. *Appl. Energy* **2019**, *195*, 1086–1099. [CrossRef]
16. McLarty, D.; Kuniba, Y.; Brouwer, J.; Samuelsen, S. Experimetnal and theoretical evidence for control requirements in solid oxide fuel cell gas turbine hybrid systems. *J. Power Sources* **2012**, *209*, 195–203. [CrossRef]
17. Mastropasqua, L.; Donazzi, A.; Campanari, S. A Multi-Scale Modelling Approach and Experimental Calibration Applied to Commercial SOFCs. *ECS Trans.* **2017**, *78*, 2645. [CrossRef]
18. Zaccaria, V.; Tucker, D.; Traverso, A. A distributed real-time model of degradation in a solid oxide fuel cell, part I—Model characterization. *J. Power Sources* **2016**, *311*, 175–181. [CrossRef]
19. Larosa, L.; Traverso, A.; Ferrari, M.K.; Zaccaria, V. Pressurized SOFC Hybrid Systems: Control System Study and Experimental Verification. *J. Eng. Gas Turbines Power* **2015**, *137*, 031602. [CrossRef]
20. Zhang, L.; Li, X.; Jiang, J.; Li, S.; Yang, J.; Li, J. Dynamic modeling and analysis of a 5-kW solide oxide fuel cell system from the perspective of cooperative control of thermal safety and high efficiency. *Int. J. Hydrogen Energy* **2015**, *40*, 456–476. [CrossRef]
21. Colombo, E.; Bisello, A.; Casalegno, A.; Baricci, A. Mitigating PEMFC degradation start-up: Locally resolved experimental analysis and transient physical modelling. *J. Electrochem. Soc.* **2021**, *168*, 054508. [CrossRef]
22. Grimaldi, A.; Baricci, A.; De Antonellis, S.; Casalegno, A. Characterization and modelling of air humidification in Fuel Cell System for transport sector. *E3S Web Conf.* **2022**, *334*, 06009. [CrossRef]
23. Cavo, M.; Gadducci, E.; Rattazzi, D.; Rivarolo, M.; Magistri, L. Dynamic analysis of PEM fuel cells and metal hydrides on a zero-emission ship: A model-based approach. *Int. J. Hydrogen Energy* **2021**, *46*, 32630–32644. [CrossRef]
24. Rabbani, A.; Rokni, M. Dynamic characteristics of an automotive fuel cell system for transitory load changes. *Sustain. Energy Technol. Assess.* **2013**, *1*, 34–43. [CrossRef]
25. Barelli, L.; Bidini, G.; Gallorini, F.; Ottaviano, A. An energetic-exergetic analysis of a residential CHP system based on PEM fuel cell. *Appl. Energy* **2011**, *88*, 4334–4342. [CrossRef]
26. Verhage, A.J.L.; Coolegem, J.F.; Mulder, M.J.J.; Yildirim, M.H.; De Bruijn, F.A. 30,000 h operation of a 70 kW stationary PEM fuel cell system using hydrogen from a chlorine factory. *Int. J. Hydrogen Energy* **2013**, *38*, 4714–4724. [CrossRef]
27. Karim, H.; Auvinen, S.; Ihonen, J.; Koski, P.; Pulkkinen, V.; Ker, T. Operational experiences of PEMFC pilot plant using low grade hydrogen from sodium chlorate production process. *Int. J. Hydrogen Energy* **2017**, *2*, 27269–27283. [CrossRef]
28. Campanari, S.; Guandalini, G.; Coolegem, J.; Ten Have, J.; Hayes, P.; Pichel, A.H. Modeling, Development, and Testing of a 2 MW Polymeric Electrolyte Membrane Fuel Cell Plant Fueled with Hydrogen from a Chlor-Alkali Industry. *J. Electrochem. Energy Convers. Storage* **2019**, *16*, 1–9. [CrossRef]
29. DEMCOPEM-2MW Project Official Website. Available online: http://www.demcopem-2mw.eu/ (accessed on 10 October 2017).
30. McBride, B.J.; Gordon, S.; Reno, M.A. *Coefficients for Calculating Thermodynamic and Transport Properties of Individual Species*; NASA Technical Memorandum 4513, no. NASA-TM-4513; NASA: Washington, DC, USA, 1993; p. 98. Available online: http://ntrs.nasa.gov/archive/nasa/casi.ntrs.nasa.gov/19940013151_1994013151.pdf (accessed on 1 March 2022).
31. Wagner, W.; Pruss, A. International Equations for the Saturation Properties of Ordinary Water Substance Revised According to the International Temperature Scale of 1990. *J. Phys. Chem. Ref. Data* **1993**, *22*, 783. [CrossRef]

32. Crespi, E.; Guandalini, G.; Gößling, S.; Campanari, S. Modelling and optimization of a flexible hydrogen-fueled pressurized PEMFC power plant for grid balancing purposes. *Int. J. Hydrogen Energy* **2021**, *46*, 13190–13205. [CrossRef]
33. Barbir, F. *PeM Fuel Cells: Theory and Practice*; Academic Press: Burlington, MA, USA, 2005.
34. Pathapati, P.R.; Xue, X.; Tang, J.Ã. A new dynamic model for predicting transient phenomena in a PEM fuel cell system. *Renew. Energy* **2005**, *30*, 1–22. [CrossRef]
35. Zenith, F.; Skogestad, S. Control of fuel cell power output. *J. Process Control* **2007**, *17*, 333–347. [CrossRef]
36. Khan, M.J.; Iqbal, M.T. Modelling and analysis of electrochemical, thermal, and recetant flow dynamics for a PEM fuel cell system. *Fuel Cells* **2005**, *5*, 463–475. [CrossRef]
37. Lee, C.; Yang, J. Modeling of the Ballard-Mark-V proton exchange membrane fuel cell with power converters for applications in autonomous underwater vehicles. *J. Power Sources* **2011**, *196*, 3810–3823. [CrossRef]
38. Malekbala, M.R.; Khodadadi Azadboni, R.; Kazempoor, P. Modeling and control of a proton exchange membrane fuel cell with the air compressor according to requested electrical current. *Therm. Sci.* **2015**, *19*, 2065–2078. [CrossRef]
39. Pukrushpan, J.T. Modeling and Control of Fuel Cell Systems and Fuel Processors. Ph.D. Thesis, University of Michigan, Ann Arbor, MI, USA, 2003.
40. Cary, M.; Ebrahimi, M.; Ffrench, K.; Sbaschnig, R. Throttle body: Modelling and identification. *Proc. Inst. Mech. Eng. Part D J. Automob. Eng.* **2001**, *215*, 813–825. [CrossRef]

energies

MDPI

Article

Modeling of Fuel Cells Characteristics in Relation to Real Driving Conditions of FCHEV Vehicles

Ireneusz Pielecha

Faculty of Civil and Transport Engineering, Poznan University of Technology, ul. Piotrowo 3, 60-965 Poznan, Poland; ireneusz.pielecha@put.poznan.pl; Tel.: +48-61-224-4502

Abstract: Fuel cells are one of the zero-emission elements of modern automotive drive systems. This article presents theoretical identification of the component parameters of indicators for the fuel cell operating conditions. Activation, ohmic, and mass transport losses were identified. Current–voltage characteristics were provided along with an analysis of typical cell losses. The actual performance characteristics of fuel cells were analyzed for Toyota Mirai I and II generation vehicles. The fuel cells operating conditions were derived and analyzed in the context of real driving conditions. Therefore, urban, rural, and motorway conditions were used. The vehicles were equipped with PEM fuel cells supplying power equal to 114 kW (1st gen.) or 128 kW (2nd gen.). The average fuel cell stack power values depend on the driving conditions: urban (about 10 kW), rural (20 kW) and motorway (about 30–40 kW) driving modes. The different power ratings of fuel cells combined with different battery generations resulted in a variation in the cells operating conditions. Analyses conducted in various traffic conditions indicated the possibility of determining losses related to the fuel cells. The analysis of fuel cell losses shows the greatest values for activation losses when the cells are under high load (for both generations)—i.e., in motorway driving conditions. The voltage of resistive losses reached its maximum in urban driving conditions when the load on the fuel cells was small.

Keywords: fuel cells; current–voltage characteristics; fuel cells modeling; fuel cell losses; road tests

Citation: Pielecha, I. Modeling of Fuel Cells Characteristics in Relation to Real Driving Conditions of FCHEV Vehicles. *Energies* **2022**, *15*, 6753. https://doi.org/10.3390/en15186753

Academic Editor: Orazio Barbera

Received: 21 August 2022
Accepted: 13 September 2022
Published: 15 September 2022

1. Introduction

The search for low-emission drive systems leads to the increasingly more common use of hybrid drive systems in transport means. Unfortunately, such systems turn out to be insufficient in terms of meeting carbon dioxide emission limits. These limits in 2021 were 95 g/km. On 14 July 2021, the European Commission proposed more ambitious goals for 2030 as part of the European Green Deal [1]. Compared to the 2021 baseline, the average CO_2 emissions from new vehicles registered between 2025 and 2030 would have to be reduced by 15% (unchanged from previous assumptions) and 55% (up from 37.5% previously), respectively. The legislation also suggests that new cars will need to be zero-emission by the year 2035. Such a drastic reduction in CO_2 emissions necessitates the use of electric drive systems or the use of fuel cell systems.

Currently, the fee for exceeding the average CO_2 emission limit (measured up to three decimal places) for a given manufacturer of passenger vehicles in accordance with the Regulation of the European Parliament and the EU Council 2019/631 amounts to 95 EUR for each newly registered vehicle [2].

As mentioned before, these regulatory proposals also suggest selling only zero-emission cars by 2035. In order to achieve a total reduction of CO_2 emissions using electric drive systems or fuel cell systems seems like the only viable solution. Such systems of production or storage of energy were shown in Figure 1, also including data on their specific power and specific energy parameters. This results in a great similarity between fuel cells and batteries in terms of specific power and energy. However, the conditions for either charging or refueling these systems are completely different. The charging frequency of an electric vehicle is several times greater than that of a vehicle powered by fuel cells. In addition, the refueling time of vehicles powered by hydrogen fuel cells is several times shorter than the charging time necessary to recharge an electric vehicle's batteries. The hydrogen refueling time is currently about 3 min—for a tank weighing about 5 kg H_2. The shortest FastCharging charging times take no less than several minutes.

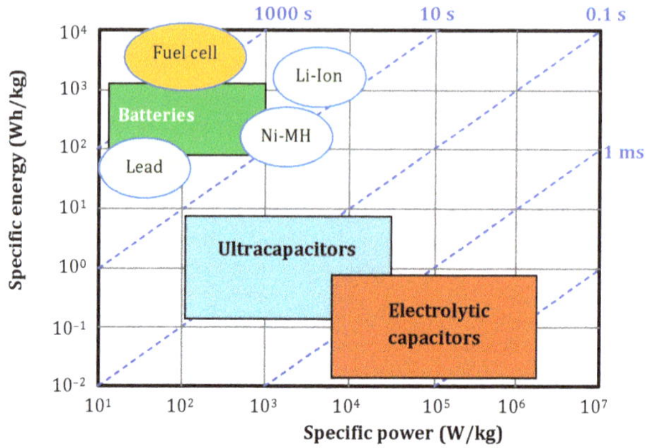

Figure 1. Specific energy vs. specific power of energy storage system [3].

The ability to model fuel cells enables their complete diagnostics, which improves the knowledge of fuel cell aging. In addition, operational information influences the prediction of the behaviour of fuel cells in real operating conditions.

2. Fuel Cell Use in Vehicle Drive Systems

Typical solutions appearing in the literature concern two vehicle propulsion solutions (shown as diagrams in Figure 2): FCEV (fuel cell electric vehicle) [3] and FCHEV (fuel cell hybrid electric vehicle) [3–5]. The first one only uses fuel cells for all vehicle propulsion. Although this solution is much simpler in technical terms, it is nevertheless more difficult to implement. The main downside of this system stems from the fact that the fuel cell is an electro-chemical device, and the maximum rate at which chemical reactions take place usually takes a certain time to be reached [6]. Therefore, it is difficult to obtain the maximum power of such a system immediately after its activation. The second solution is a typical hybrid drive system. The fuel cell is usually the primary source of energy for the vehicle.

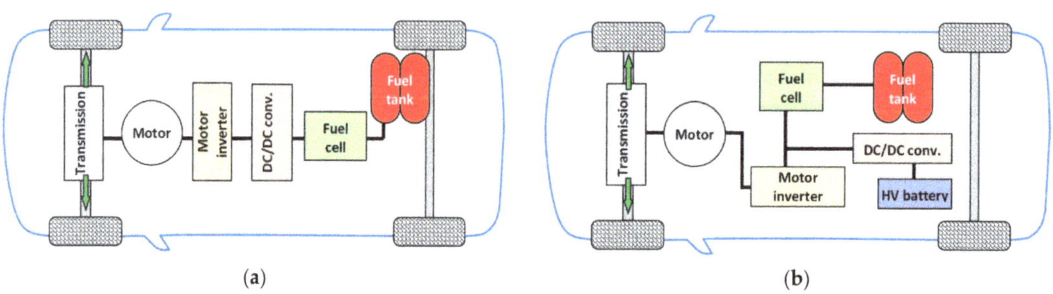

Figure 2. Drive system diagrams employing fuel cell technology: (**a**) FCEV; (**b**) FCHEV (based on [4]).

A battery with a low energy capacity of 1–3 kWh [7,8] would be used as a support device to complement the operation of the fuel cells. However, starting the vehicle and moving it in the initial period is completed largely with the use of batteries (since the fuel cell is not able to work independently immediately after activation).

In the nomenclature, hybrid vehicles are usually divided into two groups [9]: not-off-vehicle charging hybrid electric vehicles (NOVC-HEV) and off-vehicle charging hybrid electric vehicles (OVC-HEV).

Contemporary FCHEV vehicles offer relatively low battery energy capacity (NOVC-HEV): Toyota Mirai I gen.—1.59 kWh (Ni-MH); [10] Toyota Mirai II gen.—1.24 kWh (Li-Ion) [10], Hyundai Nexo—1.56 kWh (Li-Ion) [11]; Honda Clarity—1.7 kWh (Li-Ion) [12]. These small capacities are due to the fact that the fuel cell serves as the primary propulsion system of the vehicle instead of a battery. The analysis of the battery capacity and the power of the fuel cell was presented in Figure 3.

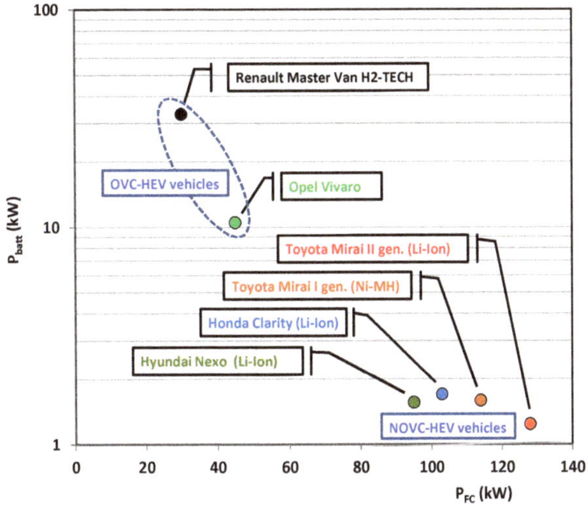

Figure 3. Comparison of cell power vs. high voltage battery capacity [10,13–15].

Figure 3 also presents some other drive systems solutions. The OVC-HEV category is used to describe plug-in vehicles. The Opel Vivaro with a fuel cell system (45 kW) includes a high voltage battery with a capacity of 10.5 kWh [16], which provides a vehicle range of 50 km in electric mode. Another example of such is the Renault Master Van H2-TECH equipped with a 30 kW fuel cell and a 33 kWh battery [17].

All the mentioned drive systems achieve a specific power density of 3.1 kW/dm^3 [10–12]. The new Toyota Mirai II achieves a specific power density of 4.4 kW/dm^3 (including end plates).

A different drive philosophy is represented by FEV proposed with the BREEZE! project [18]. It is a drive using fuel cells as a range extender. The vehicle was equipped with batteries with an electric capacity of 12 kWh and a fuel cell (PEM) with a power of 30 kW. The specific power density of the cell was 0.45 kW/dm^3, and the drive efficiency reached 40–50%. The fuel cell consisted of 150 individual cells with an active surface area of 300 cm^2.

3. Modeling Fuel Cells

Modeling of fuel cells may include electrochemical [19–21], flow [21,22] and voltage losses calculations [23].

Based on the Butler–Volmer equation [24] relating to the equilibrium potential of a fuel cell, three separate voltage loss areas can be distinguished:

- Voltage activation;
- Resistive;

- Mass transport.

The general equation that includes all types of voltage losses can be expressed as (Figure 4):

$$U_{FC} = E + U_{act} + U_{ohm} + U_{trans}, \tag{1}$$

where U_{FC}—fuel cell voltage, U_{act}—overvoltage (activation losses), U_{ohm}—ohmic voltage losses, U_{trans}—losses of tension due to mass transport.

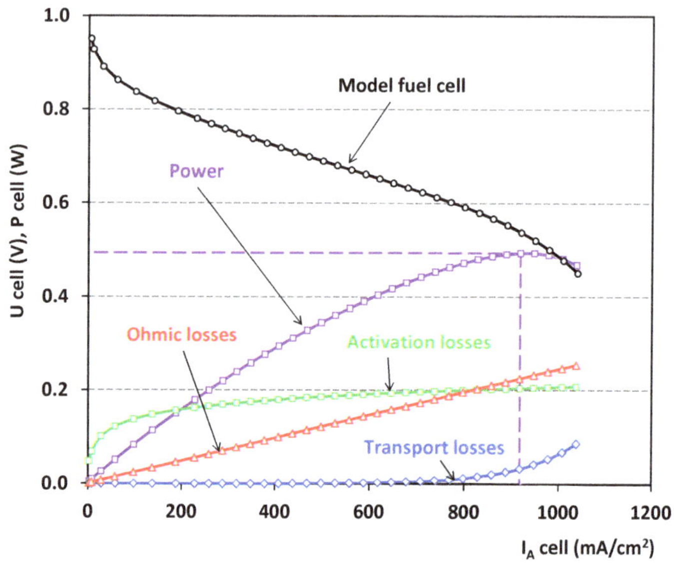

Figure 4. Current_voltage characteristics with the specification of typical losses of a fuel cell.

The voltage of the fuel cell with an open electrical circuit can be denoted as [25]:

$$E_{OC} = \frac{\Delta\overline{g_f}}{2 \cdot F'} \tag{2}$$

where $\Delta\overline{g_f}$—Gibbs free energy, F—Faraday constant.

The effect of the voltage change with an open circuit and the associated fuel cell losses were shown in Figure 5. It should be noted that the change in E_{OC} had an impact on the current–voltage characteristic values as well as on the cell power. However, it does not affect the value of the losses. The lack of impact that the E_{OC} has on the cell losses results directly from Equation (1), which means that this value is not a component of the partial elements of the fuel cell losses.

The activation loss voltage U_{act} can be expressed as:

$$U_{act} = A \cdot \ln(i), \tag{3}$$

where A—activation overvoltage, i—area-specific current (current density).

The effect of changing the A parameter on the cell characteristics was shown in Figure 6. Increasing the A value limits the cell voltage and its power. It also leads to a significant increase in activation losses.

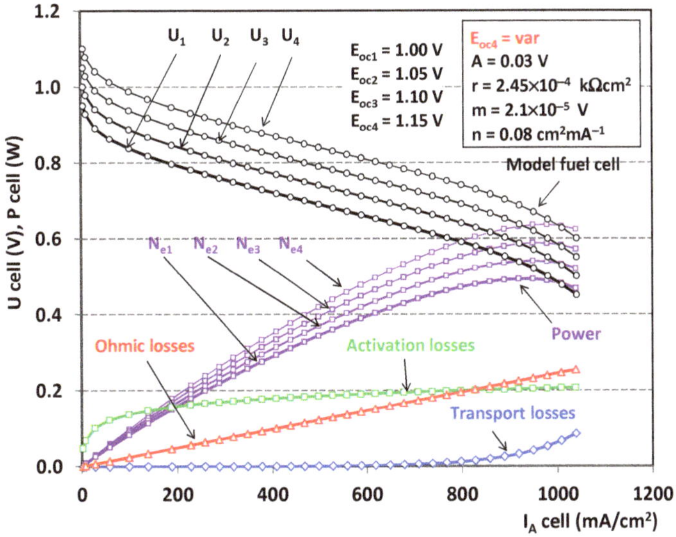

Figure 5. Influence of fuel cell E_{OC} open circuit voltage on the fuel cell characteristics and losses.

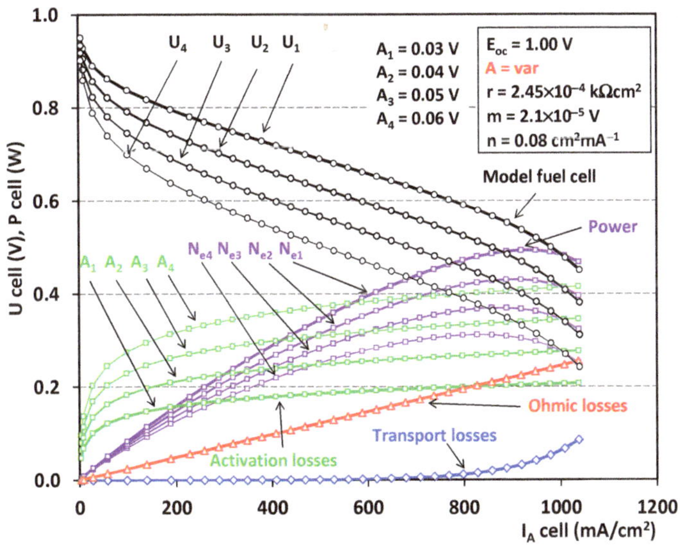

Figure 6. Impact of the changing A parameter of a fuel cell on the I–U characteristics and fuel cell losses.

Improvement in fuel cell performance through the current increase, and can be achieved by:

- Increasing the cell temperature;
- Use of effective catalysts;
- Increasing the roughness of the electrodes' surface, which increases the cell surface and leads to an increase in the current intensity;
- Increasing the concentration of the reactants, e.g., using pure oxygen instead of air;
- Increasing the reagents pressure.

Resistive (ohmic) voltage losses U_{ohm}:

$$U_{ohm} = r \cdot i, \tag{4}$$

where r—area specific resistance (internal resistance).

According to Equation (4), increasing the cell resistance r leads to an increase in resistive losses, which results in reduced cell voltage, especially in the linear range (Figure 7). The voltage drop results in a reduction of its power.

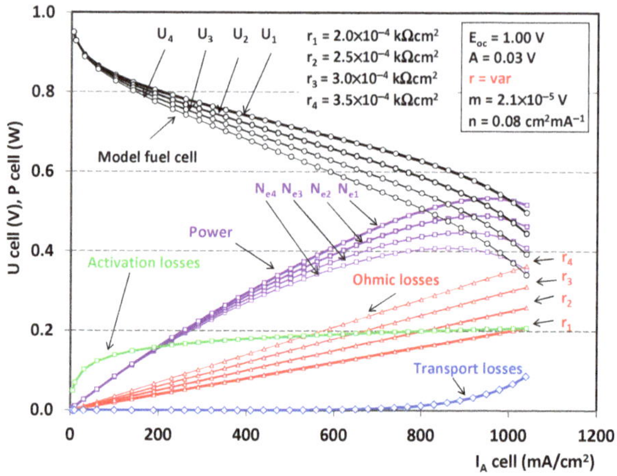

Figure 7. Impact of a changing resistance r on the I–U characteristic and fuel cell power.

There are several options for reducing resistive losses:

- Using electrodes with the highest possible conductivity;
- Dilute the electrolyte to the maximum viable values.

Mass transport voltage losses U_{trans}:

$$U_{trans} = m \cdot \exp(n \cdot i), \tag{5}$$

where m—constants in the mass-transfer overvoltage, n—constants in the mass-transfer overvoltage.

Changes in the m factor primarily affect the last fragment of the voltage–current characteristic (Figure 8). There is an increase in mass transport losses and a reduction in the voltage of the fuel cell (and thus the power of the cell) at the same time.

Using all of the component Equations (2)–(5), one can write:

$$U_{FC} = E_{OC} - A \cdot \ln(i) - r \cdot i - m \cdot \exp(n \cdot i). \tag{6}$$

Equation (6) is the basis for determining the voltage losses of a fuel cell.

The power and energy of fuel cells were determined based on the values of U_{FC}, I_{FC}, and Δt by multiplying these quantities. The battery power was determined in a similar way by multiplying the U_B and I_B values. For a complete analysis of the equations defining the energy flow analysis, see [26].

The novelty of the work in relation to the current research and analyses is the undertaken attempt to estimate the differences in the use of fuel cells in two generations of hydrogen-powered propulsion solutions. Additionally, an attempt was made to mathematically model the fuel cells using their actual operating conditions in vehicles as the basis for the calculations.

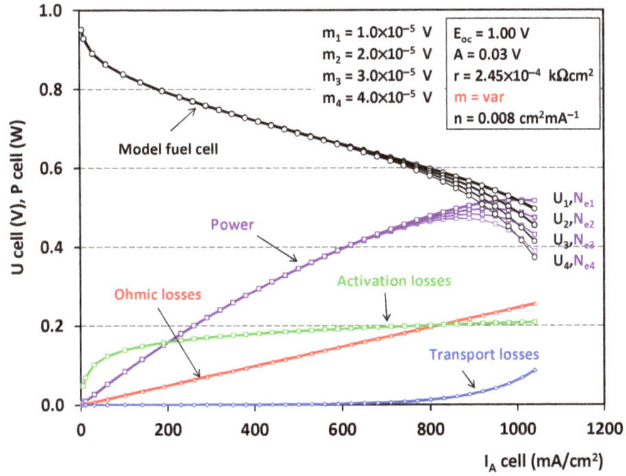

Figure 8. Impact of changes in mass m on the I–U characteristic and fuel cell power.

4. Materials and Methods

The research on fuel cell losses was carried out using as a basis two generations of the Toyota Mirai drive systems. The technical data of the drives was listed in Table 1. Both generations featured a hybrid drive system, which contained two sources of energy and propulsion. The system had the option to drive using battery power, fuel cell power, or both of these energy sources simultaneously. It was categorized as a parallel drive system as shown in Figure 5. It should be noted that, according to Table 1, the power of the fuel cell was about 4 times greater than the power provided by the battery (regardless of the energy generation in the system). This means that the fuel cell was the dominant power source in the vehicle. Both generations contain the same types of fuel cells (with slightly changed parameters) with different types of batteries. The newer generation of the drive includes Li-Ion batteries with a lower value of electric capacity but a higher value of operating voltage.

The drive system presented in Figure 9 was the same for both of the used system generations. The only differences were in the number of hydrogen storage cylinders (the newer generation of the drive had 3; while the older generation only had two) and in the drive and energy flow control method of the two generations (due to the change of system elements).

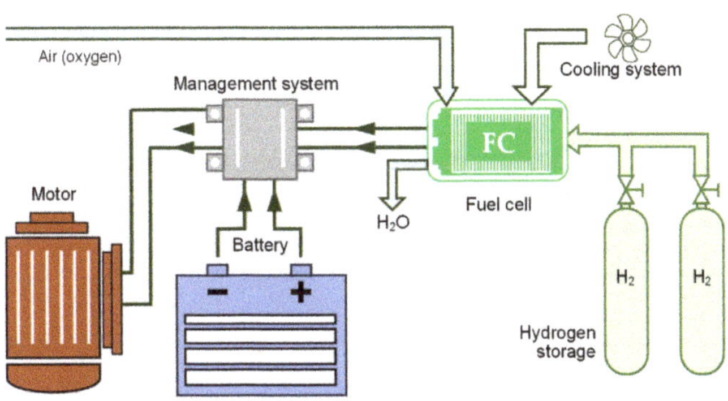

Figure 9. General schematic of the Toyota Mirai drive system.

Table 1. Toyota Mirai powertrain system [8,9,15].

Component	Parameter	Mirai I Gen.	Mirai II Gen.
Vehicle	Mass	1850 kg	2415 kg
	Top speed	179 km/h	175 km/h
	Acceleration 0 to 60 mph	9.6 s	9.2 s
	Range (homologation cycle)	Approx. 483 km	650 km
Fuel cell	Type	PEM (polymer electrolyte)	
	Power	114 kW (155 HP)	128 kW (174 HP)
	Power density	2.8 kW/kg; 3.5 kW/dm^3 (Excl. end plates)	5.4 kW/kg; 5.4 kW/dm^3 (Excl. end plates)
	Number of cells	370	330
Electric motor	Type	Permanent magnet synchronous	
	Peak power	123 kW at 4500 rpm	134 kW at 6940 rpm
	Maximum torque	335 N·m	300 N·m
	Maximum speed	13,500 rpm	16,500 rpm
Battery	Type	Nickel metal hydride (Ni-MH)	Li-Ion
	Capacity	6.5 Ah	4 Ah
	Output	25.5 kW × 10 s	31.5 kW × 10 s
	Nominal voltage	244.8 V (7.2 V × 34)	310.8 V (3.7 V × 84)
	Energy	1.59 kWh	1.24 kWh
Hydrogen storage	Internal volume	122.4 dm^3	142.2 dm^3
	Nominal pressure	70 MPa	70 MPa
	Mass	4.6 kg	5.6 kg

The research work was carried out in accordance with the RDE test requirements. Diagnostic data were obtained from a dedicated TechStream on-board diagnostic system. The data acquisition included parameters such as vehicle driving conditions (V, t), battery operating conditions (U_B, I_B), and fuel cell operating conditions (U_{FC}, I_{FC})—Figure 10.

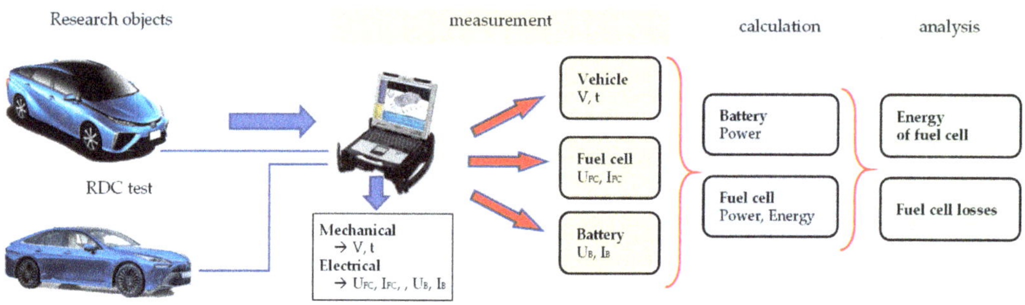

Figure 10. Methodology of testing vehicles powered by fuel cells (data sampling frequency: 1 Hz).

The mileage of the older generation Toyota Mirai (JPD10 model, production date 11 November 2015) was 40 thousand km, while the new generation (model JPD20, production date 24 November 2020)—was only 280 km. Obviously, the tested vehicles had a very different level of wear in their fuel cells. However, it was not possible to measure two generations of vehicles with the same mileage due to the production window of both these models.

Taking into account the small number of research tests, no correction factors for the difference in vehicle mileage have been introduced. Further scientific work will make it possible to identify and apply these variables to the final results of the model studies.

5. Results

5.1. Driving Test Evaluation

Road tests were carried out on a test route selected by taking into account the urban, rural, and motorway conditions in the city of Warsaw and in the vicinity. The research route fulfilled the conditions set by the RDE (real driving emissions) research test requirements [27,28]. The tests were not used to analyze exhaust emissions and therefore should be referred to as RDC (real driving conditions) tests. The main test requirements and the speed profile of both drives were shown in Figure 11. Analysis of the data in Figure 11 showed much similarity between both the driving tests. Road tests reflect the real traffic conditions to a much greater extent. However, due to the lower repeatability of such tests, the scope of requirements includes a greater possibility of variation in parameters than is normally allowed during tests performed on a chassis dynamometer. The data in Figure 11 indicated the possibility of changing the ranges of the three test phases by up to 10%. In addition, the urban phase includes a very large range of requirements in terms of stopping the vehicle. This indicates the possibility of traffic in conditions of high traffic congestion.

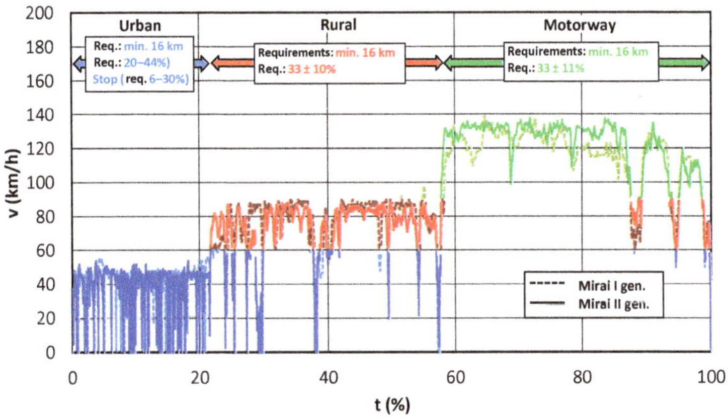

Figure 11. Research test results in accordance with the RDC requirements.

The actual test conditions, as measured, were given in Table 2. The data showed much smaller differences than the permissible values that were presented in Figure 9. Small discrepancies between the two tests indicate that further analyses are more likely to produce useful results. This means that the travel conditions were similar, and thus the chosen route meets the requirements of the RDC test.

Table 2. Selected conditions required for a valid RDC test.

Parameter	Mirai I Gen.	Mirai II Gen.	Diff. [%]	Requirements
Urban route [km]	32.93	35.08	6.1	>16
Rural route [km]	37.50	34.81	7.2	>16
Highway route [km]	42.99	44.26	2.9	>16
Total route [km]	113.43	114.15	0.6	>48
Share of urban route [%]	29.03	30.73	5.5	29–44
Share of rural route [%]	33.06	30.49	7.8	33 ± 10
Share of highway route [%]	37.90	38.77	2.2	33 ± 10
Average speed in urban route [km/h]	30.23	29.51	2.4	15–40
Share of standstill in urban route [%]	22.40	19.88	11.3	6–30
Test time [min]	114.57	119.49	4.1	90–120

The differences in the test results listed in Table 2 were designated as relative values calculated with respect to the greater value. It should be noted that despite the large permitted variations in individual travel phases, the obtained overall differences were under 10%. The maximum differences can be found in the share of parking during the urban section amounting to 11.3% (the permissible variation may be 80%).

5.2. Fuel Cells Analysis

Based on the recorded fuel cell voltage and current intensity, the cell energy consumption was determined per 100 km of the route traveled in each of the driving phases (as well as in the entire test)—Figure 12. Despite the much greater weight of the new generation vehicle, the energy consumption in the entire test was found to be lower by 5.6% (1.11 kWh/100 km). The urban phase did not show any significant difference in energy consumption (2%—corresponding to 0.1 kWh/100 km). The greatest differences were observed in the rural driving phase. Here, the differences were around 17.3% in favor of the newer vehicle generation (a change of 1.03 kWh/100 km). The last phase—motorway driving, also does not bring any differences (max 1.0%).

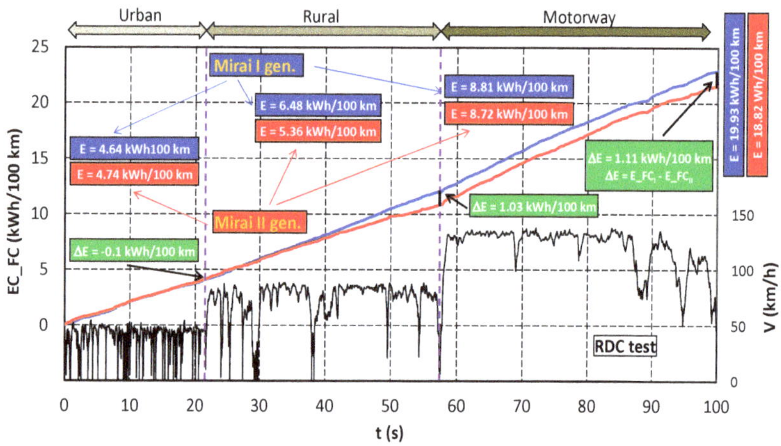

Figure 12. Changes in fuel cell energy consumption of the two generations of Toyota Mirai.

The analysis of the constituent factors made it possible to determine the fuel cells power in particular phases of the road tests (Figure 12). Greater values of energy consumption by Toyota Mirai I generation resulted from greater values of cell power in the urban and rural phases of the drive test. The average power values in all test phases were presented in the speed and power profile of the fuel cells (Figure 13).

The average fuel cell power values in the first-generation Toyota Mirai were 26% greater than the average values of the second-generation Mirai. Similarly, greater powers occurred in the rural phase of the driving test (by 35%). The motorway phase was characterized by the greater average power of the second-generation Mirai vehicle. The greater cell power was partly responsible for the observed operating conditions, but also the batteries, also with a higher power in the second generation system.

(a)

(b)

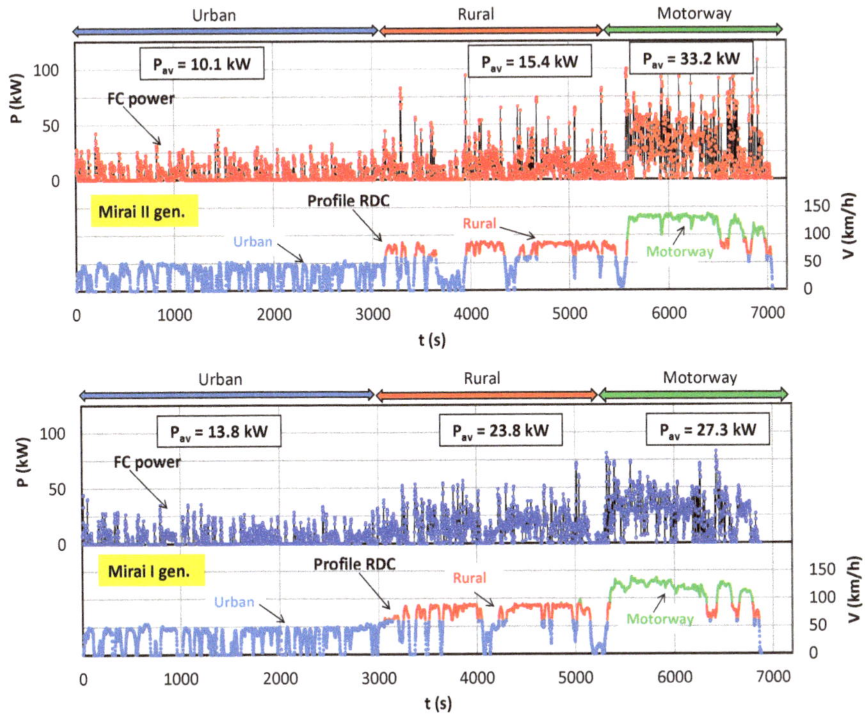

Figure 13. Fuel cell stack power and speed profiles in the RDC test: (**a**) I generation HFCHEV drive; (**b**) II generation FCHEV drive.

Analysis of the share of the cell's power in relation to the battery's power was presented in Figure 14. The data analysis showed that, regardless of the fuel cell system generation, the share of their power in the entire test was the greatest at much greater power than the battery ($P_{FC}/P_{BATT} > 90\%$). The second generation drive system also used a fairly large share of driving at the lowest system power, which means that the initial driving conditions could be achieved with battery drive alone. The transient conditions in which small shares of the cell's power and high battery power were used turned out to be small. This was shown in Figure 14 in the 20–80% P_{FC}/P_{BATT} power ratio. Summarized, the operating conditions of cells and batteries in both drive generations were within 66–69% of the total share of operating time. P_{FC}/P_{BATT} shares close to 100% do not mean the maximum power of the fuel cell. They indicate the cell's power is in the range of 90–100% greater than the battery power. This representation of the FC/BATT power ratio indicated that the system usually operates with the fuel cell providing significantly more power than that of the battery. Such a situation and vice versa were implemented in the second generation system. Cases could also be observed where the conditions were $N_{BATT} >> N_{FC}$. These are conditions labeled as $N_{FC}/N_{BATT} < 10\%$ (Figure 14b).

The analysis of the absolute power values of the fuel cell was presented in Figure 15. The fuel cell power shares were determined for each of the test phases. Although the maximum power was 114 and 128 kW, respectively, such values were not achieved in the tests, due to driving speeds required in the RDC test. The share of speeds above 145 km/h must be less than 3% of the motorway driving time (due to speed limits, the driving speed in the tests did not exceed 140 km/h).

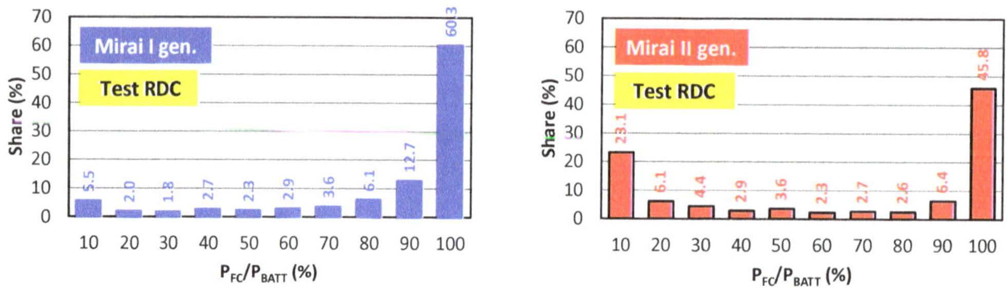

(a) (b)

Figure 14. The proportion of the fuel cell power share relative to the battery power: (**a**) for the 1st generation vehicle; (**b**) for the 2nd generation vehicle.

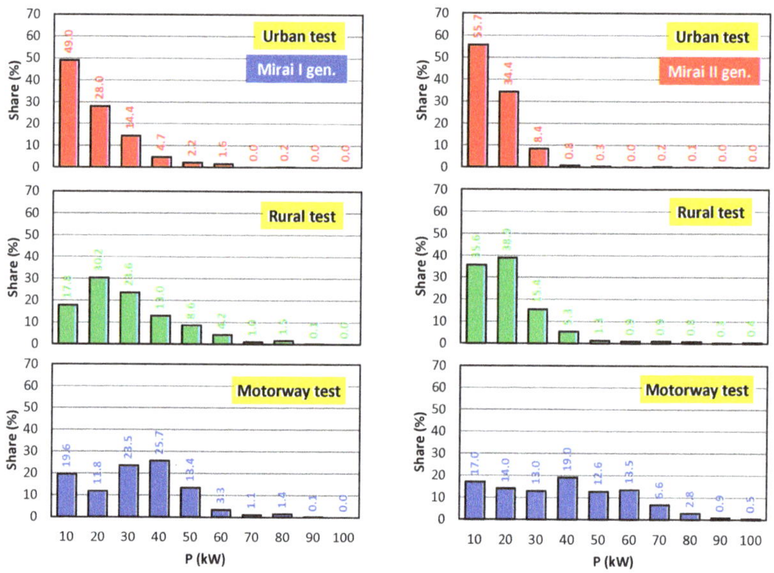

(a) (b)

Figure 15. Fuel cells power usage conditions in particular phases of drive tests: (**a**) Mirai I gen.; (**b**) Mirai II gen.

In urban driving conditions, the maximum power values were up to 30 kW, with greater values occurring sporadically. This was due to the driving speeds during the test drive. Despite the vehicle being heavier, the share of fuel cell power in the range of up to 10 kW was greater for the second-generation vehicle than for the first-generation vehicle. This share was over 50% of the share of the remaining power values in the drive test.

In rural driving conditions, the power share of up to 20 kW is most significant for both vehicle drive generations. In such driving conditions, the power values in the second-generation vehicle were much lower, which was confirmed by the values presented in Figure 13. Additionally, the energy values determined in Figure 12 were confirmed. It was found there that the energy consumed in the rural phase of the test by the first-generation system was greater by over 17%. The fuel cell power share in the highway phase was not as significant as in the other sections. The largest share of work was in the power range of

50–60 kW. In both drive generations, the shares of power above 80 kW were not used very often. This means that the drive system had a fairly large reserve of fuel cell power.

To summarize, it can be stated that the fuel cell works most frequently in the driving test phases:

- In the urban section: with power of up to 10 kW;
- In the rural section: with power of up to 20 kW;
- In the motorway section: with power of up to 30–40 kW.

As the maximum driving speed increased in the various test phases, the most frequently used power of fuel cells also increased. With each subsequent test phase (urban, rural, motorway), the most frequently used power from 10 kW in the urban phase increased by another 10 kW in each subsequent phase. The presented operating conditions of fuel cells indicate that their full power was not used, but only power of up to 70 kW was used (and that in the motorway phase only).

5.3. Modeling and Analysis of a Fuel Cell Stack Losses

Modeling of the fuel cell stack was started by modifying Equation (6) for a single fuel cell, taking into account the number of plates of the stack (n_{st}) and the surface of the current flow (a_{st}):

$$U_{st} = n_{st} \cdot U_{FC} \tag{7}$$

and

$$I = \frac{i}{a_{st}}. \tag{8}$$

Figure 16 shows the characteristics of fuel cells (cell stack current–cell stack power). Such characteristics were presented throughout the test and broken down into its individual phases. The presented characteristics already contain structured data, i.e., random data (significantly different from the typical characteristics) have been removed from the characteristics. Such conditions arise when reading data from a diagnostic tester as a result of high data sampling frequencies.

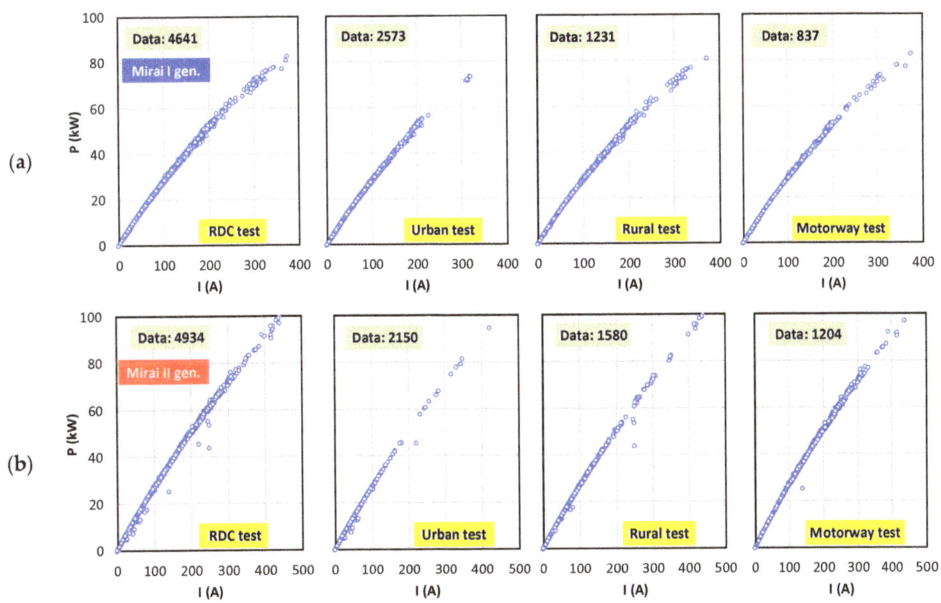

Figure 16. Characteristics of a fuel cell stack in relation to the current intensity of the fuel cell stack and the power of the fuel cell stack: (**a**) I generation FCHEV drive; (**b**) II generation FCHEV drive.

Figure 16 shows that the amount of data for both vehicle drive generations was similar (more than 4500 data points). As can be seen from the characteristics relating to individual phases, the share of higher values of the fuel cell stack power increased with the increase in the driving speed. The current values were about 20% greater for the second-generation vehicle than for the first-generation one. It is important that in both generations of cells there was no inflection of the characteristic (function local maximum), which would indicate the maximum power.

Modeling of fuel cells in order to determine their losses began with the presentation of their voltage–current characteristics. These characteristics were shown in Figure 17. Increased values of the current intensity in the second-generation system could be observed along with the decreased voltage values (by about 10%). These characteristics are presented for each of the test phases and for the entire drive test.

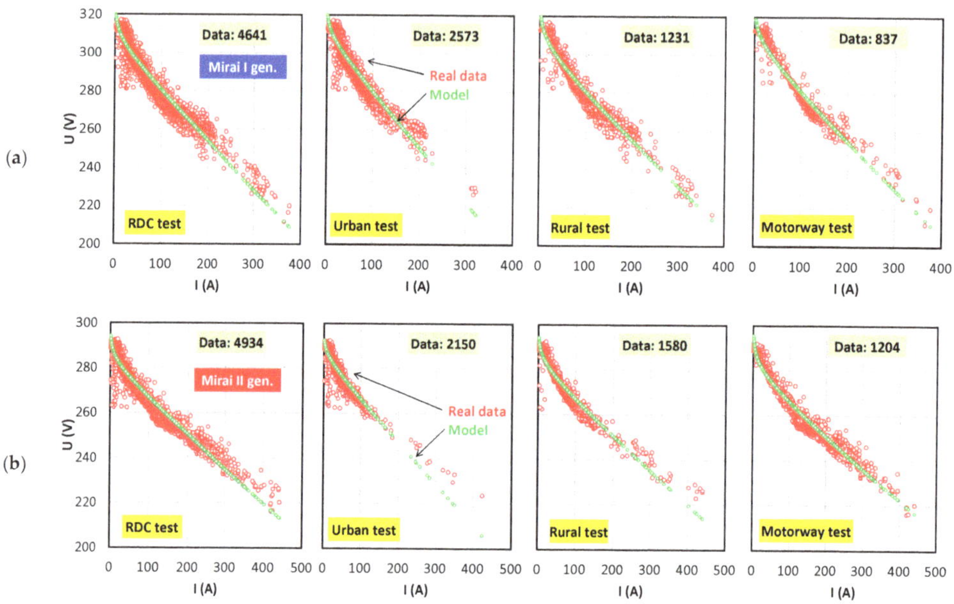

Figure 17. Current–voltage characteristics of the fuel cell stack of hydrogen-powered drives with modeling results: (**a**) I generation FCHEV drive; (**b**) II generation FCHEV drive.

The initial conditions were adopted for the modeling of the cells with values as listed in Table 3. The initial conditions were adopted based on the literature [24,25] and technical data [8,10,15]. Based on them, further mathematical analyses were performed. The differences in the starting values were in the number of fuel cells. For the first-generation drive this value was a stack of 370 fuel cells, for the second-generation it was a stack of 330 fuel cells.

Table 3. Initial conditions for modeling fuel cells.

Parameter	Symbol	Mirai I Gen.	Mirai II Gen.
Open circuit voltage	E_{oc}	0.8	0.8
Activation overvoltage	A	0.00854	0.00854
Area-specific resistance	r	22.7966	22.7966
Constants in the mass-transfer overvoltage Equation (5)	m	0.00097	0.00097
Constants in the mass-transfer overvoltage Equation (5)	n	0.62917	0.62917
Active area of stock	a_{st}	52,056.2	52,056.2
Number of cells in the fuel stack	n_{st}	370	330

The modeling of fuel cells was based on the search for the minimum value of the objective function:

$$\sum (V_{st\_x})^2 \rightarrow \min \tag{9}$$

where x were the consecutive values of current in the fuel cell stack.

The objective function is Equation (6) supplemented with changes presented in Equations (7) and (8). The result is the modeled voltage value of the fuel cell stack. Using the Solver add-in in the MS Excel package, the quantities meeting criterion (9), for which the initial values were presented in Table 3, were determined.

The non-linear generalized reduced gradient method was used for mathematical modeling [29–31]. This method allows solving the non-linear Equation (6) supplemented with Equations (7) and (8). It determines the central derivatives of the function, which leads to a solution to a certain range of variability of the initial conditions (not only in the positive direction). Taking central derivatives into account makes it possible to obtain the global minimum value of the function. Due to the iterative nature of this algorithm, the obtained data are approximate, not exact. The measure of the quality of the obtained results was the minimum sum of the squares of calculated and obtained values. It was assumed that all variables (Table 4) should be positive and non-zero (all fuel cell losses were taken into account).

Table 4. The results of the fuel cell stack modeling study.

Parameter	Mirai I Gen.				Mirai II Gen.			
	All Test	Urban	Rural	Motorway	All test	Urban	Rural	Motorway
E_{oc}	7.547×10^{-1}	7.782×10^{-1}	7.364×10^{-1}	7.469×10^{-1}	8.054×10^{-1}	8.277×10^{-1}	8.266×10^{-1}	7.919×10^{-1}
A	1.050×10^{-2}	7.883×10^{-3}	1.256×10^{-2}	1.145×10^{-2}	8.395×10^{-3}	5.610×10^{-3}	$8.923E \times 10^{-3}$	1.104×10^{-2}
r	3.370×10^{1}	3.833×10^{1}	3.098×10^{1}	3.246×10^{1}	2.292×10^{1}	2.597×10^{1}	2.308×10^{1}	2.333×10^{1}
m	1.000×10^{-7}	1.000×10^{-7}	1.000×10^{-7}	1.000×10^{-7}	9.697×10^{-4}	1.000×10^{-7}	2.112×10^{-3}	1.367×10^{-2}
n	8.654×10^{-1}	8.637×10^{-1}	8.677×10^{-1}	8.609×10^{-1}	6.292×10^{-1}	6.352×10^{-1}	6.289×10^{-1}	7.171×10^{-2}
a_{st}	5.206×10^{4}	5.206×10^{4}	5.206×10^{4}	5.206×10^{4}	5.179×10^{4}	4.794×10^{4}	5.127×10^{4}	5.927×10^{4}
n_{st}	3.712×10^{2}	3.712×10^{2}	3.710×10^{2}	3.711×10^{2}	3.284×10^{2}	3.295×10^{2}	3.195×10^{2}	3.286×10^{2}

The sums of squares for both generations of hydrogen drives in each test phase were determined using Equation (9)—Table 5. Additionally, the mean value of the square of the difference in size was also determined on this basis $\overline{(U_o - U)^2}$.

Table 5. Conditions for ending the search for the minimum of the objective function (Equation (9)).

Parameter	Mirai I Gen.				Mirai II Gen.			
	All Test	Urban	Rural	Motorway	All Test	Urban	Rural	Motorway
$\sum (U_o - U)^2$	109,827	59,591	29,249	18,317	51,588	19,616	15,967	13,878
$\overline{(U_o - U)^2}$	26.9763	23.1601	23.7609	21.8845	10.4556	9.1240	10.1062	11.5271

The data presented in Table 5 show that the results of measurements of a newer generation fuel cell drive system had a lower sum of square values. This means that a more accurate solution was found. The sum of squares as well as the mean value being lower in subsequent phases of the drive test results from the smaller number of measurement data in these phases.

The assessment of fuel cell losses was carried out using the analyses presented in Figure 18 and the data contained in Section 3.

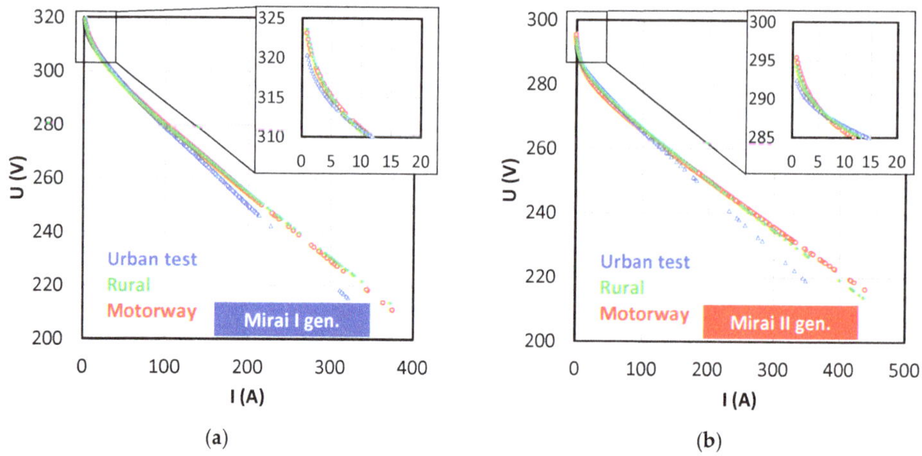

Figure 18. Comparison of model characteristics of the fuel cell stack for each phase of the road test: (**a**) I generation FCHEV drive; (**b**) II generation FCHEV drive.

The open circuit voltage changes in a similar way. For both drive system generations the lowest E_{oc} values were obtained in urban driving conditions (Figure 18). For rural conditions, the maximum E_{oc} values were obtained for the Mirai I gen. vehicle. For motorway conditions, the highest E_{oc} value was obtained for the newer generation drive—Mirai II gen. The data contained in Table 4 confirm these values. Additionally, E_{oc} in the case of the second-generation drive was about 0.5 V greater. In accordance with the previous provisions (Section 3), this voltage does not affect the values of the fuel cells losses.

The data in Table 4 indicated greater activation voltage losses in the first-generation drive. It was influenced by the A values, which were higher for the first-generation of fuel cells. For both generations, they are the smallest in urban driving conditions, and the largest in motorway driving conditions.

The resistive voltage losses were determined using the test results and data from Figure 7. This voltage turns out to be the lowest during the urban driving phase in both drive generations. In the case of higher driving speeds and the greater power of fuel cells, these losses were similar, due to the similar shape of the voltage–current curve. Increasing the resistive losses leads to a limitation of the cell voltage. These losses, according to Equation (4), increase with increasing system resistance. As shown in Table 4, these losses were significantly greater for the first-generation system.

The analysis of mass transport losses is difficult due to the discrepancy in resistive losses and the inability to analyze the characteristics at high cell powers. As only in those conditions could these losses be determined. The m and n coefficients for the second-generation cell were convergent with the first-generation cell only in urban driving conditions (Table 5). With greater power values, they also assume much greater values, which may indicate a reduction in mass transport (and at the same time an increase in these losses). The trends in fuel cell losses for each of the driving phases, taking both generations of fuel cells into account, were presented in Table 6.

Table 6. Assessment of fuel cell losses taking into account both generations of fuel cell systems.

Parameter	Mirai I Gen.			Mirai II Gen.		
	Urban	Rural	Motorway	Urban	Rural	Motorway
Open circuit voltage E_{oc}	min	max	middle	min	middle	max
Activation voltage losses	min	middle	max	middle	min	max
Resistive voltage losses	max	min	middle	max	middle	min
Mass transport voltage losses	min	min	min	min	max	max

It is planned to use the presented methodology for measuring fuel cell starts in aging conditions. Annual tests (or under specific climatic conditions) can be the basis for determining the weaknesses of fuel cells. At the same time, they can contribute to their development with regard to increasing their efficiency and reliability.

6. Conclusions

Based on the analytical work and road tests of two generations of drives with fuel cells, the following conclusions were formulated:

1. The fuel cell in the driving test phases works most often:
   - In the urban section: with a power of up to 10 kW;
   - In the rural section: with a power of up to 20 kW;
   - In the motorway section: with a power of up to 30–40 kW.
2. The share of fuel cells' power use increased as the maximum driving speed increased in the various phases of the driving test. In the urban phase, this power was on average 10 kW. Each subsequent phase of the test (rural and motorway) increased the power demand by an average of 10 kW in each phase. The presented operating conditions of fuel cells indicated that their full power was not utilized, and the maximum power used reached up to 70 kW (in the motorway phase). This was about 70% of the maximum power.
3. The presented research methodology shows the possibilities of estimating the fuel cells' voltage losses. It is possible to determine each component of these losses, but the mass transport losses require the fuel cells to be under maximum load to be measured properly (which is difficult to achieve in the RDC drive test due to its limitations).
4. The analysis of fuel cell losses showed the maximum values of activation voltage losses when cells were placed under high load (for both generations). The voltage of resistive losses reached its maximum in urban driving conditions, when the fuel cells were under a small load.

Funding: This research received no external funding.

Institutional Review Board Statement: Not applicable.

Informed Consent Statement: Not applicable.

Data Availability Statement: The data presented in this study are available on request from the corresponding author.

Acknowledgments: The author wish to thank Andrzej Szałek and Toyota Central Europe sp. z o.o. from Warsaw for providing the vehicles for the road tests.

Conflicts of Interest: The author declare no conflict of interest.

References

1. Proposal for a Directive of the European Parliament and of the Council Amending Directive (EU) 2018/2001 of the European Parliament and of the Council, Regulation (EU) 2018/1999 of the European Parliament and of the Council and Directive 98/70/EC of the European Parliament and of the Council as Regards the Promotion of Energy from Renewable Sources, and Repealing Council Directive (EU) 2015/652. Available online: https://eur-lex.europa.eu/legal-content/EN/TXT/?uri=CELEX:52021PC0557 (accessed on 8 May 2022).
2. Regulation (EU) 2019/631 of the European Parliament and of the Council of 17 April 2019 Setting CO2 Emission Performance Standards for New Passenger Cars and for New Light Commercial Vehicles, and Repealing Regulations (EC) No 443/2009 and (EU) No 510/2011. Available online: http://data.europa.eu/eli/reg/2019/631/oj (accessed on 4 May 2022).
3. Zhang, W.; Wang, J.; Du, S.; Ma, H.; Zhao, W.; Li, H. Energy management strategies for hybrid construction machinery: Evolution, classification, comparison and future trends. *Energies* **2019**, *12*, 2024. [CrossRef]
4. Das, H.S.; Tan, C.W.; Yatim, A.H.M. Fuel cell hybrid electric vehicles: A review on power conditioning units and topologies. *Renew. Sustain. Energy Rev.* **2017**, *76*, 268–291. [CrossRef]
5. Zhou, S.; Fan, L.; Zhang, G.; Gao, J.; Lu, Y.; Zhao, P.; Wen, C.; Shi, L.; Hu, Z. A review on proton exchange membrane multi-stack fuel cell systems: Architecture, performance, and power management. *Appl. Energy* **2022**, *310*, 118555. [CrossRef]

6. Nakagaki, N. The Newly Developed Components for the Fuel Cell Vehicle, Mirai. In *SAE Technical*; SAE International: Warrendale, PA, USA, 2015. [CrossRef]
7. Yoshizumi, T.; Kubo, H.; Okumura, M. Development of High-Performance FC Stack for the New MIRAI. In *SAE Technical*; SAE International: Warrendale, PA, USA, 2021. [CrossRef]
8. Konno, N.; Mizuno, S.; Nakaji, H.; Ishikawa, Y. Development of Compact and High-Performance Fuel Cell Stack. *SAE Int. J. Altern. Powertrains* **2015**, *4*, 123–129. [CrossRef]
9. Commission Implementing Regulation (EU) 2019/1840 of 31 October 2019 Amending Implementing Regulation (EU) 2017/1153 as Regards the Reporting of WLTP CO2 Values for Certain Categories of New Passenger Cars and Adjusting the Input Data for the Correlation Tool. Available online: http://data.europa.eu/eli/reg_impl/2019/1840/oj (accessed on 8 May 2022).
10. Toyota Mirai. Available online: https://www.techdoc-toyota.com (accessed on 6 May 2022).
11. Hyundai Nexo Blue. Available online: https://www.hyundai.com/content/dam/hyundai/ww/en/images/footer/downloads/eco/e-brochure/nexo-fe-lhd-ebrochure-2019.pdf (accessed on 6 May 2022).
12. Honda Leverages Older Name for New FCV: Clarity Fuel Cell Makes Its Debut at Tokyo; 435-Mile Range on JC08. Available online: https://www.greencarcongress.com/2015/10/20151027-clarity.html (accessed on 7 May 2022).
13. Luo, Y.; Wu, Y.; Li, B.; Mo, T.; Li, Y.; Feng, S.-P.; Qu, J.; Chu, P.K. Development and application of fuel cells in the automobile industry. *J. Energy Storage* **2021**, *42*, 103124. [CrossRef]
14. Hong, B.K.; Kim, S.H. Recent advances in fuel cell electric vehicle technologies of Hyundai. *ECS Trans.* **2018**, *86*, 3–11. [CrossRef]
15. Enomoto, K.; Ida, A.; Harada, T.; Takeuchi, H. Development of Safety Performance for FC Stack in the New Toyota FCEV. In *SAE Technical*; SAE International: Warrendale, PA, USA, 2022. [CrossRef]
16. Opel Begins Production and Deliveries of Hydrogen Vivaro van. Available online: https://www.greencarcongress.com (accessed on 6 May 2022).
17. Bernier, A. Hyvia Exhibits the Renault Master Van H2-TECH. Available online: https://www.world-hydrogen-summit.com (accessed on 3 August 2022).
18. Walters, M.; Wick, M.; Tinz, S.; Ogrzewalla, J.; Sehr, A.; Pischinger, S. Fuel cell system development: A strong influence on FCEV performance. *SAE Int. J. Altern. Powertrains* **2018**, *7*, 335–350. [CrossRef]
19. Wurzenberger, J.C.; Rašić, D.; Tavcar, G.; Glatz, T.; Mele, I.; Katrasnik, T. FCEV Performance Assessment—Electrochemical Fuel Cell and Battery Modelling on Vehicle Level. In *SAE Technical*; SAE International: Warrendale, PA, USA, 2020. [CrossRef]
20. Bagherabadi, K.M.; Skjong, S.; Pedersen, E. Dynamic modelling of PEM fuel cell system for simulation and sizing of marine power systems. *Int. J. Hydrog. Energy* **2022**, *47*, 17699–17712. [CrossRef]
21. Zhang, G.; Wu, L.; Qin, Z.; Wu, J.; Xi, F.; Mou, G.; Wang, Y.; Jiao, K. A comprehensive three-dimensional model coupling channel multi-phase flow and electrochemical reactions in proton exchange membrane fuel cell. *Adv. Appl. Energy* **2021**, *2*, 100033. [CrossRef]
22. Liu, Y.; Dirkes, S.; Kohrn, M.; Wick, M.; Pischinger, S. A high-fidelity real-time capable dynamic discretized model of proton exchange membrane fuel cells for the development of control strategies. *J. Power Sources* **2022**, *537*, 231394. [CrossRef]
23. Bressel, M.; Hilairet, M.; Hissel, D.; Bouamama, B.O. Model-based aging tolerant control with power loss prediction of Proton Exchange Membrane Fuel Cell. *Int. J. Hydrog. Energy* **2018**, *45*, 11242–11254. [CrossRef]
24. Barbir, F. Chapter 3—Fuel Cell Electrochemistry. In *PEM Fuel Cells*; Barbir, F., Ed.; Academic Press: Cambridge, MA, USA, 2005; pp. 33–72. [CrossRef]
25. Larminie, J.; Dicks, A. *Fuel Cell Systems Explained*, 3rd ed.; John Wiley & Sons Ltd.: Hoboken, NJ, USA, 2018. [CrossRef]
26. Pielecha, I.; Szałek, A.; Tchorek, G. Two generations of hydrogen powertrain—an analysis of the operational indicators in Real Driving Conditions (RDC). *Energies* **2022**, *15*, 4734. [CrossRef]
27. Pielecha, J.; Skobiej, K.; Kurtyka, K. Exhaust emissions and energy consumption analysis of conventional, hybrid, and electric vehicles in Real Driving Cycles. *Energies* **2020**, *13*, 6423. [CrossRef]
28. Andrych-Zalewska, M.; Chlopek, Z.; Merkisz, J.; Pielecha, J. research on exhaust emissions in dynamic operating states of a combustion engine in a Real Driving Emissions test. *Energies* **2021**, *14*, 5684. [CrossRef]
29. Ferreira, M.E.; Oliveira, M.C.; Menezes, L.F.; Andrade-Campos, A. 3–Numerical Optimization Strategies for Springback Compensation in Sheet Metal Forming. In *Woodhead Publishing Reviews: Mechanical Engineering Series, Computational Methods and Production Engineering*; Davim, J.P., Ed.; Woodhead Publishing: Sawston, UK, 2017; pp. 51–82. [CrossRef]
30. Arora, J.S. *Introduction to Optimum Design*, 4th ed.; Academic Press: Cambridge, MA, USA, 2017. [CrossRef]
31. Facó, J.L.D. A generalized reduced gradient algorithm for solving large-scale discrete-time nonlinear optimal control problems. *IFAC Proc. Vol.* **1989**, *22*, 45–50. [CrossRef]

energies

Article

A Multi-Fluid Model for Water and Methanol Transport in a Direct Methanol Fuel Cell

Anders Christian Olesen, Søren Knudsen Kær * and Torsten Berning *

Department of Energy Technology, Aalborg University, 9220 Aalborg, Denmark
* Correspondence: skk@europeanenergy.dk (S.K.K.); tbe@et.aau.dk (T.B.)

Abstract: Direct-methanol fuel cell (DMFC) systems are comparatively simple, sometimes just requiring a fuel cartridge and a fuel cell stack with appropriate control devices. The key challenge in these systems is the accurate determination and control of the flow rates and the appropriate mixture of methanol and water, and fundamental understanding can be gained by computational fluid dynamics. In this work, a three-dimensional, steady-state, two-phase, multi-component and non-isothermal DMFC model is presented. The model is based on the Eulerian approach, and it can account for gas and liquid transport in porous media subject to mixed wettability, i.e., the simultaneous presence of hydrophilic and hydrophobic pores. Other phenomena considered are variations in surface tension due to water–methanol mixing and the capillary pressure at the gas diffusion layer–channel interface. Another important aspect of DMFC modeling is the transport of methanol and water across the membrane. In this model, non-equilibrium sorption–desorption, diffusion and electro-osmotic drag of both species are included. The DMFC model is validated against experimental measurements, and it is used to study the interaction between volume porosity of the anode gas diffusion layer and the capillary pressure boundary condition at the anode, and how it affects performance and limiting current density.

Keywords: direct-methanol fuel cells (DMFC); multi-phase flow; fuel cells; computational fluid dynamics (CFD); limiting current density; reactant cross-over; fuel cell heat and mass transfer; anode gas diffusion layer; porous media; water-methanol mixture

Citation: Olesen, A.C.; Kær, S.K.; Berning, T. A Multi-Fluid Model for Water and Methanol Transport in a Direct Methanol Fuel Cell. *Energies* **2022**, *15*, 6869. https://doi.org/10.3390/en15196869

Academic Editor: Orazio Barbera

Received: 19 August 2022
Accepted: 12 September 2022
Published: 20 September 2022

1. Introduction

During the last decades, several types of fuel cells (FCs) have been subject to intensive research, both on component and system-level. While many types of FCs require large volumetric fuel storage or complicated reforming and gas cleaning systems, liquid-fed direct methanol fuel cells (DMFCs) circumvent this by directly oxidizing a fuel consisting of a methanol–water solution. Hence, this type not only offers a high electrical efficiency, it has a high energy density, fast start-up characteristics, and nearly zero recharge time. These characteristics make DMFCs suitable for backup systems and portable power applications in direct competition to batteries. A recent review of the different methanol producing technologies was published by Araya et al. [1].

The system around a direct-methanol fuel cell is very simple, and this is one of the reasons that they have been successfully commercialized already, especially in the enthusiast and leisure market such as camping and sailboats. Figure 1 depicts such an EFOY system by the manufacturer SFC Energy that can provide a maximum charge capacity of 250 Ah/day and a maximum power of 125 W to recharge common types of 12 V or 24 V on-board batteries. According to the manufacturer, a 10 L container filled with pure methanol affords autonomy for up to four weeks, and it can be easily replaced and refilled. While the system of a direct-methanol fuel cell is simple compared to other fuel cell types, the heat and mass transfer processes that occur in these fuel cells are substantially more complex, and this gives rise to the need for detailed computational fluid dynamics analysis.

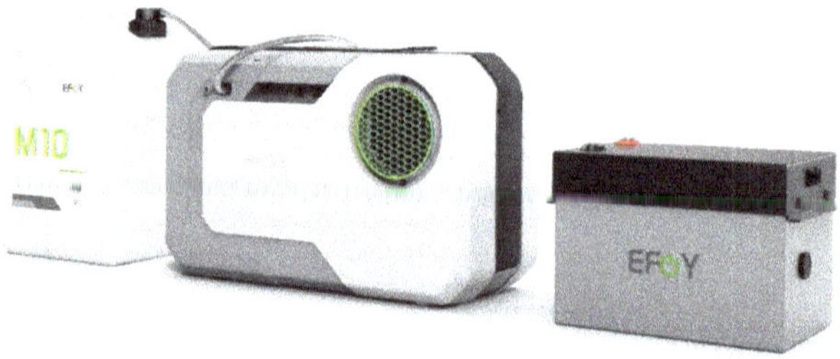

Figure 1. Commercial direct methanol fuel cell system. Photo credit: SFC Energy AG.

A common problem of DMFCs is the high permeation rate of water and methanol through the polymer electrolyte membrane. While methanol crossover results in a loss of fuel and a mixed potential at the cathode [2,3], water crossover causes a flooding of the cathode-backing layer and channels and consequently inhibits oxygen mass transport [4]. The extent of methanol crossover is typically reduced by operating DMFCs with thicker membranes and dilute methanol solutions, whereas the extent of water flooding can be controlled by the air stoichiometry and dew point diagrams [5]. Another two-phase flow phenomenon affecting DMFC operation is gas phase blockage of the anode backing layers and channels. Just as the liquid phase at the cathode can obstruct air mass transport, the gas phase at the anode can obstruct methanol mass transport in the liquid phase [6,7].

An aid in the investigation of these macroscopic transport phenomena governing the electrolyte membrane and backing layers is mathematical modeling. With the ongoing progress in experimental measurements of transport, thermodynamic and electro-chemical properties, the development of detailed mechanistic models of physiochemical processes and increased computational power, incremental improvements are continuously added to mathematical models of DMFCs. In recent years, this has entailed a substantial number of publications in the area of DMFC modeling. In the beginning, these models predominantly covered one-dimensional, multi-component, isothermal and single-phase transport, and were aimed at predicting concentration distributions, methanol crossover and polarization curves [8]. More recently, models have been developed that account for two-phase flow, thermal effects and higher dimensional transport effects as well as detailed membrane transport phenomena. In the following, a short review is given on the progress within these latter areas.

In an early attempt, Wang and Wang [9] investigated the coupling between two-phase flow, methanol crossover and the resulting mixed potential in a two-dimensional framework. This was done while ignoring charge transport and thus assuming a uniform overpotential in the anode and cathode catalyst layer (CL) as well as treating the CL as an interface without thickness. Their results underlined the importance of keeping the methanol concentration below 2 M in order to avoid excessive methanol crossover and performance loss. In their work, two-phase flow in the porous media was described using the multiphase mixture formulation, where only one set of governing conservation equations is solved for both thermodynamic phases, as opposed to the two-fluid method where two sets of governing equations are solved. Another feature of the developed DMFC model was the formulation of the capillary pressure in the momentum equations which was described by the dimensionless Leverett *J*-function [10]. This conveniently expresses the mathematical relationship between capillary pressure, surface tension, viscous permeability, porosity and contact angle of a porous medium in contact with two immiscible fluids as a function of wetting phase saturation. It does so by assuming an idealized hydrophilic or hydrophobic porous medium, consisting of non-connected capillary tubes. One merit of this approach is the characterization of capillary pressure through easily accessible data. In later work, Berning et al. [11] proposed to correlate the steepness of the Leverett curve with the pore-size distribution, concluding that a wider pore-size distribution of the porous medium results in a steeper capillary pressure curve and is thus desirable. Nonetheless, it should be noted that the validity of the Leverett function is strongly debated due to its limited ability to predict capillary characteristics of the porous media found in FC [12,13].

In the two-dimensional, two-phase and multi-component model by Divisek et al. [14] published around the same time as Wang and Wang [9], a different approach was taken to two-phase modeling. Their model was instead based on the two-fluid approach, which enabled them to account for mixed wettability as well as irreducible saturation in the formulation of the capillary pressure equation. Although a comprehensive model was developed accounting for spatial distributions in the CL, detailed electrochemical reactions, charge transport, dissolved species transport and mixed wettability, the effect of mixed potential was not addressed and validation could not be obtained.

Ge and Liu [15] presented a three-dimensional, two-phase and multi-component liquid-fed DMFC model. The two-phase model was based on the multiphase mixture model, capillary pressure was described by the Leverett *J*-Function and thermodynamic equilibrium was assumed between phases. Although the model accounted for the spatial resolution of the CL and one-dimensional methanol crossover due to electro-osmotic drag and diffusion, it neglected methanol transport in the CL and thus assumed a uniform distributed parasitic current density across the CL. Their study underlined the improved model predictability of switching from a single-phase to a two-phase flow model at high current densities. In particular, it was shown that the predictability of methanol crossover was improved. Shortly after, Liu and Wang [16] also presented a three-dimensional, isothermal, two-phase model of a liquid feed DMFC, which included proton transport. This model was to some extent a continuation of the work by Wang and Wang [9]. Unlike the previous model, this model included a spatially resolved CL and a non-uniform overpotential across the CL. Compared with Liu and Wang [16], the authors did not account for species transport inside the membrane, but rather formulated source terms at the CL–membrane interfaces assuming a one-dimensional, linear diffusive and convective transport across the membrane. Their model highlighted the effect of the land area on the non-uniform distribution of methanol and current density.

Yang and Zhao [17] presented a two-dimensional, isothermal and two-phase mass transport model for liquid-fed DMFCs. The proposed model accounted for two-phase flow in the porous layers using the two-fluid method and the Leverett *J*-function. Two different approaches were taken for modeling two-phase flow in the anode and cathode channels. In the cathode, a homogeneous flow model was used, whereas a modified one-dimensional drift–flux model was applied at the anode to account for a difference in phase velocities. The model was later updated by Yang and Zhao [18] to account for species transport and non-equilibrium phase change. This enhancement was used for highlighting the improved predictability by switching from a thermodynamic phase equilibrium condition to a non-equilibrium phase condition while accounting for methanol vapor transport. The developed model was further expanded to a three-dimensional domain by Yang et al. [19]. With this improvement, the effect of the liquid phase saturation at the interface between the channel and the gas diffusion layer was investigated. It was shown that the higher the saturation, the lower was the mass transport resistance for methanol transport. In the work by Xu et al. [20], the issue of non-equilibrium sorption/desorption of water between its dissolved state in the electrolyte phase and a gas–liquid mixture was addressed. This was done using a one-dimensional, isothermal, two-phase model, similar to the mass transport model presented in [17,18,20].

Miao et al. [21] developed a two-dimensional, isothermal, two-phase model of a liquid feed DMFC including electron and proton transport. The employed two-phase model was based on the two-fluid method, the Leverett *J*-function and non-equilibrium phase change, similar as in [18,19]. Their model, however, accounted for a non-uniform overpotential and parasitic current density, as well as a detailed agglomerate model. The model assumed a fully hydrated membrane and that membrane transport occurred through the liquid phase, similar as in references [9,14]. A paramount issue their model addressed was the treatment of the liquid saturation boundary condition at the GDL–channel interface at the anode. In the majority of DMFC models, a constant liquid saturation interface condition of nearly one is assumed independent of the operation condition [4,16,18–20,22,23]. The underlying basis of this specification has neither been substantiated experimentally nor mechanistically. Miao et al. formulated the GDL–channel interface based on an average liquid saturation between inlet and outlet [21]. Their work highlighted that a significant nonuniform parasitic current density and overpotential distribution can be observed. Later, Miao et al. [22] improved the previously presented model by accounting for homogeneous heat transport, thermal and electrical contact resistance, inhomogeneous compression of the GDL and a more detailed reaction mechanism at the anode.

Yang et al. [24] re-examined the existence of a fixed liquid saturation boundary condition at the GDL/channel interface, which had been used in their previous work [17–20,23]. It was attempted to remove the GDL/channel interface boundary condition by introducing a liquid water saturation boundary in the CL, and thereby quantifying the liquid distribution in the porous media. The thus obtained boundary condition was current-density-dependent.

In a study by Garvin and Meyers [25], a detailed thermodynamic, multicomponent and two-phase model of the anode backing layer was developed. Although only one-dimensional, the model combined in detail a non-ideal thermodynamic description of species saturation pressures and a bundle-of-capillaries model of the two-phase flow in porous media. Their model was used for analyzing the promotion of gas removal and a lowering of the methanol concentration in the catalyst layer by tailoring the backing layer properties such as contact angle, permeability, etc., and it was found that the backing layer has to be tailored for specific applications.

He et al. [4] presented a two-dimensional, non-isotherm, two-phase DMFC model including current and proton transport. The developed model is fundamentally similar to the one presented by Miao et al. [21], with two exceptions: the electrolyte model and the anode reaction mechanism. Although the electrolyte model accounts for saturation-weighted equilibrium water content in the CL (rather than assuming a fully hydrated membrane as in references [9,14,17–19,21–23]), it does neglect a significant difference in the non-equilibrium sorption–desorption kinetic rate of liquids and gases.

In 2013, Bahrami and Faghri [26] published a detailed literature review and highlighted the differences between the various modeling approaches.

Based on the literature review, it can be concluded that within the field of DMFC modeling, the simultaneous presence of hydrophilic and hydrophobic pores has not been taken into account in three-dimensional studies, even though the fraction of hydrophilic pores for SGL-type GDLs ranges between 15–60%, depending on the PTFE content [13,27], and hence should not be neglected. It will be shown below that this is of particular importance for DMFCs where at the anode side the water/methanol mixture reaches the catalyst layer primarily using the hydrophilic pores in the MPL and combination of hydrophilic and hydrophobic pores in the GDL, while the product gas only reaches the channel via the hydrophobic pores in any layer. Hence, in a DMFC model it is of utmost importance to account for this fractional wettability. Moreover, it appears that most modeling attempts for DMFCs that include detailed membrane transport and sorption/desorption in conjunction with detailed two-phase flow in porous media are limited to one-dimensional and sometimes two-dimensional models.

Likewise, most models neglect detailed two-phase flow transport in channels of the anode, even though the amount of gas exiting the anode can be substantial and therefore should not be disregarded. Hence, in the present effort, a boundary condition for the GDL/channel interface based on the channel pressure is prescribed. The anode pressure loss ranges up to several thousand Pascals, and hence constitutes a significant pressure that can force the liquid phase into the hydrophobic pores of the GDL [28]. However, it is doubtful if it is high enough to substantiate a liquid saturation of nearly one at the GDL–channel interface. At the present moment, in order to keep this boundary condition as simple as possible, it is limited to a constant average pressure value for a given channel length.

In order to study the importance of a pressure-based boundary condition, two simulation cases with different anode GDL volume porosities are compared and examined in detail with regard to liquid saturation and methanol distribution in the anode. Moreover, focus is put on how these distributions are impacted by fractional wettability of each layer as well as variations in the surface tension.

2. Mathematical Model

The three-dimensional, steady-state, two-fluid, multicomponent and non-isothermal DMFC model presented here consists of a membrane, anode and cathode electrodes along with adjacent channels. The DMFC model was implemented in the commercial CFD software ANSYS CFX. The two-phase flow model employed is based on the multi-fluid approach, i.e., the model solves one set of transport equations for each phase. The multi-fluid approach uses the notion of interpenetrating continua, and hence inherently only captures statistical macroscopic phenomena.

Two phases are assumed in the channels, a liquid and a gas phase. In the porous media an additional solid phase is present, and heat transfer in all three phases is included. Moreover, charge transport is added to the porous media. In the membrane, only ion transport, dissolved species and heat transport are modeled.

2.1. Computational Grid

The geometry used for this study is depicted in Figure 2. The mesh comprises approximately 52,000 hexahedral elements. The computational domain consists of a membrane electrode assembly (MEA) and two flow channels. Each electrode has a catalyst layer (CL), micro porous layer (MPL) and gas diffusion layer (GDL). The bipolar plates (BP's) are currently neglected. Instead, a temperature and electrical potential boundary condition is applied to channel walls and GDL surface area in contact with BP.

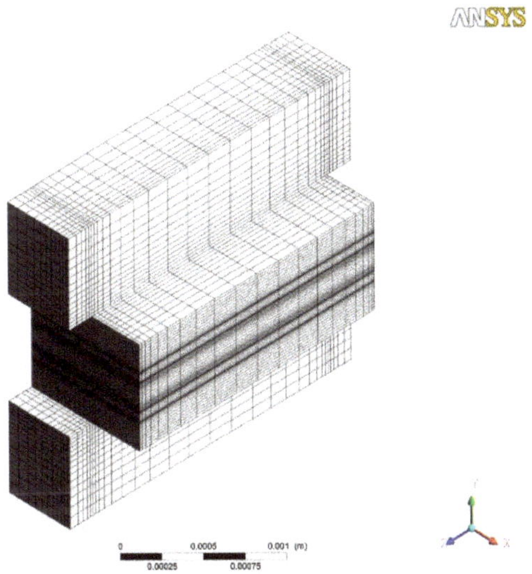

Figure 2. Computational mesh used in solving the mathematical model.

2.2. Assumptions

Because of the inherent complexity of the model, a number of simplifying assumptions were made to ensure satisfactory convergence. Care was taken that none of these simplifications had a strong impact on the results discussed below.

- On the cathode side, the gas phase consists of nitrogen, oxygen and water vapor, whereas the liquid phase only comprises water. The anode gas phase consists of water vapor, methanol and carbon dioxide. The liquid phase is modeled as a binary solution consisting of water and methanol;

- Only water and methanol undergo phase change. Species from one phase cannot be dissolved in the other phases and transported;
- The cross-over of carbon dioxide through the membrane is currently neglected;
- The oxygen reduction reaction (ORR) at the cathode is assumed to produce liquid water. However, when the relative humidity of the adjacent gas phase is below 100%, this water will evaporate immediately;
- Gases are assumed to behave ideally, since pressures are low;
- In general, all phases share the same pressure field. However, in the CL, MPL and GDL, capillary forces are accounted for that act as so-called body force terms in the momentum equations;
- Membrane swelling is currently neglected.

2.3. General Governing Equations

The general set of conservation equations for mass, momentum and species are depicted in Equations (1)–(3), respectively.

$$\nabla \varepsilon s_\alpha \rho_\alpha \mathbf{U}_\alpha = \varepsilon s_\alpha \mathbf{S}_\alpha + (\dot{m}_{\alpha\beta} - \dot{m}_{\beta\alpha}), \tag{1}$$

$$\nabla \varepsilon s_\alpha (\rho_\alpha \mathbf{U}_\alpha \otimes \mathbf{U}_\alpha - \nabla \boldsymbol{\tau}) = \varepsilon s_\alpha (\mathbf{S}_\alpha - \nabla p_\alpha) + \varepsilon \mathbf{M}_\alpha + (\dot{m}_{\alpha\beta} \mathbf{U}_\beta - \dot{m}_{\beta\alpha} \mathbf{U}_\alpha), \tag{2}$$

$$\nabla \varepsilon s_\alpha (\rho_\alpha \mathbf{U}_\alpha Y_{A\alpha} - \rho_\alpha D_{A\alpha} \nabla Y_{A\alpha}) = \varepsilon s_\alpha S_A + (\dot{m}_{\alpha\beta} Y_{A\beta} - \dot{m}_{\beta\alpha} Y_{A\alpha}), \tag{3}$$

where the subscripts α and β denote the phases, s the saturation, ε the porosity, ρ the density, $\dot{m}_{\alpha\beta}$ and $\dot{m}_{\beta\alpha}$ the directionally dependent interfacial mass flows, $Y_{A\alpha}$ the mass fraction of component A in phase α, τ the stress tensor, p the pressure, and $\mathbf{M}_\alpha$ the interfacial forces acting on phase α due to the presence of other phases. Moreover, $D_{A\alpha}$ is the diffusion coefficient of species A in the phase α, and $\mathbf{S}$ is the source term. The true phase velocity or intrinsic velocity, $\mathbf{U}_\alpha$, is related to the superficial velocity using the expression

$$\mathbf{u}_\alpha = \varepsilon s_\alpha \mathbf{U}_\alpha, \tag{4}$$

In addition to mass and momentum, the transport of energy is modeled. However, in the present case, energy transport is simplified by neglecting kinetic energy changes and by solving a homogeneous temperature field. Thus, the energy equation can simply be written as:

$$\nabla \varepsilon s_\alpha (\rho_\alpha \mathbf{U}_\alpha H_\alpha - \lambda_\alpha \nabla T_\alpha) = \varepsilon s_\alpha \mathbf{S}_\alpha + \nabla \varepsilon s_\alpha \sum_{i=1}^{N_c} \Gamma_i h_i \nabla Y_i, \tag{5}$$

where H_α denotes the static enthalpy, λ_α is the thermal conductivity, and N_C is the number of species. The molar enthalpy of species i is denoted by h_i. The source term $\mathbf{S}$ results from all the local overpotentials, both activation and ohmic.

These are the general transport equations. Depending on the region of the cell, the general sink and source terms require different expressions which will now be described.

2.4. Flow Channels

Inside the channels, momentum is transferred between the phases by an interphase drag force. The current implementation uses the following mixture formulation:

$$M_{D,l} = C_D A_{lg} \rho_{lg} (u_g - u_l) |u_g - u_l|, \tag{6}$$

where A_{lg} and ρ_{lg} denote interfacial area density and the mixture density, respectively. For two-phase channel flow, the interfacial area density is defined relative to the phase volume fraction and particle size. In this work, a distinction is made between anode and cathode channel due to a difference in the flow morphology. The anode is modeled as a continuous–continuous flow, whereas the cathode is modeled as dispersed–continuous flow. For large gas volume fractions, the gas–liquid flow may approach a continuous–continuous regime, and hence cause a violation of the dispersed gas assumption in the calculation of the interfacial area. Not accounting for this may overpredict interfacial drag in the channel and have a detrimental effect on the convergence behavior. The interfacial area densities of these two types of flows are modeled as follows, where Equation (7) applies to the anode and Equation (8) to the cathode:

$$A = s(1-s)/d, \tag{7}$$

$$A = 6s/d. \tag{8}$$

Here, d is the average particle size, which in our case is assumed to be 20 µm, and s is again the saturation. It should be noted that the abovementioned distinction is only made with regard to interfacial area density.

2.5. Porous Media

In the porous media of a fuel cell, the momentum equation is dominated by viscous and capillary forces. One way to describe the effect of viscous forces is through Darcy's law. This equation effectively relates a phase pressure gradient to the phase velocity as follows:

$$\nabla p_g = -\varepsilon(1-s)\mu_g/\left(k_{rel,g} \times \overline{\mathbf{K}}\right) \times \mathbf{U}_g, \tag{9}$$

$$\nabla p_l = -\varepsilon s \mu_l/\left(k_{rel,l} \times \overline{\mathbf{K}}\right) \times \mathbf{U}_l, \tag{10}$$

where $\overline{\mathbf{K}}$ and k_{rel} denote permeability tensor and relative permeability, respectively, while μ denotes the dynamic viscosity. In general, the relative permeability depends on pore structure, wettability, capillary forces and saturation history [29] and thus is rather complex in nature. However, it is often modeled as being dependent only on saturation. A common approach in FC modeling is a power law correction [9,11,30]:

$$k_{rel,g} = (1-s)^q \;\vee\; k_{rel,l} = S_e{}^q, \tag{11}$$

The coefficient q is an empirical constant in the range of 2–5, hence implying a strong interaction with local saturation, and S_e is the effective saturation. The effective saturation is a function used to account for the presence of irreducible saturation s_{irr} in a porous medium, i.e., the wetting phase is immobile below a saturation value of $s_w = s_{w,irr}$ [31]. The effective saturation S_e is defined as follows [32]:

$$S_e = \left(s_w - s_{w,irr}\right)/\left(1 - s_{w,irr}\right). \tag{12}$$

With the introduction of Darcy's equation, the presence of capillary forces can be easily accounted for. The capillary pressure is defined as the difference in phase pressure according to:

$$\nabla p_{cap} = \nabla p_g - \nabla p_l. \tag{13}$$

In order to introduce Equations (9), (10) and (13) in the momentum equations, they have to be rewritten as a source terms or so-called body force terms [11,33]. Moreover, if the capillary pressure gradient is known, only one pressure field has to be solved for, which either could be the liquid or gas phase pressure. Assuming that the gas phase pressure is solved for, the resulting momentum source terms are as follows:

$$\mathbf{S}_{g,mom} = -\varepsilon(1-s)\mu_g/\left(k_{rel,g} \times \overline{\mathbf{K}}\right) \times \mathbf{U}_g. \tag{14}$$

$$\mathbf{S}_{l,mom} = \nabla p_{cap} - \varepsilon s \mu_l/\left(k_{rel,l} \times \overline{\mathbf{K}}\right) \times \mathbf{U}_l. \tag{15}$$

When modeling the capillary pressure p_{cap} on a macroscopic scale, it is convenient to describe it in terms of the Leverett's function, Equation (16a,b), which combines the fluid properties, material properties and liquid saturation. This approach was first proposed by Udell [34]. However, the Leverett function in its original form can only handle a mixed wettability and no fractional wettability. Hence, in the current study the standard Leverett function is altered by introducing an effective wetting saturation based on the hydrophilic pore fraction, i.e., Equation (17). This approach assumes that the hydrophilic pores are initially filled, and subsequently the hydrophobic pores. Hence, the capillary pressure behaves as follows: $p_{cap} < 0$ for $s < f_{hi}$, $p_{cap} = 0$ for $s = f_{hi}$, and $p_{cap} > 0$ for $s > f_{hi}$, where f_{hi} denotes the hydrophilic pore fraction. A similar implementation has been used in other studies [11,32,35,36]. In references [11,32], it was assumed that only the hydrophobic phase was mobile.

The complete set of equations used to describe the capillary pressure is as follows:

$$p_{cap} = \sigma \cos\theta (\varepsilon/K)^{1/2} J(S) \tag{16a}$$

$$J(S) = 1.417S - 2.120S^2 + 1.263S^3 \tag{16b}$$

$$S = \begin{cases} (s - f_{hi})/(1 - f_{hi}) & s > f_{hi} \\ (1 - s - f_{ho})/(1 - f_{ho}) = (f_{hi} - s)/f_{hi} & s < f_{hi} \end{cases} \tag{17}$$

where σ denotes surface tension, θ the contact angle, ε the porosity, K the effective permeability and $J(S)$ the Leverett function. In the Leverett function, the term $\sqrt{K/\varepsilon}$ represents the inverse of the characteristic pore radius, $1/r_c$. This length scale is helpful in evaluating the effect of the capillary pressure.

For methanol–water mixtures, it is important to account for the variation in surface tension due to the methanol molar fraction. The mixture surface tension is described using the following equation [37]:

$$\sigma = \sigma_{H_2O} - [1 + a(1 - x_{MeOH})/(1 - b(1 - x_{MeOH}))] \times x_{MeOH}(\sigma_{H_2O} - \sigma_{MeOH}) \tag{18}$$

where a and b are empirical coefficients that exhibit a linear dependency on the temperature in the range between 20 °C and 50 °C [37]. In our study, it was assumed that these values are valid at higher temperatures as well, since the temperature dependency is small and linear. The mixture surface tension behaves strongly nonlinear with changes in methanol molar fraction [37], and hence is very important to incorporate.

In the present model, both Fickian and Knudsen diffusion are accounted for, similar to the model presented in [38]. The model also accounts for effective properties due to multicomponent diffusion and porous media obstruction. The implemented equations are shown in Table 1.

Table 1. Diffusivity constitutive relations.

Parameter	Symbol	Unit	Equation
Gas phase multicomponent correction	$D_{A\alpha}(T_0, p_0)$	cm^2/s	$(1 - x_A)\left(\sum_{i \neq A}^{n} x_i/D_{Ai}\right)^{-1}$
Temperature and pressure correction of the Fickian diffusivity	$D_{A\alpha}(T, p)$	cm^2/s	$D_{A\alpha}(T_0, p_0) \times (p_0/p)(T/T_0)^{3/2}$
Porous media correction	$D_{A\alpha}^{eff}$	cm^2/s	$\frac{1}{\tau}(1 - s)^p D_{A\alpha}(T, p)$
Knudsen diffusivity	$D_{A\alpha}^{K}$	cm^2/s	$\frac{D_{pore}}{3}\sqrt{8RT/\pi MW_g}$
Combined Fickian and Knudsen diffusivity	$D_{A\alpha}$	cm^2/s	$D_{A\alpha}^{eff} D_{A\alpha}^{K} / \left(D_{A\alpha}^{eff} + D_{A\alpha}^{K}\right)$

Phase Change

The phase change rate of a planar surface can be expressed according to the Hertz–Knudsen–Langmuir equation:

$$\dot{m}_{\alpha\beta} = A_{lg}\sqrt{MW/2\pi R}\left(\xi_c p_g/\sqrt{T_g} - \xi_e p_l/\sqrt{T_l}\right) \tag{19}$$

where ξ denotes the condensation/evaporation coefficient, and MW is the molecular weight of the species undergoing phase change. Equation (19) exists in various forms. In the following, a simplified version is used where an average temperature is assumed and the liquid phase pressure is given by the saturation pressure. This leads to the following implementation [11,29]:

$$\dot{m}_{\alpha\beta} = k_{xm} A_{lg} MW_i\left[p_i - p_{i,mix}^{sat}(T_{avg})\right]/RT_{avg} \tag{20}$$

where k_{xm} denotes the convective mass transfer coefficient in [m/s]. The convective mass transfer coefficient was originally derived based on kinetic theory and the assumption of ideal gas behavior; however, the diffusion process needs to be corrected due to the presence of other species by accounting for intermolecular collisions. Further, the ability of the droplet to absorb water molecules needs to be accounted for [32]:

$$k_{xm} = \Gamma_u \overline{u}_m = \Gamma_u\sqrt{8RT/\pi MW} \tag{21}$$

where $\overline{u}_m$ denotes the mean molecular speed and Γ the dimensionless uptake coefficient, which accounts for the presence of non-condensable species at the gas–liquid interface.

The interfacial area density depends on the phase change mechanism. In contrast to boiling, evaporation requires a liquid–gas interfacial area in order to occur. This is also in contrast to condensation, which can occur either on a pre-existing liquid layer or on a hydrophilic surface [32]. Inside porous media, it is reasonable to assume that the interfacial area density depends on the pore surface area [32]. Furthermore, the mass transfer rate is corrected for local liquid saturation and porosity:

$$A_{lg} = \begin{cases} \epsilon(1-s)y_{g,i}\Gamma_s A_{pore}, & p_i > p_{i,mix}^{sat} \\ \epsilon s(1-s)y_{l,i}\Gamma_s A_{pore}, & p_i < p_{i,mix}^{sat} \end{cases} \tag{22}$$

where A_{pore} denotes specific pore surface area, and Γ_s the surface accommodation coefficient. Fundamentally, it has to be true in the case of evaporation that $A_{lg} \to 0$ for $s \to 1 \vee s \to 0$ and in the case of condensation that $A_{lg} \to 0$ for $s \to 1$. Moreover, the surface area exposed to evaporation is only covered by a fraction of a given species. This fraction is assumed equivalent to the species liquid molar fraction.

Since methanol and water together make up a non-ideal vapor–liquid mixture, the equilibrium saturation pressure of each species needs to be corrected. At equilibrium, the fugacity of each component in each phase is equal. By introducing an activity coefficient and assuming that the liquid fugacity of component i at standard state is equal to the saturation pressure of its pure component, and assuming an ideal vapor phase (i.e., vapor fugacity coefficient of 1), the vapor pressure of component i can be stated as follows:

$$f_i^l = f_i^g \tag{23}$$

$$x_i\gamma_i P_i^{sat} = y_i P \tag{24}$$

The activity of any species in the gas or liquid phase can be defined relative to pure standard conditions as:

$$a_{i,l} = f_i^l/P_i^{sat} = x_i\gamma_i \tag{25}$$

$$a_{i,g} = f_i^g/P_i^{sat} = y_i P/P_i^{sat} \tag{26}$$

It is important to correct the saturation pressure at low molar fractions of methanol, since it behaves highly non-ideal in this range. The activity coefficients of a vapor–liquid mixture can be estimated via the empirical Non-Random Two Liquid (NRTL) model. Moreover, the saturation pressures of pure water and methanol have a strong non-linear dependence on temperature. To improve numerical robustness, a curve fit to tabulated saturation pressures in an interval between 0 °C and 100 °C is implemented.

2.6. Catalyst Layers

The current and ion density distributions are modelled using Ohm's law:

$$\nabla\left(-\sigma_{eff}\nabla\Phi_s\right) = R \tag{27}$$

$$\nabla\left(-\kappa_{eff}\nabla\Phi_m\right) = -R \tag{28}$$

$$\sigma_{eff} = \varepsilon^{1.5}\sigma \tag{29}$$

where Φ_s and Φ_m denote the solid potential and membrane potential, respectively, σ the electron conductivity, κ the ion conductivity, and R the charge production or consumption rate. The ion density vector is given as $\mathbf{i} = -\sigma\nabla\Phi_s$ and the electric current density vector as $\mathbf{j} = -\kappa_{eff}\nabla\Phi_m$. These currents generated in the membrane and solid phase also give reason for ohmic heating, implemented as the source term in the energy equation, Equation (5), where in the regions in question, the source term is $S = i^2/\sigma$ and/or $S = j^2/\kappa$.

At the interface between the GDL and BP, a voltage drop exists due to contact resistance between the carbon fibers and BP. This contact resistance depends on clamping pressure and PTFE content. The voltage drop is modeled as follows:

$$I = (V_{GDL} - V_{BP})/R_C \tag{30}$$

where R_C denotes contact resistance.

The rate of charge production and consumption at the cathode is given based on the oxygen reduction reaction (ORR) and the direct methanol oxidation reaction (DMOR), respectively:

$$O_2 + 4H^+ + 4e^- \leftrightarrow 2H_2O \tag{31}$$

$$CH_3OH + 1.5O_2 \rightarrow CO_2 + 2H_2O \tag{32}$$

In order to determine the volumetric current density at the cathode due the ORR and DMOR, the Butler–Volmer expression is used. This approach assumes a single rate-determining step.

$$R = j_0^{eff}\left[exp(\alpha F/RT \times \eta) - exp(-(1-\alpha)F/RT \times \eta)\right] \tag{33}$$

$$j_0^{eff} = (1-s)aj_0^{T_0}\left(\xi \times C_{O_2}/C_{O_2,ref}\right)\exp\left[-E_{act,i}/RT\left(1 - T/T_{ref}\right)\right] \tag{34}$$

$$j_0^{eff} = aj_0^{T_0}\left(C_{CH_3OH}/C_{CH_3OH_{ref}}\right)^{\gamma}\exp\left[-E_{act,i}/RT\left(1 - T/T_{ref}\right)\right] \tag{35}$$

where $\eta = -(\Phi_s - \Phi_m - U^0)$ $\eta_c = -(\Phi_s - \Phi_m - U_c^0)$ is the cathode overpotential and $\xi = RT/H_0$ is an uptake coefficient based on Henry's law [12], accounting for the difference in the oxygen concentration in the gas and the ionomer phase. Moreover, the exchange current density is corrected for its temperature dependence and active sites being blocked.

The rate of charge production and consumption at the anode is given based on the methanol oxidation reaction:

$$CH_3OH_{(l,g)} + H_2O \leftrightarrow CO_2 + 6H^+ + 6e^- \tag{36}$$

This reaction can be considered to consist of the following elementary reaction steps, as proposed by Gasteiger et al. [39]:

$$CH_3OH + M_1 \rightarrow CH_3OH_{ad,1} \quad (37)$$

$$CH_3OH_{ad,1} \leftrightarrow CO_{ad,1} + 4H^+ + 4e^- \quad (38)$$

$$H_2O + M_2 \leftrightarrow OH_{ad,2} + H^+ + e^- \quad (39)$$

$$CO_{ad,1} + OH_{ad,2} \rightarrow CO_2 + H^+ + e^- + M_1 + M_2 \quad (40)$$

The volumetric current density is determined by an expression derived by Meyer and Newman [40] based on the previously shown reaction mechanism, neglecting the reaction in the cathode direction:

$$R = j_0^{eff} \times \left[\widetilde{C}_{CH_3OH} \exp(\alpha F/RT \times \eta)\right] / \left[\widetilde{C}_{CH_3OH} + K \exp(\alpha F/RT \times \eta)\right] \quad (41)$$

$$j_0^{eff} = a j_0^{T_0} \exp\left[-E_{act,i}/RT\left(1 - T/T_{ref}\right)\right] \quad (42)$$

where K is an equilibrium constant, $\eta_a = \Phi_s - \Phi_m - U_a^0$ is the anode overpotential and $\widetilde{C}_{CH_3OH} = (1-s)C_{CH_3OH,l} + sC_{CH_3OH,g}$ is the corrected methanol concentration. The electrochemical reactions in the anode and cathode CL introduce sink and source terms in the continuity, charge and energy equations given as functions of current density drawn from the DMFC and the number of electrons involved in the consumption or formation of a given species, as listed in Equations (31) and (36).

The simultaneous presence of ORR and DMOR at the cathode electrode gives rise to a mixed potential. By subtracting the virtual parasitic current density from the cathode current density in the charge source terms, the mixed potential is accounted for.

The anode and cathode current densities are defined as follows:

$$I_{An} = \frac{1}{A}\int_{V_{ACL}} R_{An} dV \quad (43)$$

$$I_{Cat} = \frac{1}{A}\int_{V_{CCL}} R_{Cat} dV \quad (44)$$

$$I_P = \frac{1}{A}\int_{V_{CCL}} R_P dV \quad (45)$$

$$I_{Cell} = I_{An} = I_{Cat} - I_P \quad (46)$$

where A denotes the cross-section area. The remaining properties used in the electrochemical model are found in Table 2. These values have been assumed and adjusted so that the calculated performance curve provided a good match to experimental results as shown below. It has been assured that these values are within a reasonable range.

Table 2. Electrochemical properties.

Parameter	Symbol	Unit	Value
Anode reference exchange current density	aj_0	A/m^3	1.0×10^5
Cathode reference exchange current density	aj_0	A/m^3	2.8×10^3
Anode transfer coefficient	α	-	0.55
Cathode transfer coefficient	α	-	0.50
Anode activation energy	E_{act}	kJ/mol	35.57
Cathode activation energy	E_{act}	kJ/mol	66
Anode reference concentration	C_{ref}	mol/m^3	1.0×10^2
Cathode reference concentration	C_{ref}	mol/m^3	9.6

2.7. Membrane Model

Water and methanol transport in the ionomer is described using dilute solution theory (i.e., Fickian diffusion transport and an electro-osmotic drag term) [19,33,41,42] and including the absorption/desorption term S_A:

$$\nabla\left(\rho_{mem} D_{eff}/EW \times \nabla\lambda_A\right) = \nabla(n_d/F \times i) + S_A \tag{47}$$

where λ denotes the content of component A in the membrane phase, ρ_{mem} is the dry membrane density (ρ_{mem} = 2000 kg/m^3), and EW is the equivalent weight of the membrane (EW = 1.1 kg/mol) [43], while F denotes Faraday's constant (96,485 C/mol). The species content is defined relative to the number of sulfonic acid groups:

$$\lambda_j = n_j/n_{SO_3^- H^+} \tag{48}$$

The species content of Nafion equilibrated with a water–methanol mixture is known to depend non-linearly on species activities and temperature. A few thermodynamic and mechanistic models have been proposed in the literature describing this dependency [40,44,45]. However, none of these models can differentiate between gas and liquid phase equilibration, or what is known as Schroeder's paradox. Essentially, this phenomenon describes how a difference in water content is observed for vapor and liquid at equal activity (i.e., equal chemical potential), even though from a thermodynamic point of view, it should not occur. Some research groups have proposed models explaining Schroeder's paradox based on capillary forces [46] or a difference in surface energy of vapor and liquid water on Nafion [47]. Meanwhile, recent studies have shown the absence of Schroeder's paradox [48,49], if care is put into the thermal history and the equilibration time and activity. However, these observations have not been confirmed by other authors. Thus, in the present modeling effort, two-phase sorption/desorption is accounted for by volume fraction weighting the equilibrium liquid and vapor content.

In the current model, a distinction is made between anode and cathode sorption/desorption. At the cathode, only water exists in both liquid and vapor states; i.e., all methanol that crosses the membrane immediately reacts. Thus, equilibrium sorption of water from liquid and vapor states is given as follows [50]:

$$\lambda_{equi,H_2O,l} = 22 \tag{49}$$

$$\lambda_{equi,H_2O,g} = 0.043 + 17.81 \times a_{H_2O,g} - 39.85 \times a^2_{H_2O,g} + 36 \times a^3_{H_2O,g} \tag{50}$$

At the anode, a two-phase mixture of methanol and water exists. For the limiting case of dilute liquid mixtures of water and methanol (i.e., $x_{CH_3OH} < 0.2$), simplified relations can be given. Equilibrium sorption of liquid methanol–water solution has been measured by several authors [44,51,52]. Based on these measurements, a constant liquid water sorption is implemented, similar to the one at the cathode. For liquid methanol sorption, a second-order polynomial fit of the measurements by Ren et al. [52] is used. The methanol content is given as function of methanol activity:

$$\lambda_{equi,CH_3OH,l} = 5.635 a_{CH_3OH,l} + 30,3259 a_{CH_3OH,l}{}^2 \tag{51}$$

To our best knowledge, no sorption data for gaseous mixtures of methanol and water vapor exist in the literature. Thus, the same dilute behavior of methanol sorption is assumed as in the liquid case, however scaled down relative to the difference in maximum sorption. For water sorption, the equation by Zawodzinski et al. [50] is used, neglecting the effect of methanol on water sorption.

$$\lambda_{equi,CH_3OH,g} = 3.585 a_{CH_3OH,g} + 19.2983 a_{CH_3OH,g}{}^2 \tag{52}$$

where a denotes activity and the subscript *equi* denotes the equilibrium state.

The sorption/desorption source terms are implemented according to Equations (53)–(55). A similar non-equilibrium formulation of the uptake rate has previously been implemented by [12,33,42]. Moreover, in the membrane phase of the cathode CL, only water exists, since methanol becomes oxidized. Hence, sorption/desorption only occurs for water. However, in the anode CL, sorption/desorption can occur for both species:

$$S_{i,\lambda} = \alpha\rho_{mem}/EW\left[sk_{i,l}\left(\lambda - \lambda_{equi,l}\right) + (1-s)k_{i,g}\left(\lambda - \lambda_{equi,g}\right)\right] \tag{53}$$

$$\dot{m}_{i,l} = s\alpha k_{i,l}M_i\rho_{mem}/EW \times \left(\lambda - \lambda_{equi,l}\right) \tag{54}$$

$$\dot{m}_{i,g} = (1-s)\alpha k_{i,g}M_i\rho_{mem}/EW \times \left(\lambda - \lambda_{equi,g}\right) \tag{55}$$

where α denotes area density and k the sorption/desorption kinetic coefficient.

The kinetics describing interfacial mass transport of polar vapors and liquids are significantly different in Nafion. Liquid water and methanol sorb approximately ten times faster than their vapor counterpart. Moreover, one needs to differentiate between desorption and sorption kinetics. Desorption is approximately two times faster than sorption for both methanol and water [53–55]. The sorption/desorption mechanisms of water and methanol appear to occur similarly [56]. Hence, similar uptake kinetics of methanol and water are assumed. The effective uptake kinetic coefficient is modeled similar to Ge et al. [54]:

$$k_{s,j,i} = k^{T_0}_{s,j,i}f_{v,i}\exp[2416 \times (1/303 - 1/T)] \tag{56}$$

$$k_{des,j,i} = k^{T_0}_{des,j,I}f_{v,i}\exp[2416 \times (1/303 - 1/T)] \tag{57}$$

$$f_{v,i} = \sum_{i=1}^{N}\lambda_i V_i \Bigg/ \left(V_m + \sum_{i=1}^{N}\lambda_i V_i\right) \tag{58}$$

where $V_m = EW/\rho_m$ is the partial molar volume of dry membrane, $V_i = MW_i/\rho_{i,l}$ is the partial volume of species i, and j denotes the phase facing the membrane phase.

The diffusivity of water and methanol has during the last decade been subject to intensive discussions due to inconsistencies in the reported values which varied over three orders of magnitude, depending on the measurement technique employed (i.e., mass uptake, permeation and NMR-relaxation) [53]. It was discussed by Majzstrik et al. [53] that these differences were caused by not correctly accounting for membrane swelling and the sorption/desorption phenomenon.

Moreover, local maxima in the Fickian diffusivity as a function of water content have been reported [57,58]. However, it was then shown by our group [59] that the spike in the Fickian diffusivity most likely is a mathematical artifact due to the conversion of the chemical diffusivity into a Fickian diffusivity. We further showed that the diffusivity instead exhibits a transition regime with a sudden increase in diffusivity and afterwards stabilization where a low second-order change with water content is observed, depending on whether membrane swelling is accounted for or not:

$$D_{H_2O} = 5.39e^{-2}\left(1.0 + 2.7e^{-3}\lambda^2\right)\left(1.0 + \tanh\left(\left(\lambda - \lambda_{tp}\right)/\delta_{ti}\right)\right)\exp[-3343/T] \tag{59}$$

where $\lambda_{tp} = 2.6225$ denotes the transition point and $\delta_{ti} = 0.8758$ the transition interval. This expression reflects the experimental observation by Benziger et al. [60], who used PSGE NMR measurements at long delay times. They showed that at a low water activity or water content, the connection between the hydrophilic pores is low, causing a high tortuosity. Increasing the species content facilitates more interconnected hydrophilic pores with a resulting higher diffusivity. Furthermore, it was shown in [56] that the diffusivity of methanol in Nafion follows the same tendency as water, although diffusing slower than water.

$$D_{CH_3OH} = P(\lambda)\exp[-3343/T] \tag{60}$$

The EOD coefficient of water in Nafion has been shown to depict the same discontinuous behavior as the water content. For vapor-equilibrated and pre-dried Nafion, a constant EOD coefficient $n_d = 1$ is observed [61]. Whereas for pre-boiled and liquid-equilibrated Nafion, an EOD coefficient of circa $n_d = 2.5$ has been reported numerous times [61–63]. To account for a transition between the two states the following function is used:

$$n_d = \max(2.5\lambda/22, 1.0) \tag{61}$$

It was noted in previous work that the application of a constant value for the EOD coefficient leads to a diffusion-only transport mechanism of water inside the membrane [64,65].

In the case that the membrane is equilibrated with a mixture containing water and methanol, it has been shown that both EOD coefficients depend on the molar fraction of methanol present in the bulk mixture [62,63]. At low methanol molar fractions, the EOD coefficient of methanol becomes marginal. It is thought that methanol has a tendency to stick to the polymer backbone structure at low molar fractions. As the methanol molar fraction increases, the total EOD coefficient increases likewise. This phenomenon occurs since the EOD of methanol is more strongly dependent on methanol molar fraction than on the EOD of water. The following equations were fitted to experimental data from [62] for $x_{CH_3OH} \in [0, 0.3]$ normalized:

$$n_{d,H_2O,mix} = n_{d,H_2O}(1.0 - 0.72x_{CH_3OH}) \tag{62}$$

$$n_{d,CH_3OH,mix} = n_{d,H_2O}\left(x_{CH_3OH} + 1.058x^2_{CH_3OH}\right) \tag{63}$$

2.8. Boundary Conditions

In order to solve the governing transport equations, a set of boundary conditions is needed for each modeling domain. The used mesh is split up into three sub-domains: anode, membrane and cathode. In addition to interfaces between the sub-domains, extra boundaries are found. All boundaries are marked with a dashed line in Figure 3. The following specifications are given for these boundaries:

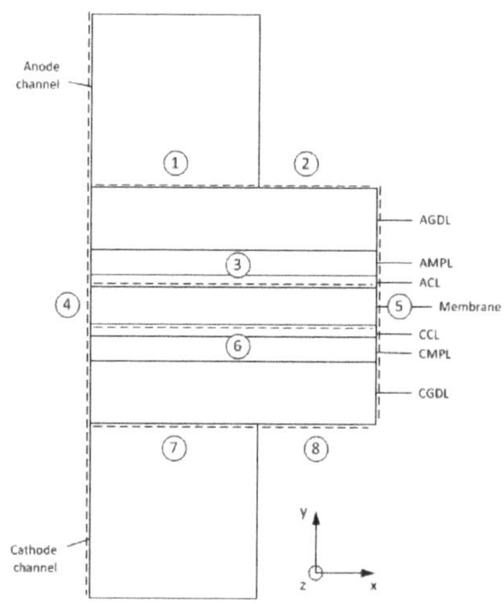

Figure 3. Schematics of the simulation domain and boundary condition regions.

- Boundary 1: This boundary constitutes the interface between the anode channel and anode GDL. It is here that the liquid methanol solution enters and a gas consisting of carbon dioxide, saturated with methanol vapor and water vapor, leaves the GDL. For this boundary, the following is specified: $p_{l,GDL} = p_{l,chan}$, $p_{g,GDL} = p_0$, $N_{e^-} = 0$.

- Boundary 2: This boundary defines the land area in contact with anode GDL, through which electrons and heat is transported: $u_g = 0$, $u_l = 0$, $V = V_{Cell}$, $T = T_{wall}$.
- Boundary 3: This boundary separates the anode CL from the membrane and is impermeable to gas and electrons: $N_{e^-} = 0$, $N_{g,CH_3OH} = 0$, $N_{g,H_2O} = 0$, $N_{g,CO_2} = 0$.
- Boundaries 4 and 5: Both boundaries are treated as mirror planes, meaning that any change of any variable in the x-direction has to be equal to zero: $d\varphi/dx = 0$.
- Boundary 6: Similar to boundary 3, this boundary separates the CL from the membrane at the cathode, and is impermeable to gas and electrons: $N_{e^-} = 0$, $N_{g,N_2} = 0$, $N_{g,H_2O} = 0$, $N_{g,CO_2} = 0$, $N_{g,O_2} = 0$.
- Boundary 7: At this boundary, the liquid phase exits the GDL and enters the channel as droplets. Simultaneously, air enters the GDL from the channel. For this boundary, the following is specified: $p_{l,GDL} = p_{l,chan}$, $p_{g,GDL} = p_0$, $N_{e^-} = 0$.
- Boundary 8: This boundary defines the cathode land area in contact with GDL, through which electrons and heat is transported: $u_g = 0$, $u_l = 0$, $V = 0$, $T = T_{wall}$.

It should be noted that the implemented pressure boundary condition at boundary 1 and 7 implies that the GDL–channel interface capillary pressure is directly specified as a pressure difference between the liquid phase channel and GDL gas phase pressure, as follows:

$$P_c = P_{l,chan} - P_{g,GDL} \tag{64}$$

This approach is in contrast to earlier attempts where either a constant liquid saturation or a channel-averaged liquid saturation has been specified. The reason for altering this boundary condition is to obtain a more physically correct boundary condition that is based on pressure rather than saturation. Interestingly, what this boundary condition effectively enforces is that the liquid phase becomes pushed into the GDL due to an overpressure in the channel relative to the gas phase pressure inside the GDL.

3. Results and Discussion

In this section, the mathematical model is validated and detailed results are presented, including the distributions of liquid methanol concentration, liquid saturation, fluid temperature and dissolved methanol and water content in the electrolyte phase.

In the one-dimensional variable distributions, attention is on the impact of the anode capillary pressure boundary condition at the channel–GDL interface. In order to investigate the importance of this boundary condition, two cases are depicted in each distribution. Each case represents a specific porosity of the anode GDL. The porosity affects both the relative permeability as well as the characteristic pore radius used in the capillary pressure function [66].

In Case 1, a porosity of 0.75 is specified and in Case 2, a porosity of 0.8. This is equivalent to a permeability of 1.2×10^{-11} m^2 and 3.2×10^{-11} m^2, and a characteristic pore size of 8 μm and 10 μm, respectively.

In the three-dimensional plots, emphasis is solely on discussing some of the phenomena the model is able to capture and how they affect some variable distributions. The various input parameters for this model are listed in Tables 3 and 4. Moreover, the electrical contact resistance between the GDL and the bipolar plate was 20 mΩ-cm^2 [67], and the thermal contact resistances were 1.5×10^{-4} K-m^2/W [68].

Table 3. Operating Conditions.

Parameter	Symbol	Value	Unit
Anode inlet methanol concentration	$C_{MeOH,in}$	1000	mol/m^3
Cathode inlet air relative humidity	$RH_{Cat,in}$	75%	-
Anode inlet temperature	$T_{An,in}$	60	°C
Cathode inlet temperature	$T_{Cat,in}$	40	°C
Bipolar plates temperature	T_{Wall}	75	°C
Anode relative inlet pressure	$P_{An,in}$	2500	Pa
Cathode relative outlet pressure	$P_{Ca,out}$	0	Pa

Table 4. Porous medium parameters for the base case.

Parameter	Symbol	Unit	GDL	MPL	CL
Permeability	K	m^2/s	2.0×10^{-14}	1.0×10^{-14}	3.0×10^{-14}
Porosity	ε	-	0.75	0.55	0.55
Tortuosity	τ	-	3.5	4	5
Irreducible saturation	s_{irr}	-	0.10	0.10	0.10
Hydrophilic fraction	f_{Hi}	-	0.35	0.30	0.40
Spec. interfacial area	α	m^2/m^3	5.0×10^5	2.5×10^6	1.0×10^6
Electric Conductivity	σ	S-m	4.0×10^3	7.0×10^3	7.0×10^3

3.1. Predicted Polarization Curve

The presented model with the base case parameters has been validated against experimental measurements of the voltage to current density relation. Figure 4 shows the comparison for a bipolar plate temperature of 348 K (i.e., 75 °C at the bipolar plates, see Table 3) The electrochemical parameters in Table 2 were curve-fitted to match these experimental results.

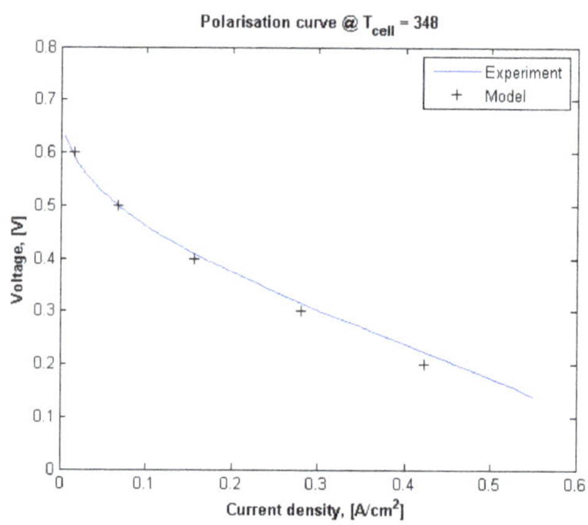

Figure 4. Comparison of calculated performance curve and experiments.

With this, the predicted performance is in excellent agreement with the experimental data. At low current densities, in the activation overpotential region, the model captures the change in voltage, especially in the important low current density regime that is dominated by cross-over of methanol. This initial change in voltage is dominated by a change in anode overpotential. The cathode overpotential loss is already high due to methanol crossover. As the current density increases, the change in voltage becomes linear, indicating the beginning of the ohmic region. It seems that the predicted curve falls steeper than the experimentally measured one. This suggests a slightly higher resistance due to either contact resistance, ionic or electron transport. However, it could also be due to an underestimation of the anode overpotential loss. Further improvements may be obtained by slightly adjusting these abovementioned parameters, but overall, the agreement is deemed very satisfactory.

3.2. Comparison of Two Cases

In the following, the two cases with different GDL porosities and different characteristic pore radii as mentioned above are examined in detail. The distribution of liquid saturation in the anode electrode is shown in Figure 5a. It is evident that the two cases nearly coincide in the CL and MPL, whereas a clear distinction is possible in the GDL. Even though a large difference in liquid saturation level appears, the same trend of a gas build-up under the land area is visible. This build-up of gas is not only a reflection of transport resistance, but also due to a difference in surface tension. Under the land, the molar fraction of methanol is lower than under the channel, as seen in Figure 5b. This difference in molar fraction imposes a difference in surface tension under land as this is highly dependent on methanol molar fraction. Moreover, this results in a difference in liquid saturation in order to balance the capillary pressure.

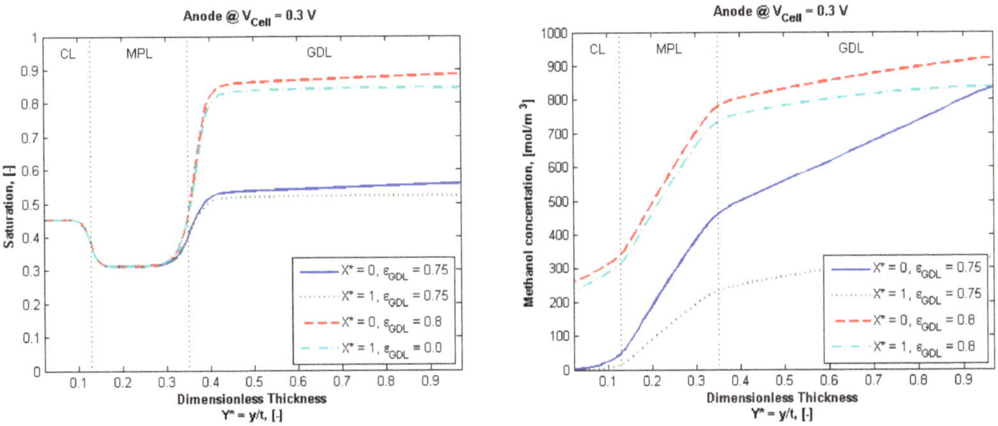

(a) (b)

Figure 5. (**a**) Liquid saturation distribution in the anode electrode for two cases of GDL porosity at a cell voltage 0.3 V; (**b**) methanol concentration distribution in the anode electrode for two cases of GDL porosity at cell voltage 0.3 V. A dimensionless distance of $X^* = 0$ and $X^* = 1$ are equivalent to under the channel and land, respectively.

Another phenomenon that the model can capture is the saturation jump condition at the CL–MPL and MPL–GDL interface. This jump condition in liquid saturation is caused by an abrupt change in porous media properties such as porosity, permeability and hydrophilic pore fraction. By having a highly hydrophobic MPL with small pores, the liquid saturation level can be kept fairly low. This effectively decreases methanol diffusivity, thereby minimizing the methanol crossover rate. Unfortunately, this also affects the extent of the anode overpotential loss when the limiting current sets in.

The use of micro-pores has another interesting benefit. In the CL and MPL, this pushes the liquid saturation close to the hydrophilic pore fraction. By having a small characteristic pore size of 0.163 µm and 0.426 µm, respectively, it takes a large pressure for the liquid phase to overcome the repelling capillary forces and intrude the hydrophobic pores. Meanwhile, the characteristic pore size of the GDL is much larger. Consequently, the pressure requirement for intruding these hydrophobic pores is less. As it appears from Figure 5, by decreasing the porosity of the GDL slightly, the saturation level is pushed considerably closer to the hydrophilic pore fraction. Again, this occurs due to the increasing capillary forces, which the liquid phase has to overcome. It is evident from these two cases that a too small characteristic pore size of the GDL significantly lowers the liquid saturation level. In turn, this leads to a decrease in the diffusivity of methanol in the liquid phase, hence limiting current density and performance. For Case I and Case II, this gives a current density of 0.17 A/cm^2 and 0.28 A/cm^2, respectively, a surprisingly large difference for such a comparatively small change in the physical properties.

A clear distinction between the two cases can also be seen in the liquid phase methanol distributions, shown in Figure 5b. As discussed above, a lower liquid saturation level in the GDL causes a decreased methanol transport rate. Hence, a much steeper methanol concentration gradient is visible in the through-plane direction of the GDL for Case I. The same applies to the in-plane direction towards the middle of the land. However, not only a decrease in saturation plays a role here. A decrease in porosity and a corresponding increase in tortuosity decrease the effective diffusivity of methanol. Meanwhile, it should be noted that the distribution of methanol in the liquid phase is furthermore dependent on the rate of methanol evaporation and condensation. Even though the net transport of gas phase is in the direction of the channel, that of methanol vapor may move toward the CL and help improve performance or the limiting current density. However, the extent of this effect on performance is fairly low, as the mass transport rate in the gas phase is lower compared to that of the liquid phase.

The fluid temperature distribution in the anode electrode is shown in Figure 6a. Again, a clear difference can be seen between the two cases. In Case II, where the higher current density is produced, the temperature is also highest; this difference is due to the rate of heat production associated with catalytic burning of crossover methanol, the ORR and MOR reactions, and electron and ion transport. Importantly the methanol crossover generates a high production of heat and temperature rise. This difference in the temperature distribution is also interesting since it affects the sorption/desorption in the electrolyte phase of the CL, species crossover, and phase change.

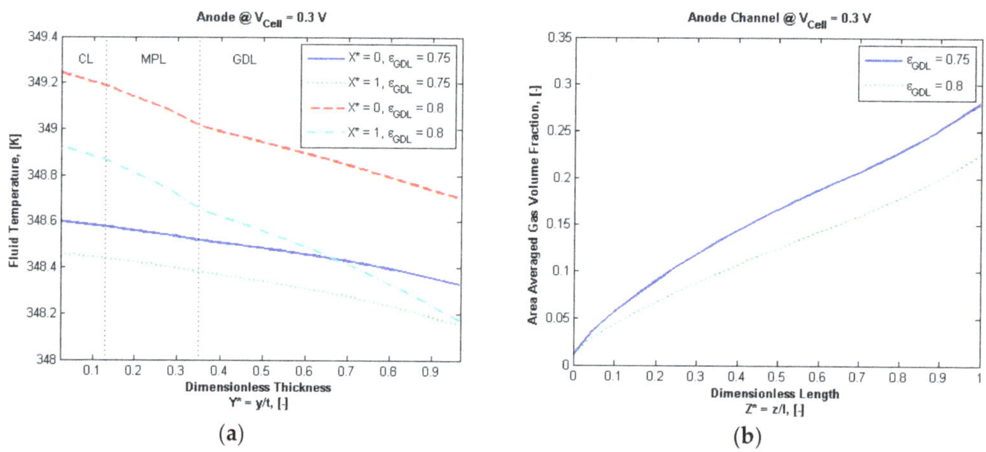

Figure 6. (**a**) The area-averaged liquid saturation distribution along the anode channel for two cases of GDL porosity at a cell voltage 0.3 V; (**b**) fluid temperature concentration distribution in the anode electrode for two cases of GDL porosity at cell voltage 0.3 V.

The decrease in performance between cases with a different porosity is visible from cross sectional area-averaged gas volume fraction in the channel, depicted in Figure 6b. For Case II, a much larger amount of gas is predicted in the channel. Inherently, this occurs since a higher current density means a higher production rate of carbon dioxide. However, the exact amount of gas leaving the GDL is a complex matter to predict, since it depends on the rate of methanol and water evaporation, which in turn depends on temperature.

The three-dimensional distribution of the gas volume fraction in the anode gas diffusion electrode and channel is shown in Figure 7a. A fairly even distribution is seen inside the electrode along the channel length. This occurs in opposition to the channel height, where a gradual increase is visible. This increase was also observed in Figure 5, and it occurs because gas continuously enters the channel from the electrode. When taking a closer look at the middle of the channel, it can be observed that the gas phase is pushed away from the GDL–channel interface and is concentrated at a short distance from the GDL–channel interface. This trend is quite different near the channel corner facing the GDL. Here, a large gas pocket is forming, which gradually increases along the channel length, obstructing more and more liquid phase transport. This phenomenon occurs due to the velocity distribution around the corner rather than surface tension forces, as they are not included at the present stage.

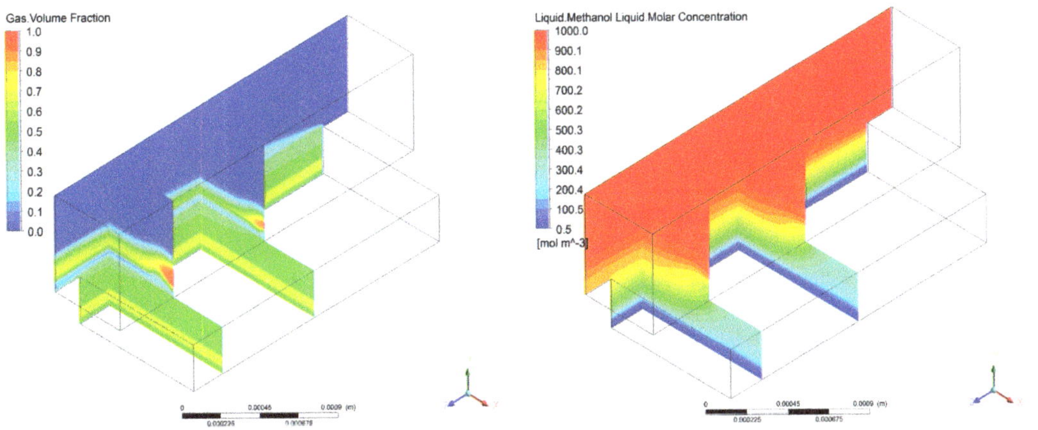

(**a**) (**b**)

Figure 7. (**a**) Three-dimensional liquid saturation distribution and (**b**) three-dimensional liquid phase methanol concentration distribution in the anode channel and electrode at a GDL porosity of 0.75, cell voltage of 0.3 V and a current density of 0.17 A/cm^2. The flow direction is from right, back, to left, front.

In Figure 7b, an overview is given of the methanol concentration distribution along the channel for Case 1. Again, a fairly even distribution of methanol is seen along the channel, which results from the high stoichiometry. It is only close to the GDL–channel interface that a gradient in the methanol concentration is visible. This thinning out of the methanol concentration inside the catalyst layer causes a high concentration overpotential, but it also reduces the methanol crossover effect, which is one of the biggest loss mechanisms in DMFCs.

In Figure 8, the methanol and water content distributions of the electrolyte phase are shown. In both figures, a steeper gradient can be seen in the CL, caused by a decrease in ionomer content in comparison to the membrane. Decreasing the ionomer content in the CL effectively lowers the diffusivity of methanol and, unfortunately, also the ion conductivity. However, this gradient is not only caused by a decrease in the ionomer content, it is partially also due to EOD and non-equilibrium uptake that balances diffusion. The EOD is quite large in the CL, since the current density changes in the through-plane direction, imposing a large potential field gradient.

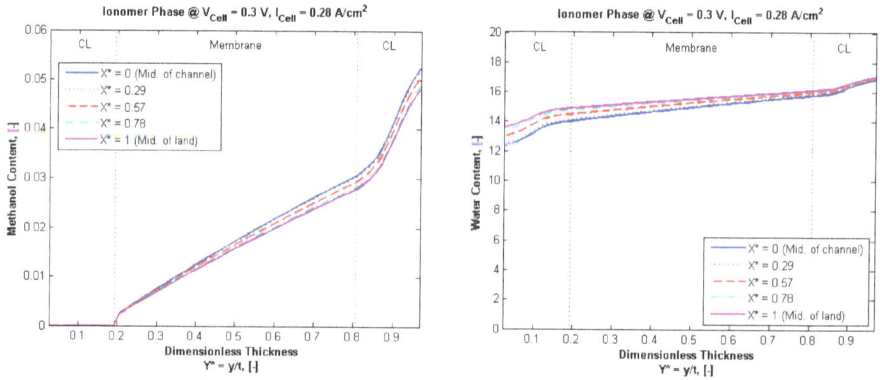

(a) (b)

Figure 8. (**a**) Membrane methanol content for GDL porosity of 0.75 at cell voltage of 0.3 V and current density of 0.17 A/cm^2; (**b**) membrane water content distribution for GDL porosity of 0.75 at cell voltage of 0.3 V and current density of 0.17 A/cm^2.

Inherently, the distributions of methanol and water are rather different in the membrane. As seen from Figure 7a, the methanol content approaches zero near the interface between the cathode CL and membrane, whereas the distribution of water content is again fairly even. The methanol content has to approach zero at the interface, since methanol is catalytically burned when getting in contact with air. The water content level is dependent on the air's relative humidity of the cathode, which is a function of water crossover, the water production rate due to ORR and DMOR, as well as air stoichiometry.

Finally, it should be noted that the methanol and water content are highly dependent on the hydrophilic pore fraction of the anode CL. Since the liquid saturation is nearly identical to the hydrophilic pore fraction, and the equilibrium methanol and water content are highly dependent on the liquid saturation, and since the liquid phase has a higher equilibrium content and sorption rate, the hydrophilic pore fraction effectively determines the anode water and methanol content.

4. Conclusions

Direct methanol fuel cell systems are typically simple in comparison with other types of fuel cells. However, the underlying heat and mass transfer mechanisms are exceedingly complex, and computational fluid dynamics modelling can help to shed light into how to improve system performance. In this work, a steady-state, three-dimensional, two-fluid, multi-component and non-isothermal liquid-fed direct methanol fuel cell model was presented and compared against experimental measurements in good agreement. The developed model was used for investigating the interaction between volume porosity of the anode GDL and the capillary pressure boundary condition on fuel cell performance. The simulation results indicate that while keeping the force of liquid phase intrusion into the GDL constant, the change in GDL porosity plays a significant role in performance. This would not have been observed for a fixed saturation condition as it implicitly adjusts the force required for intruding hydrophobic pores when changing the characteristic pore size radius of the GDL and hence its capillary pressure. It was further shown that accounting for variations in surface tension due to methanol decreases the prediction of liquid saturation under the land as opposed to under the channel, which pronounces the resistance to methanol diffusion under the land area. Moreover, the importance of accounting for the hydrophilic pore fraction was underlined. The smaller the characteristic pore size becomes for highly flooded porous media, the closer the liquid saturation is pushed to the hydrophilic pore fraction, making methanol transport losses more prominent.

Author Contributions: Conceptualization, A.C.O., S.K.K. and T.B.; methodology, A.C.O., S.K.K. and T.B.; software, A.C.O., S.K.K. and T.B.; experimental, A.C.O.; formal analysis, A.C.O., S.K.K. and T.B.; resources, S.K.K.; writing—original draft preparation, A.C.O.; writing—review and editing, A.C.O. and T.B.; supervision, S.K.K. and T.B.; project administration, S.K.K.; funding acquisition, S.K.K. All authors have read and agreed to the published version of the manuscript.

Funding: This work was partially supported by the Danish Energy program, 2008-1-10076, Aalborg University, and the HyFC Research School.

Acknowledgments: A.C.O. would like to thank IRD A/S for the support to obtain experimental data.

Conflicts of Interest: The authors declare no conflict of interest.

References

1. Araya, S.S.; Liso, V.; Cui, X.; Li, N.; Zhu, J.; Sahlin, S.L.; Jensen, S.H.; Nielsen, M.P.; Kær, S.K. A Review of The Methanol Economoy: The Fuel Cell Route. *Energies* **2020**, *13*, 596. [CrossRef]
2. Heinzel, A.; Barragán, V.M. A review of the state-of-the-art of the methanol crossover in direct methanol fuel cells. *J. Power Sources* **1999**, *84*, 70–74. [CrossRef]
3. Neburchilov, V.; Martin, J.; Wang, H.; Zhang, J. A review of polymer electrolyte membranes for direct methanol fuel cells. *J. Power Sources* **2007**, *169*, 221–238. [CrossRef]
4. He, Y.; Miao, Z.; Zhao, T.; Yang, W. Numerical study of the effect of the GDL structure on water crossover in a direct methanol fuel cell. *Int. J. Hydrogen Energy* **2012**, *37*, 4422–4438. [CrossRef]
5. Berning, T. The dew point temperature as a criterion for optimizing operating conditions of proton exchange membrane fuel cells. *Int. J. Hydrogen Energy* **2012**, *37*, 10265–10275. [CrossRef]
6. Argyropoulos, P.; Scott, K.; Taama, W.M. Gas evolution and power performance in direct methanol fuel cells. *J. Appl. Electrochem.* **1999**, *29*, 661–669. [CrossRef]
7. Lu, G.Q.; Wang, C.Y. Electrochemical and flow characterization of a direct methanol fuel cell. *J. Power Sources* **2004**, *134*, 33–40. [CrossRef]
8. Oliveira, V.; Falcao, D.; Rangel, C.; Pinto, A. A comparative study of approaches to direct methanol fuel cells modelling. *Int. J. Hydrogen Energy* **2007**, *32*, 415–424. [CrossRef]
9. Wang, Z.H.; Wang, C.Y. Mathematical Modeling of Liquid-Feed Direct Methanol Fuel Cells. *J. Electrochem. Soc.* **2003**, *150*, A508–A519. [CrossRef]
10. Leverett, M.C. Capillary behavior in porous solids. *Trans. AIME* **1976**, *142*, 152. [CrossRef]
11. Berning, T.; Odgaard, M.; Kær, S. A computational analysis of multi-phase flow through the porous media of a PEMFC cathode using the multi-fluid approach. *J. Electrochem. Soc.* **2009**, *156*, B1301. [CrossRef]
12. Wu, H.; Li, X.; Berg, P. On the modeling of water transport in polymer electrolyte membrane fuel cells. *Electrochim. Acta* **2009**, *54*, 6913–6927. [CrossRef]
13. Kumbur, E.C.; Sharp, K.V.; Mench, M.M. Validated leverett approach for multiphase flow in PEFC diffusion media. *J. Electrochem. Soc.* **2007**, *154*, B1295–B1304. [CrossRef]
14. Divisek, J.; Fuhrmann, J.; Gärtner, K.; Jung, R. Performance Modeling of a Direct Methanol Fuel. *J. Electrochem. Soc.* **2003**, *150*, A811. [CrossRef]
15. Ge, J.; Liu, H. A three-dimensional two-phase flow model for a liquid-fed direct methanol fuel cell. *J. Power Sources* **2007**, *163*, 907–915. [CrossRef]
16. Liu, W.; Wang, C.Y. Modeling water transport in liquid feed direct methanol fuel cells. *J. Power Sources* **2007**, *164*, 189–195. [CrossRef]
17. Yang, W.W.; Zhao, T.S. A two-dimensional, two-phase mass transport model for liquid-feed DMFCs. *Electrochim. Acta* **2007**, *52*, 6125–6140. [CrossRef]
18. Yang, W.W.; Zhao, T.S. Two-phase, mass-transport model for direct methanol fuel cells with effect of non-equilibrium evaporation and condensation. *J. Power Sources* **2007**, *174*, 136–147. [CrossRef]
19. Yang, W.W.; Zhao, T.S.; Xu, C. Three-dimensional two-phase mass transport model for direct methanol fuel cells. *Electrochim. Acta* **2007**, *53*, 853–862. [CrossRef]
20. Xu, C.; Zhao, T.S.; Yang, W.W. Modeling of water transport through the membrane electrode assembly for direct methanol fuel cells. *J. Power Sources* **2008**, *178*, 291–308. [CrossRef]
21. Miao, Z.; He, Y.; Li, X.; Zou, J. A two-dimensional two-phase mass transport model for direct methanol fuel cells adopting a modified agglomerate approach. *J. Power Sources* **2008**, *185*, 1233–1246. [CrossRef]
22. Miao, Z.; He, Y.; Zou, J. Modeling the effect of anisotropy of gas diffusion layer on transport phenomena in a direct methanol fuel cell. *J. Power Sources* **2010**, *195*, 3693–3708. [CrossRef]
23. Liu, W.; Wang, C.Y. Three-dimensional simulations of liquid feed direct methanol fuel cells. *J. Electrochem. Soc.* **2007**, *154*, B352. [CrossRef]

24. Yang, W.W.; Zhao, T.S.; Chen, R.; Xu, C. An approach for determining the liquid water distribution in a liquid-feed direct methanol fuel cell. *J. Power Sources* **2009**, *190*, 216–222. [CrossRef]
25. Garvin, J.J.; Meyers, J.P. Modeling of Coupled Multiphase Transport in Direct Methanol Fuel Cell Diffusion Layers. *J. Electrochem. Soc.* **2011**, *158*, B1119. [CrossRef]
26. Bahrami, H.; Faghri, A. Review and advances of direct methanol fuel cells: Part II: Modeling and numerical simulation. *J. Power Sources* **2013**, *230*, 303–320. [CrossRef]
27. Gostick, J.T.; Fowler, M.W.; Ioannidis, M.A.; Pritzker, M.D.; Volfkovich, Y.M.; Sakars, A. Capillary pressure and hydrophilic porosity in gas diffusion layers for polymer electrolyte fuel cells. *J. Power Sources* **2006**, *156*, 375–387. [CrossRef]
28. Argyropoulos, P.; Scott, K.; Taama, W.M. Modelling pressure distribution and anode/cathode streams vapour–liquid equilibrium composition in liquid feed direct methanol fuel cells. *Chem. Eng. J.* **2000**, *78*, 29–41. [CrossRef]
29. Crowe, C.T. *Multiphase Flow Handbook*; CRC Press: Boca Raton, FL, USA, 2005.
30. Mazumder, S.; Cole, J.V. Rigorous 3-D Mathematical Modeling of PEM Fuel Cells. *J. Electrochem. Soc.* **2003**, *150*, A1510–A1517. [CrossRef]
31. Bear, J. *Dynamics of Fluids in Porous Media;* Courier Corporation: Chelmsford, MA, USA, 2013.
32. Nam, J.H.; Kaviany, M. Effective diffusivity and water-saturation distribution in single- and two-layer PEMFC diffusion medium. *Int. J. Heat Mass Transf.* **2003**, *46*, 4595–4611. [CrossRef]
33. Gurau, V.; Zawodzinski, T.A.; Mann, J.A. Two-phase trans.sport in PEM fuel cell cathodes. *J. Fuel Cell Sci. Technol.* **2008**, *5*, 021009. [CrossRef]
34. Udell, K.S. Heat transfer in porous media considering phase change and capillarity-the heat pipe effect. *Int. J. Heat Mass Transf.* **1985**, *28*, 485–495. [CrossRef]
35. Grötsch, M.; Mangold, M. A two-phase PEMFC model for process control purposes. *Chem. Eng. Sci.* **2008**, *63*, 434–447. [CrossRef]
36. Lemoine-Nava, R.; Hanke-Rauschenbach, R.; Mangold, M.; Sundmacher, K. The gas diffusion layer in polymer electrolyte membrane fuel cells: A process model of the two-phase flow. *Int. J. Hydrogen Energy* **2011**, *36*, 1637–1653. [CrossRef]
37. Vazquez, G.; Alvarez, E.; Navaza, J.M. Surface Tension of Alcohol Water + Water from 20 to 50 °C. *J. Chem. Eng. Data* **1995**, *40*, 611–614. [CrossRef]
38. Olesen, A.C.; Berning, T.; Kær, S.K. The Effect of Inhomogeneous Compression on Water Transport in the Cathode of a PEM. *Fuel Cell* **2011**, *54693*, 839–850.
39. Gasteiger, H.A.; Markovic, N.; Ross, P.N.; Cairns, E.J. Methanol electrooxidation on well-characterized platinum-ruthenium bulk alloys. *J. Phys. Chem.* **1993**, *97*, 12020–12029. [CrossRef]
40. Meyers, J.P.; Newman, J. Simulation of the Direct Methanol Fuel Cell. *J. Electrochem. Soc.* **2002**, *149*, A718. [CrossRef]
41. Bahrami, H.; Faghri, A. Water Management in a Passive DMFC Using Highly Concentrated Methanol Solution. *J. Fuel Cell Sci. Technol.* **2010**, *8*, 021011. [CrossRef]
42. Berning, T.; Odgaard, M.; Kær, S.K. Water balance simulations of a polymer-electrolyte membrane fuel cell using a two-fluid model. *J. Power Sources* **2011**, *196*, 6305–6317. [CrossRef]
43. Ju, H.; Luo, G.; Wang, C.Y. Probing liquid water saturation in diffusion media of polymer electrolyte fuel cells. *J. Electrochem. Soc.* **2007**, *154*, B218–B228. [CrossRef]
44. Gates, C.M.; Newman, J. Equilibrium and diffusion of methanol and water in a nafion 117 membrane. *AICHE J.* **2000**, *46*, 2076–2085. [CrossRef]
45. Schultz, T.; Sundmacher, K. Mass, charge and energy transport phenomena in a polymer electrolyte membrane (PEM) used in a direct methanol fuel cell (DMFC): Modelling and experimental validation of fluxes. *J. Membr. Sci.* **2006**, *276*, 272–285. [CrossRef]
46. Choi, P.; Datta, R. Sorption in Proton-Exchange Membranes an Explanation of Schroeder's Paradox. *J. Electrochem. Soc.* **2003**, *150*, E601–E607. [CrossRef]
47. Freger, V. Hydration of ionomers and schroeder's paradox in nafion. *J. Phys. Chem. B* **2008**, *113*, 24–36. [CrossRef]
48. Onishi, L.M.; Prausnitz, J.M.; Newman, J. Water and nafion equilibria. absence of schroeder's paradox. *J. Phys. Chem. B* **2007**, *111*, 10166–10173. [CrossRef]
49. Jeck, S.; Scharfer, P.; Kind, M. Water sorption in physically crosslinked poly (vinyl alcohol) membranes: An experimental investigation of Schroeder's paradox. *J. Membr. Sci.* **2009**, *337*, 291–296. [CrossRef]
50. Zawodzinski, T.A.; Neeman, M.; Sillerud, L.O.; Gottesfeld, S. Determination of Water Diffusion Coefficients in Perfluorosulfonate Ionomeric Membranes. *J. Phys. Chem.* **1991**, *95*, 6040–6044. [CrossRef]
51. Skou, E.; Kauranen, P.; Hentschel, J. Water and methanol uptake in proton conducting Nafion®membranes. *Solid State Ion.* **1997**, *97*, 333–337. [CrossRef]
52. Ren, X.; Springer, T.E.; Zawodzinski, T.A.; Gottesfeld, S. Methanol transport through nafion membranes. Electro-osmotic drag effects on potential step measurements. *J. Electrochem. Soc.* **2000**, *147*, 466–474. [CrossRef]
53. Majsztrik, P.W.; Satterfield, M.B.; Bocarsly, A.B.; Benziger, J.B. Water sorption, desorption and transport in Nafion membranes. *J. Membr. Sci.* **2007**, *301*, 93–106. [CrossRef]
54. Ge, S.; Li, X.; Yi, B.; Hsing, I. Absorption, Desorption, and Transport of Water in Polymer Electrolyte Membranes for Fuel Cells. *J. Electrochem. Soc.* **2005**, *152*, A1149–A1157. [CrossRef]
55. Zhao, Q.; Majsztrik, P.; Benziger, J. Diffusion and Interfacial Transport of Water in Nafion. *J. Phys. Chem. B* **2011**, *115*, 2717–2727. [CrossRef] [PubMed]

56. Zhao, Q.; Carro, N.; Ryu, H.Y.; Benziger, J. Sorption and transport of methanol and ethanol in H+-nafion. *Polymer* **2012**, *53*, 1267–1276. [CrossRef]
57. Springer, T.E.; Zawodzinski, T.A.; Gottesfeld, S. Polymer Electrolyte Fuel Cell Model. *J. Electrochem. Soc.* **1991**, *138*, 2334–2342. [CrossRef]
58. Motupally, S.; Becker, A.J.; Weidner, J.W. Diffusion of Water in Nafion 115 Membranes. *J. Electrochem. Soc.* **2000**, *147*, 3171–3177. [CrossRef]
59. Olesen, A.C.; Berning, T.; Kær, S.K. On the Diffusion Coefficient of Water in Polymer Electrolyte Membranes. *ECS Trans.* **2012**, *50*, 979–991. [CrossRef]
60. Benziger, J.; Bocarsly, A.; Cheah, M.; Majsztrik, P.; Satterfield, B.; Zhao, Q. Mechanical and transport properties of nafion: Effects of temperature and water activity. In *Fuel Cells and Hydrogen Storage*; Bocarsly, A., Mingos, D.M.P., Eds.; Springer: Berlin/Heidelberg, Germany, 2011; Volume 141, pp. 85–113.
61. Zawodzinski, T.A.; Davey, J.; Valerio, J.; Gottesfeld, S. The water content dependence of electro-osmotic drag in proton-conducting polymer electrolytes. *Electrochim. Acta* **1995**, *40*, 297–302. [CrossRef]
62. Hallberg, F.; Vernersson, T.; Pettersson, E.T.; Dvinskikh, S.V.; Lindbergh, G.; Furó, I. Electrokinetic transport of water and methanol in Nafion membranes as observed by NMR spectroscopy. *Electrochim. Acta* **2010**, *55*, 3542–3549. [CrossRef]
63. Tschinder, T.; Schaffer, T.; Fraser, S.; Hacker, V. Electro-osmotic drag of methanol in proton exchange membranes. *J. Appl. Electrochem.* **2007**, *37*, 711–716. [CrossRef]
64. Berning, T. On water transport in polymer electrolyte membranes during the passage of current. *Int. J. Hydrogen Energy* **2011**, *36*, 9341–9344. [CrossRef]
65. Berning, T. On the Nature of Electro-Osmotic Drag. *Energies* **2020**, *13*, 4726. [CrossRef]
66. Tomadakis, M.M.; Robertson, T.J. Viscous Permeability of Random Fiber Structures: Comparison of Electrical and Diffusional Estimates with Experimental and Analytical Results. *J. Compos. Mater.* **2005**, *39*, 163–188. [CrossRef]
67. Ismail, M.S.; Damjanovic, T.; Ingham, D.B.; Pourkashanian, M.; Westwood, A. Effect of polytetrafluoroethylene-treatment and microporous layer-coating on the electrical conductivity of gas diffusion layers used in proton exchange membrane fuel cells. *J. Power Sources* **2010**, *195*, 2700–2708. [CrossRef]
68. Nitta, I.; Himanen, O.; Mikkola, M. Thermal conductivity and contact resistance of compressed gas diffusion layer of PEM fuel cell. *Fuel Cells* **2008**, *8*, 111. [CrossRef]

energies

Article

Modelling and Performance Analysis of an Autonomous Marine Vehicle Powered by a Fuel Cell Hybrid Powertrain

Giuseppe De Lorenzo [1], Francesco Piraino [1], Francesco Longo [1], Giovanni Tinè [2], Valeria Boscaino [2], Nicola Panzavecchia [2], Massimo Caccia [3] and Petronilla Fragiacomo [1,*]

1 Department of Mechanical, Energy and Management Engineering, University of Calabria, Via P. Bucci, Arcavacata di Rende, 87036 Cosenza, Italy
2 Institute of Marine Engineering, National Research Council of Italy, Via Ugo La Malfa, 153, 90146 Palermo, Italy
3 Institute of Marine Engineering, National Research Council, Via De Marini 6, 16149 Genoa, Italy
* Correspondence: petronilla.fragiacomo@unical.it

Abstract: This paper describes the implementation of a hydrogen-based system for an autonomous surface vehicle in an effort to reduce environmental impact and increase driving range. In a suitable computational environment, the dynamic electrical model of the entire hybrid powertrain, consisting of a proton exchange membrane fuel cell, a hydrogen metal hydride storage system, a lithium battery, two brushless DC motors, and two control subsystems, is implemented. The developed calculation tool is used to perform the dynamic analysis of the hybrid propulsion system during four different operating journeys, investigating the performance achieved to examine the obtained performance, determine the feasibility of the work runs and highlight the critical points. During the trips, the engine shows fluctuating performance trends while the energy consumption reaches 1087 Wh for the fuel cell (corresponding to 71 g of hydrogen) and 370 Wh for the battery, consuming almost all the energy stored on board.

Keywords: dynamic analysis; proton exchange membrane fuel cells; hybrid electric propulsion system; li-ion battery; control system; autonomous marine vehicle

Citation: De Lorenzo, G.; Piraino, F.; Longo, F.; Tinè, G.; Boscaino, V.; Panzavecchia, N.; Caccia, M.; Fragiacomo, P. Modelling and Performance Analysis of an Autonomous Marine Vehicle Powered by a Fuel Cell Hybrid Powertrain. *Energies* **2022**, *15*, 6926. https://doi.org/10.3390/en15196926

Academic Editor: Nicu Bizon

Received: 30 August 2022
Accepted: 18 September 2022
Published: 21 September 2022

1. Introduction

Autonomous Surface Vehicles (ASVs) are becoming consolidated robotic tools for marine, coastal and inland surveys. ASVs are usually equipped with electronic instruments to perform autonomous geo-morphological, biological, chemical, and physical analyses, as well as data collection.

Autonomous Unmanned Surface Vehicles (USVs), also known as unmanned surface ships or (in some cases) as autonomous surface vehicles (ASVs), are boats that operate on the surface of the water unmanned [1].

Remotely operated USVs were used as early as the end of World War II in mine clearance operations [2]. Since then, advancements in USV control systems and navigation technology have resulted in USVs with partially autonomous control and USVs (ASVs) with full autonomy [2]. Modern applications and research areas for USVs and ASVs include environmental and climate monitoring, seabed mapping [3], surveillance [4], maintenance and inspection of infrastructures [5], and military and naval operations [2]. USVs are valuable in oceanography [6] too, as they are more capable than moored or drifting weather buoys but are much cheaper than equivalent weather and research vessels and more flexible than the contributions of a commercial vessel.

The National Research Council, formerly the Institute on Intelligent Systems for Automation (ISSIA) and now the Institute of Marine Engineering (INM), has developed numerous autonomous marine electric vehicles with batteries, both submarine (PROTEUS) [7] and surface vehicles (CHARLIE [8], SWAMP [9]).

For autonomous vehicles, electric propulsion is preferred for its efficiency, reliability, low cost, low thermal and acoustic impact, and low vibration levels [10]. Generally, the battery system supplies electric energy, although the specific energy of current batteries limits their duration and range. Conventional lithium polymer (LiPo) batteries have a specific energy of 150–250 Wh/kg [11] and provide a typical runtime of 60 to 90 min [12]. It is used in different applications, starting from chip to electric vehicle [13,14]. However, to extend the runtime to many hours, a power system with higher specific energy than that of batteries must be used. The fuel cell-based systems fed by hydrogen are becoming the new energy system for future automotive [15] and road freight transportation [16]. They can provide higher specific energy in different vehicles: autobus [17], bicycles and motorcycles [18], trains [19,20], unmanned underwater [21] and ground [22] vehicles and ship [23]. Combining a fuel cell with a reservoir of compressed hydrogen with a weight fraction of 6%, for example, results in specific energy greater than 800–1000 Wh/kg [24].

The commercially available electric motors for AUV and ASV applications are:

1. permanent magnet DC brush electric motors;
2. brushless DC electric motors;
3. stepper electric motors.

For marine applications that require greater thrust power, alternating current motors with inverters can be used, which can be synchronous or asynchronous.

The synchronous motor is a type of alternating current electric motor, whose rotation speed is synchronized with the electric frequency and the synchronous motor is also called vector motor or Rowan motor.

In recent years, the use of power electronics has drastically simplified the start-up operation; in fact, it is possible to regulate both the voltage (and therefore the current) of the power supply and the frequency. Thus, starting from zero frequency and making it grow very gradually, a torque is continuously activated, which can accelerate the motor from a standstill. The components (inverters or cyclo-converters), which allow this mode, are made with semiconductor components such as the thyristor or the IGBT transistor (Insulated Gate Bipolar Transistor) and allow the creation of electronic speed control systems.

The power levels that can be developed by synchronous and asynchronous electric motors are usually up to 12 kW (synchronous motor) and up to 1000 kW (asynchronous motor). At low power, the synchronous motor has a higher efficiency (0.95 for a rated power between 2 and 12 kW) than the asynchronous motor (from 0.85 for small 4-kW motors to 0.95 for large 45-kW motors) [25], while the inverter that controls the motor has an efficiency that is generally between 0.8 and almost 1. On the other hand, the synchronous electric motor is often used to drive variable-speed loads when it is powered by a static converter (inverter), as is the case in most electric vehicles, for example.

In this paper, the dynamic electric model of the entire PEM fuel cell-based hybrid electric propulsion system for an autonomous surface vehicle and the electric model of its control system are formulated and implemented in the Matlab-Simulink environment. The entire PEM fuel cell-based hybrid electric propulsion system consists of the power-controlled PEM fuel cell subsystem with its hydrogen metal hydride storage subsystem, the Li-ion battery subsystem, and the electric motors.

The calculation tool is reliable and robust, it can simulate long-term missions and is flexible since it can consider different types of PEM fuel cells or Li-ion batteries.

The dynamic analyses of the entire fuel cell-based hybrid-electric propulsion system with its control subsystem operating in real long-term tests are performed with the developed calculation tool for four different working missions. The goal of these analyses is to monitor the charge states of the battery subsystem and hydrogen storage system during the missions, determine the feasibility of the working missions, and highlight the critical points before the propulsion system is built and installed on board the vehicle.

2. Modeling

The heart of the paper concerns the powertrain modeling of an autonomous marine vehicle. All the main components are modeled, using electrical, chemical, thermal, mechanical, and logical systems of equations. After the formalization phase, the dynamic models are implemented in the Matlab-Simulink environment, building dedicated blocks for each component. Once the modeling phase is completed, a validation analysis is performed to evaluate the behavior of the components compared to real components available on the market. Three different sections are defined: the vehicle presentation, the power, and the propulsive systems modeling.

2.1. Autonomous Marine Vehicle Powertrain

The introduction of an innovative fuel cell-based powertrain in an autonomous marine vehicle (AMV) is the main goal of this paper. The AMV discussed is denominated SWAMP (Shallow Water Autonomous Multipurpose Platform) [9], and it is electric, modular, portable, lightweight, and highly controllable catamaran; the energy required is provided by a battery pack, while the propulsion system is constituted by four azimuth Pump-Jet thrusters, specifically designed for this vehicle. Regarding the physical parameters, the vehicle is 1.23 m long, 1.1 m high and 1.1 m wide, but this last value is variable between 0.7 m and 1.25 m, given by the sliding structure used; in addition, its weight is 38 kg, but the catamaran allows embarking up to 100 kg.

Considering these features, useful for calculating power consumption, the introduction of a fuel-cell system is evaluated, working together with the battery, with the aim of extending the vehicle's range. The conceptual scheme of this new layout is illustrated in Figure 1. As shown, the main components can be gathered into two macro-systems:

1. The power system, which provides power and energy demands of the AMV, is useful for defining a specific power-sharing, and is composed of a fuel cell, battery, DC/DC converter, and power-sharing control systems;
2. The propulsive system, that transforms the power system outputs into parameters useful for the AMV traction, is constituted by two BLDC motors, the propulsive and azimuth ones, formed by an electric motor, an inverter, and a motor control subsystem.

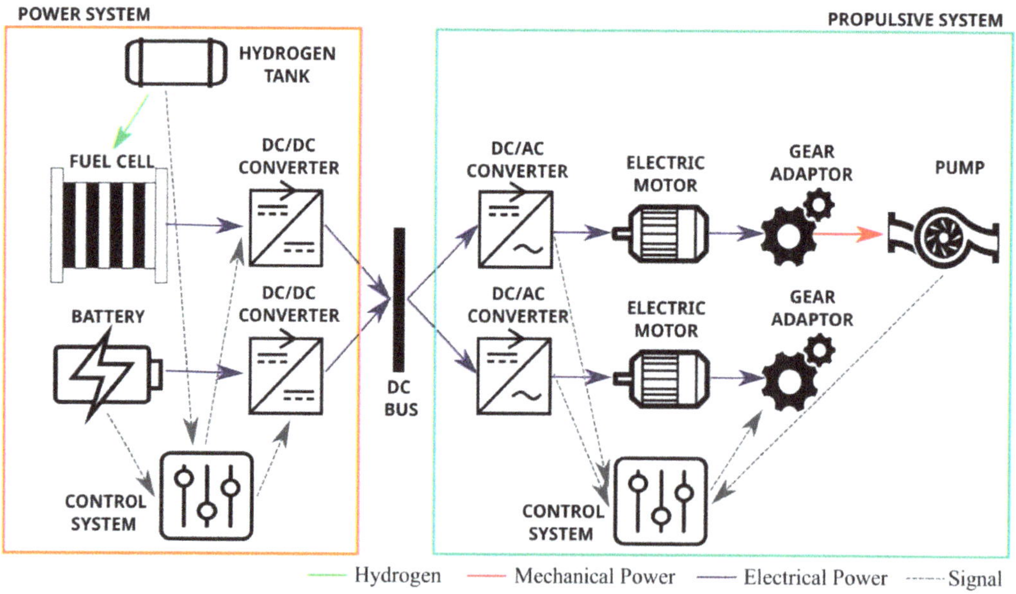

Figure 1. Simplified scheme of the AMV innovative powertrain.

2.2. Power System

Electrochemically, fuel cells convert reactant chemical energy directly into electrical energy [17,20]. An external storage system continuously supplies reagents, unlike batteries. The fuel cell system and its auxiliary systems can be sized to meet the required operation's energy needs, while the external hydrogen storage system is sized to meet the fuel cell system's energy needs for the same mission. Since the energy density of the hydrogen storage system is much higher than that of batteries, fuel cell systems' energy density can potentially exceed the batteries' energy density.

Among the various types of fuel cells [26–29], low-temperature proton exchange membrane fuel cells (PEMFCs) are best suited for use in power generation systems aboard an ASV or AUV for the following reasons:

1. the thermal energy requirements of an ASV and an AUV are minimal and strictly necessary to prevent the vehicle temperature from dropping below the minimum recommended value for the proper functioning of the on-board instrumentation;
2. their weight must be as low as possible to maintain good vehicle handling;
3. the compatibility of the devices with the marine environment, in which they operate, must be guaranteed;
4. PEMFCs are the most mature from a technical and commercial point of view among the different types of low-temperature fuel cells.

The fuel cell power generation system with fuel cells and battery and without super-capacitors is the best configuration for the ASV propulsion system since the power demands of the considered ASV vehicle do not have a power peak request of such an extent as to justify the use of super-capacitors, which would unnecessarily burden the same vehicles.

The AMV Power System consists of a power-controlled PEM fuel cell subsystem, a Li-ion battery subsystem, and a metal hydride hydrogen storage subsystem. The PEM fuel cell power system is designed to be controlled by the powertrain control system and integrates an innovative and more efficient PEM stack. The powertrain system is complemented by an innovative low-pressure metal hydride hydrogen storage system, which is more compact than a conventional tank at the same pressure, and the Li-ion battery subsystem. The DC−DC converter models of the PEM fuel cell subsystem and the Li-ion battery subsystem have been simplified to reduce software calculation time.

The power system consists of three different blocks: the power-controlled PEM fuel cell power subsystem, the Li-ion battery subsystem, and the metal hydride hydrogen storage subsystem. After the implementation of each subsystem, their validation tests will be performed.

2.2.1. Power-Controlled PEM Fuel Cells Subsystem

The power-controlled PEM Fuel Cell subsystem consists of the PEM stack with all its ancillaries (measuring instruments, actuators, air blower, etc.), the DC−DC converter, and the control subsystem.

The main equations of the PEM Fuel Cells stack are Equations (1)–(6):

$$V_s(t) = OCV_s - V_{an,s}(t) - V_{cat,s}(t) - R_{el,s}(t) \cdot i_s(t) \tag{1}$$

$$i_s(t) = i_{an,s,1}(t) + i_{an,s,2}(t) = i_{cat,s,1}(t) + i_{cat,s,2}(t) \tag{2}$$

$$\frac{dV_{an,s}(t)}{dt} \cdot C_{an,s}(t) = i_{an,s,1}(t) \tag{3}$$

$$\frac{dV_{cat,s}(t)}{dt} \cdot C_{cat,s}(t) = i_{cat,s,1}(t) \tag{4}$$

$$V_{cat,s}(t) = R_{cat,s}(t) \cdot i_{cat,s,2}(t) \tag{5}$$

$$V_{an,s}(t) = R_{an,s}(t) \cdot i_{an,s,2}(t) \tag{6}$$

Equation (1) is the equation for the calculation of the real voltage of the PEM fuel cell stack, Equation (2) represents the current balance equation at the circuit main nodes, Equations (3) and (4) are the current equations for the capacitive anodic and cathodic activation phenomena, while Equations (5) and (6) are the voltage equations for the anodic and cathodic ohmic phenomena.

The main equations of the simplified DC−DC converter are Equations (7) and (8):

$$\eta_{DC-DC_{conv}} \cdot V_s(t) \cdot i_s(t) = V_{out} \cdot i_{out}(t) \tag{7}$$

$$V_{out} = D(t) \cdot V_s(t) \tag{8}$$

In Equations (7) and (8), $\eta_{DC-DC_{conv}}$ and $D(t)$ are respectively the DC−DC converter efficiency and the DC−DC converter duty ratio.

The equations for the electric loads of ancillaries and user, which are assumed as pure resistive loads, are Equations (9) and (10):

$$i_{out}(t) = i_{user}(t) + i_{aux}(t) \tag{9}$$

$$V_{out} \cdot i_{out}(t) = R_{user}(t) \cdot i_{user}^2(t) + R_{aux}(t) \cdot i_{aux}^2(t) \tag{10}$$

Equations (9) and (10) represent the currents and power balance equations for the PEM fuel cells subsystem.

The control subsystem regulates the DC−DC converter duty ratio, $D(t)$, which influences electrochemical phenomena (associated with parameters such as $R_{el,s}(t)$, $R_{cat,s}(t)$, $R_{an,s}(t)$, $C_{an,s}(t)$, $C_{cat,s}(t)$), in a way that the electric power delivered by the power subsystem is equal to the electric power required by the control system of the entire hybrid electric propulsion system.

Regarding the validtion, the PEM stack simulated is the PEM stack model H300 type C, produced and marketed by the company H2Planet by Hydro2Power s.r.l., whose main technical data are reported in Table 1.

Table 1. Main PEM stack technical data.

Technical Parameter	Unit	Value
Fuel cells number	-	60
Nominal electric power	W	300
Operative temperature	°C	5–30
H_2 flow rate consumption at the nominal electric power	Nl min^{-1}	3.9
Electric efficiency (referred to Hydrogen HHV)	-	40
Start-up time	s	≤30

Figure 2 shows the comparison between the PEM stack experimental data supplied by the manufacturer and the polarization and electric power curves produced by the simulation model. Figure 2 shows that there is a good agreement between the simulation model results and the experimental data.

2.2.2. Li-Ion Battery Subsystem

The Li-ion battery subsystem consists of the Li-ion battery pack and the simplified DC−DC converter subsystem for the management of the charging and discharging phases.

In the discharging ($i^* > 0$) and charging phases ($i^* < 0$), the main equations of the Li-ion battery pack model are Equations (11) and (12) [30–32]:

$$V_{batt} = E - R \cdot i - K \cdot \frac{Q}{Q - Q_{ext}} \cdot i^* - K \cdot \frac{Q}{Q - Q_{ext}} \cdot Q_{ext} + A \cdot e^{-B \cdot Q_{ext}} \tag{11}$$

$$V_{batt} = E - R \cdot i - K \cdot \frac{Q}{Q_{ext} + 0.1 \cdot Q} \cdot i^* - K \cdot \frac{Q}{Q - Q_{ext}} \cdot Q_{ext} + A \cdot e^{-B \cdot Q_{ext}} \tag{12}$$

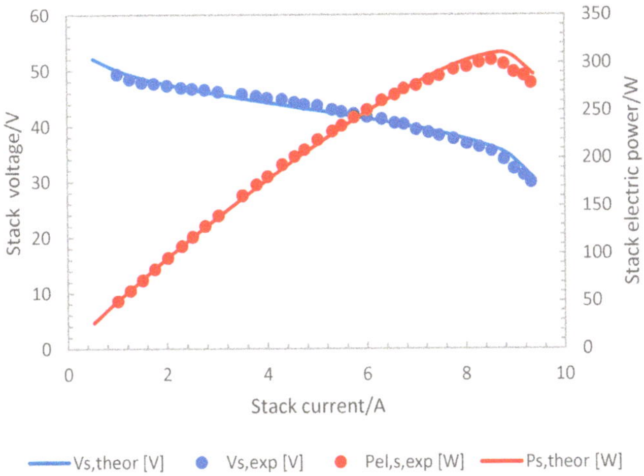

Figure 2. Comparison between the PEM stack theoretical polarization (blue line) and electric power (red line) curves and PEM stack experimental data supplied by the manufacturer.

In Equations (11) and (12) V_{batt}, i, R, E, K, i^*, Q_{ext}, Q, A and B are respectively the voltage, the current, the internal resistance, the constant voltage, the polarization constant or the polarization resistance, the low frequency filtered current, the extracted capacity, the capacity, the exponential voltage, and the exponential zone time constant.

The voltage of the battery pack at a fully charged state is calculated by adopting Equation (13) [30–32]:

$$V_{full} = E - R \cdot i + A \tag{13}$$

The battery pack voltage at the exponential section and at the nominal zone are calculated by Equations (14) and (15) [30–32]:

$$V_{exp} = E - R \cdot i - K \cdot \frac{Q}{Q - Q_{exp}} \cdot (Q_{exp} + i) + A \cdot e^{-\frac{3}{Q_{exp}} \cdot Q_{exp}} \tag{14}$$

$$V_{nom} = E - R \cdot i - K \cdot \frac{Q}{Q - Q_{nom}} \cdot (Q_{nom} + i) + A \cdot e^{-\frac{3}{Q_{exp}} \cdot Q_{nom}} \tag{15}$$

In Equations (14) and (15), Q_{exp} and Q_{nom} are the exponential and nominal capacities.

The equivalent circuit of the Li-ion battery pack is discussed somewhere else [28–30]. The parameters of the equivalent circuit can be modified to represent the Li-ion battery, based on its discharge characteristics (exponential voltage drop, normal discharge, and total discharge sections). The Simulink battery block implements a Li-ion battery pack dynamic model parameterized to represent the Li-ion battery pack installed on board the autonomous marine vehicle.

The model uses the SOC as a state variable [33], and the open circuit voltage is calculated using a nonlinear equation based on the analysis of the state of charge.

Regarding the battery validation, the Lithium-ion battery (NCM) pack simulated is the model IS36V13, which is produced by the company Map batteries by FAM batteries, and it has a nominal voltage and capacity equal to 37 V and 13 Ah. All other technical specifications can be found in [34].

Figure 3 shows the comparison between the Li-ion battery pack experimental data supplied by the manufacturer and the Li-ion battery pack battery theoretical discharge curve produced by the simulation model at different typical operating discharge currents (6.5, 13, 26 A). Figure 3 shows that there is a good agreement between the simulation model results and the experimental data.

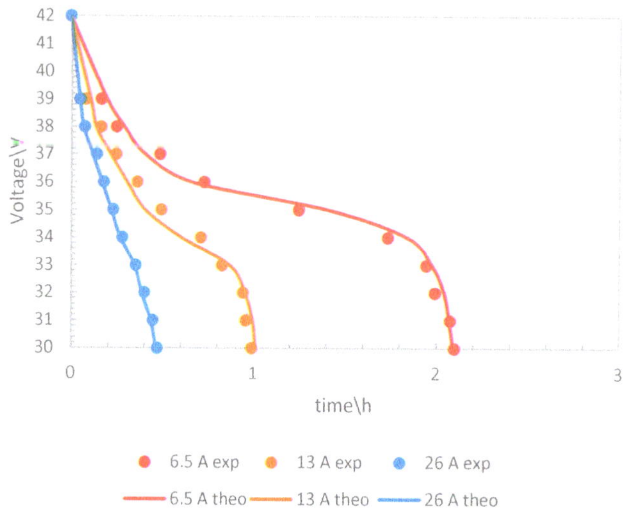

Figure 3. Comparison between the theoretical Li-ion battery pack discharge theoretical (theo) curves and the Li-ion battery pack experimental (exp) data, supplied by the manufacturer at different typical operating currents (6.5, 13, and 26 A).

The amount of energy available and accumulated instant by instant are functions of the battery SOC. Improving the available capacity has been necessary to carefully assess the depth of discharge, the charging voltage, and the currents in the processes of charging and discharging [35].

2.2.3. Metal Hydride Hydrogen Storage Subsystem

The hydrogen storage system based on metal hydrides (MH) operates in discharging mode, and the main equation of the hydrogen storage system is Equation (16):

$$SOC_{H_2}(t) = \frac{mol_{H_2}(t)}{mol_{H_2,max}} = SOC_{H_2}(t_i) - \frac{n_{fc}}{k}\int_{t_i}^{t} i_s(t)\cdot dt \tag{16}$$

where $mol_{H_2}(t)$, $mol_{H_2,max}$, $SOC_{H_2}(t_i)$, n_{fc}, $i_s(t)$ and k are, respectively, the instantaneous and maximum hydrogen moles in the storage system, the state of charge at the initial instant t_i, the number of fuel cells, the instantaneous stack current, and the conversion parameter for the calculation of the state of charge variation from the stack charge variation.

The MH H2 storage system simulated is the My-H2 900, produced and marketed by the company H2Planet by Hydro2Power s.r.l., whose main technical data are reported in Table 2.

Table 2. Main MH-H2 tank technical data.

Technical Parameter	Unit	Value
Storage capacity	Nl	900
Volume	l	1.7
Weight	kg	6.9
Dimensions (Diameter × Height)	cm	10 × 38
Charge pressure range	bar	5–12
Maximum pressure	bar	30

The PCT diagram of the metal hydride shows the equilibrium relationship between the hydrogen pressure and the percentage of hydrogen mass, $\%_{m,H2}$, as a function of the amount of alloy present in the system. Under dynamic conditions, i.e., during charging or discharging, the respective curve deviates from this equilibrium curve to a greater or lesser extent, depending on the cooling/heating parameters chosen, in particular the temperature.

Figure 4 shows the comparison between the theoretical desorption curve MH PCT and the experimental data from MH provided by the manufacturer at the typical operating temperature (20 °C). Figure 4 shows that there is good agreement between the results of the simulation model and the experimental data.

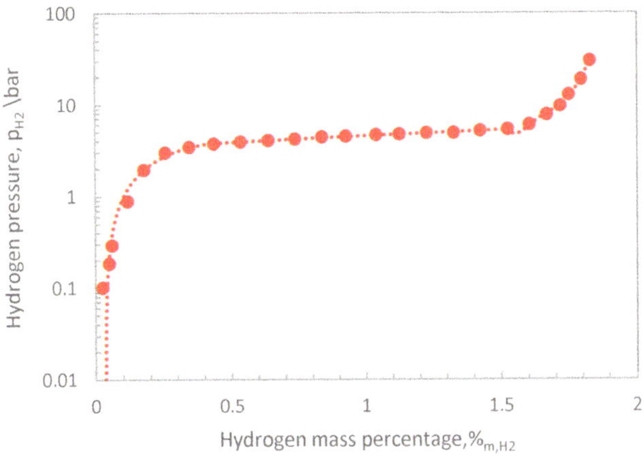

Figure 4. Comparison between the theoretical Metal Hydride PCT desorption curve and experimental data, supplied by the manufacturer at the typical operating temperature (20 °C).

2.2.4. Power System Control Strategy

Control strategies are fundamental to allow a proper management of the energy sources within the powertrains [36,37]. Between the numerous strategies, the state machine control strategy is based on different states useful for selecting the operating power level of the Fuel Cell (FC) subsystem [38]. This control technique aims to reduce the FC power fluctuations to improve efficiency and lifetime. For this reason, the fuel cell power fluctuations occur only when the battery state of charge (SOC) exceeds the predetermined thresholds.

In general, the control technique tries to keep the fuel cell in the optimal state, i.e., the value corresponding to the maximum efficiency, which is its only output parameter. Nevertheless, the only controlled parameter, the fuel cell power, is changed according to the battery SOC between the maximum power output and the minimum value. Therefore, the fuel cell is never turned off to avoid the long turn-on times and low efficiency at low power rates. Two input parameters are considered, namely the battery SOC and the fuel cell subsystem power at the previous time step. The battery operates in a wide SOC range between 20% and 100%, which is set as the operating limit, but four intermediate levels of SOC are considered, namely 30%, 40%, 80%, and 90%, which are useful for appropriate charging or discharging operations. Finally, to ensure the proper behavior of the fuel cell subsystem under the extreme conditions of the battery subsystem, the performance of the fuel cell at the previous time is studied, considering two threshold values, namely the maximum and minimum performance of the fuel cell. The limited values of the input and output parameters are summarized in Table 3.

Table 3. Limits imposed on input and output parameters of the state machine control.

Battery SOC	FC Power
$SOC_L = 30\%$ $SOC_{ML} = 40\%$ $SOC_{MH} = 80\%$ $SOC_H = 90\%$	$P_{FC,min} = 60$ W $P_{FC,opt} = 204$ W $P_{FC,max} = 250$ W

Evaluating the suitable combinations of the input limits, six states are considered, listed in Table 4.

Table 4. Control rules of the state machine control.

State	Battery SOC	FC Power (t − 1)	FC Power (t)
1	$SOC \leq SOC_L$	-	$P_{FC} = P_{FC,max}$
2	$SOC \leq SOC_{ML}$	P_{FC} (t − 1) = $P_{FC,max}$	$P_{FC} = P_{FC,max}$
3	$SOC > SOC_L$	-	$P_{FC} = P_{FC,opt}$
4	$SOC < SOC_H$	-	$P_{FC} = P_{FC,opt}$
5	$SOC \geq SOC_{MH}$	P_{FC} (t − 1) = $P_{FC,min}$	$P_{FC} = P_{FC,min}$
6	$SOC \geq SOC_H$	-	$P_{FC} = P_{FC,min}$

Following these rules, the control strategy is composed of a hysteresis cycle for each SOC extreme condition, as shown in Figure 5: in detail, a hysteresis cycle between 20% and 40% of battery SOC is implemented (Figure 5a), while another one between 80% and 100% of battery SOC (Figure 5b).

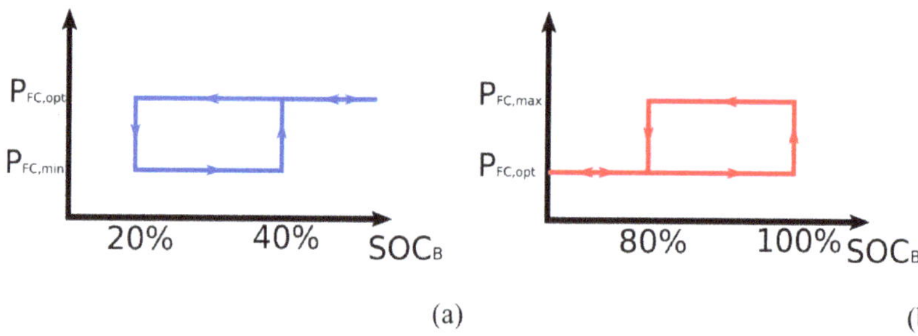

Figure 5. Hysteresis cycle of the state machine control system: (**a**) low battery SOC levels; (**b**) high battery SOC levels.

A further feature implemented in the power-sharing strategy concerns the simulation stops; when one of the two SOCs reaches the minimum value allowed (11% for the SOCH2 and 20% for the SOCB), the vehicle is stopped and the drive cycle is not completed.

2.3. *Propulsive System*

The propulsive system of the tested AMV consists of two brushless DC (BLDC) motors, which are named motor system, for each propulsive pump; during the test, 2 or 4 propulsive pumps are used. The BLDC motors used are three-phase synchronous motors with permanent magnets on the rotor and electronic commutators instead of brush gears and mechanical commutators. In this way, the system takes advantage of three-phase motors, behaves like a DC system, and avoids limitations of electromagnetic interference, frame, speed, and noise [39]. Compared to brushed DC and induction motors, they offer satisfactory performances such as high efficiency, long life, noiseless operation, wide speed ranges, high torque relative to frame size, excellent dynamic behavior, and good speed–torque characteristics.

Generally, BLDC systems consist of three different blocks: the electric motor, the inverter, and the control system. After implementation, the validation tests are performed.

2.3.1. Electric Motor

Brushless DC motors are synchronous motors, namely the two magnetic fields, belonging to the stator and rotor, have the same frequency. They can be single-phase, two-phase, and three-phase, according to the stator windings. Generally, three-phase motors are the most used, thanks to their high efficiency and accurate control. In this case, permanent magnets' synchronous motors with trapezoidal back electromotive force, rather than sinusoidal ones, are the most used in BLDC systems [40]. The model implemented is based on the block present in the Matlab/Simulink environment [41,42]. The main assumption made is the linear magnetic circuit, without stator and rotor saturation.

The electric motor behavior is described by Equations (17)–(19), which are useful for calculating the three-phase currents (i_a, i_b, and i_c), considering the phase voltages (V_{ab} and V_{bc}), the flux amplitude induced by the rotor magnet to the stator (λ), and the electromotive forces in per-unit (φ_a, φ_b, φ_c):

$$\frac{di_a}{dt} = \frac{1}{3L_{sw}}\left[2V_{ab} + V_{bc} - 3\,R_{sw}\,i_a + \lambda p\omega_m\,(2\phi_a + \phi_b + \phi_c)\,\right] \tag{17}$$

$$\frac{di_b}{dt} = \frac{1}{3L_{sw}}\left[-V_{ab} + V_{bc} - 3\,R_{sw}\,i_b + \lambda p\omega_m\,(\phi_a - 2\phi_b + \phi_c)\,\right] \tag{18}$$

$$\frac{di_a}{dt} + \frac{di_b}{dt} + \frac{di_c}{dt} = 0 \tag{19}$$

2.3.2. Inverter

The inverter is fundamental in BLDC systems. It is needed to convert the single-phase direct current supplied by the power grid into the three-phase alternating current. In addition, it is critical for detailed control of the electric motor by determining the required mechanical torque and speed of the motor. The detailed modeling of the inverter, which consists of six switches, requires a high computational effort that is suitable for simulating a detailed behavior of the device but for a limited simulation time. However, the scope of this work is to analyze an autonomous marine vehicle in different simulation campaigns, and a simplified model is more appropriate in terms of low computational cost and sufficient in terms of data accuracy.

The inverter model used is based on the block available in the Matlab Simulink environment [42,43]. It consists of a DC-controlled current source, whose behavior is described in Equation (20), and two AC-controlled voltage sources that manage the trapezoidal inputs provided by the electric motor model:

$$I_{DC} = \frac{P_{AC,out} + P_{loss}}{V_{in}} \tag{20}$$

The AC reference currents, calculated through the control system, the back electromotive force voltage, and the current are the inputs needed to compute the inverter voltages. The main assumption is the rate limiter, which regulates the output voltage during transitions, employing a saturation degree coefficient.

2.3.3. Motor Control Strategy

Once the main components of the BLDC motor have been defined, it is important to analyze the electronic commutations, capable of controlling the interaction between the inverter and electric motor and achieving the outputs needed, as shown in Figure 6.

In the present work, among the different commutation strategies, the communication with Hall sensors is performed, given by the BLDC motor used [44,45]. In brief, there are three Hall sensors (H_1, H_2, and H_3 in Figure 6) on the motor at an angle of 120°, which detect the magnetic field of the control magnet on the shaft. In this way, the signal received, as the combination of the Hall sensors, is univocal for each 60° of rotation. Therefore, according to the rotor position and the outputs needed, it is possible to change the stator magnetic field, acting on the three-phase voltages and, hence, on the duty-cycle of the inverter switches ($S_{1'}$, $S_{1''}$, $S_{2'}$, $S_{2''}$, $S_{3'}$, $S_{3''}$).

The control strategy can be divided into two subsystems, namely a commutation system and a speed controller system. The former is a rule-based strategy, capable of providing a vector that contains phase voltage details, starting from the hall sensor signals. For example, a rule is shown in the following sentence:

"If Hall sensor signal is [1 0 1] then phase voltage vector is [+1 −1 0]"

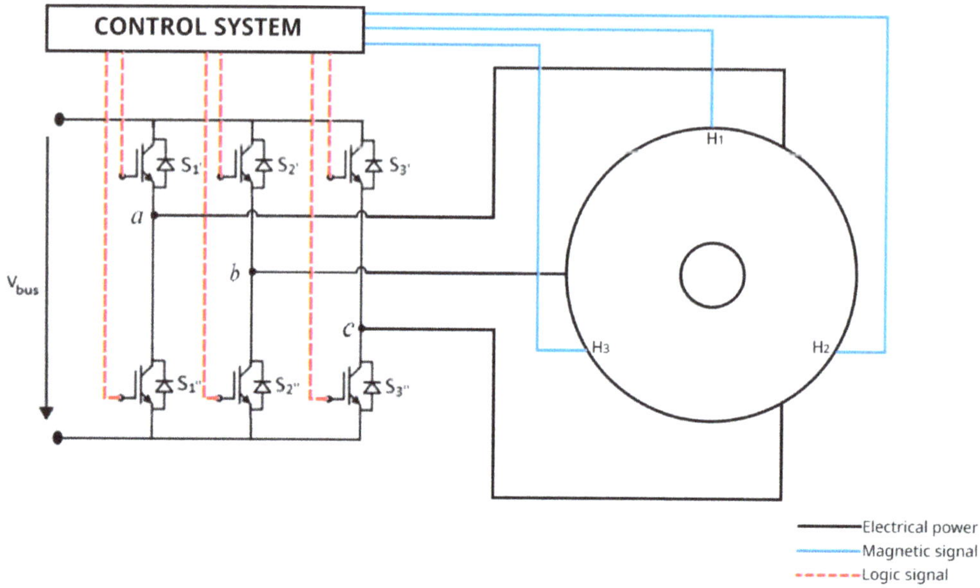

Figure 6. Motor system: electric motor, inverter, and control subsystems.

The speed controller is useful for calculating the torque reference value, namely the torque imposed to perform a specific journey. A proportional-integral (PI) structure is used, considering as a controlled value the error between the reference and real values of the motor rpm.

By combining the outputs of the two subsystems with a suitable motor factor, a control vector can be created that specifies a suitable duty cycle for each inverter switch.

2.3.4. Validation

As shown in the simplified schematic of the AMV in Figure 1, the propulsion system consists of two brushless DC motors, one for propulsion and one for directional control, the azimuth motor. Due to these tasks, two motors with different characteristics are tested in the numerical simulations, which are listed in Table 5. Both are connected to the DC bus and work together depending on the requirements of the vehicle. The drive motor is a Maxon Ec-4pole, a motor with a rated speed of 16,800 rpm and a rated torque of 54 mNm, coupled with a 14:1 reduction gearbox [46]. The azimuth motor, on the other hand, is a Faulhaber 2232-BX, characterized by a rated speed of 4930 rpm and a rated torque of 14.3 mNm, with a reduction ratio of 59:1 [47].

Table 5. Performance features of the two electric motors.

	Propulsion Motor	Azimuth Motor
Manufacturing Company	Maxon	Faulhaber
Company's Headquarter	Sachseln (Switzerland)	Schönaich (Germany)
Model	Ec-4pole	2232-BX
Nominal torque	54 mNm	14.3 mNm
Nominal speed	16,800 rpm	4930 rpm
Nominal voltage	36 V	24 V
Maximum efficiency	90%	74%
Rotor inertia	5.1 gcm^2	8.91 gcm^2
Gear ratio coupled	14:1	59:1

Starting from these values, and considering the torque trends present on the data sheets provided by the manufacturing company, validation tests have been performed, obtaining as results the plots shown in Figure 7; the blue line represents the model behavior, while the red dots are the real motor data. The error achieved during the simulations is less than 1%; therefore, the model turns out to be appropriate for the studied application.

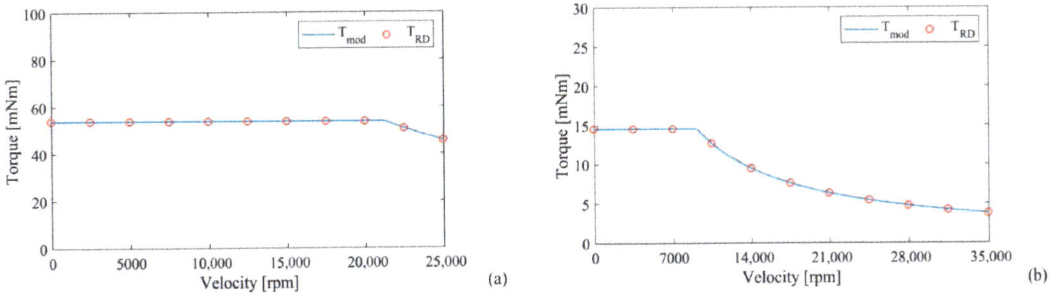

Figure 7. Validation tests of (**a**) the propulsion motor and (**b**) the directional motor.

3. Simulation and Results

The hydrogen-based marine vehicle was numerically tested in four case studies to evaluate the performance and range of the vehicle and to discuss the SOC of the fuel cell and battery. The energy parameters of each drive cycle are presented, and the response of the powertrain is mentioned.

3.1. Case Studies

In the simulation campaigns, four journeys with different power levels and time intervals are considered, which are useful to test the AMV under different conditions. Figure 8 shows the power and energy requirements in blue and red, respectively, measured at the DC bus connecting the power and motor systems. They are the sum of the motor demand, i.e., the electrical power requested by the propulsion and azimuth motors, and the auxiliary power, i.e., the theoretical consumption of the instruments on board.

As shown in the figure, all trips have a minimum power of 176 W, which is the instrument consumption considered as a constant power for the whole drive-cycle time.

In Figure 8a, the power and energy demands of Journey 1 are shown. The trip lasts 5 h and has numerous power fluctuations, with a maximum value of 370 W and an energy consumption of 1100 Wh. The second drive cycle, shown in Figure 8b, achieves higher power but lower energy consumption compared to Journey 1, about 585 W and 425 Wh, respectively, due to less frequent power fluctuations and a shorter drive time of 85 min. Journey 3 (Figure 8c), unlike the other case studies, uses only two propulsive pumps, instead of four, namely, two motor systems (composed of an azimuth and a propulsive motor) are used. Consequently, the maximum power value is 385 W, with very frequent and faster power fluctuations, more than the other journeys. Although the power range is limited, the energy consumption of Journey 3 is the highest, about 1615 Wh, since the trip is about 6 h long. Finally, Journey 4 (Figure 8d) is almost 3 h long, with high power, 585 W, the same value as Journey 2, but the total energy is less than 1 kWh, a value lower than that of Journey 3.

3.2. Result Analysis

The four drive-cycles shown in Figure 8 are used to numerically test the autonomous marine vehicle and analyze the behavior of the powertrain under different operating conditions without changing the size of the energy storage device. The energy consumed by the fuel cell and the battery varies greatly depending on the trip tested; therefore, the final values of the hydrogen SOC and the battery SOC are different. A separate section is dedicated to each journey.

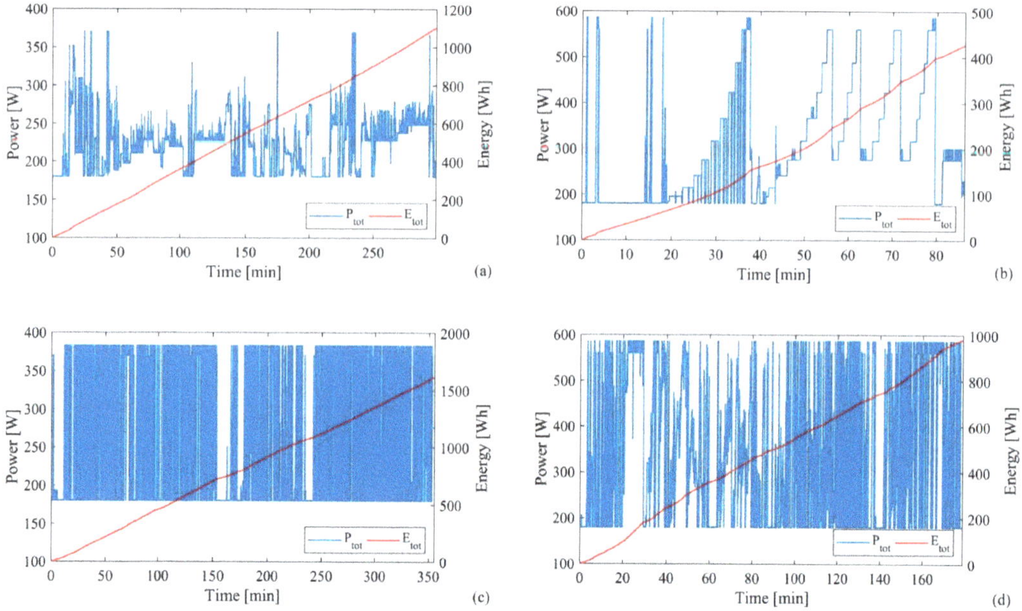

Figure 8. Power (blue line) and energy (red line) trends of the four analyzed journeys: (**a**) Journey 1; (**b**) Journey 2; (**c**) Journey 3; (**d**) Journey 4.

3.2.1. Journey 1

The first Journey is almost 5 h long, with a smaller power interval compared to the other journeys, between 176 W and 370 W, even though four motor systems are used. The simulation results are shown in Figure 9. The right part (Figure 9a) shows the mechanical power variations of the propulsion motor and the azimuth motor, while the left part (Figure 9b) illustrates the power system performances, namely the fuel cell power, the hydrogen SOC and the battery SOC. The power of the propulsion motor (blue line in Figure 9a) exhibits numerous and sudden fluctuations between 0 W and 160 W, while the azimuth motor is turned on a few times during the trip, and its power (red line in Figure 9a) assumes an impulse trend in a range between 2 W and 20 W. As in the other runs, the fuel cell power (blue line on the left axis in Figure 9b) exhibits a step-like trend starting from the minimum power (60 W) as the battery is fully charged and varying to the optimal value (about 200 W) when one of the control strategy thresholds is exceeded. The maximum FC power value is not reached as the battery SOC remains at a high level. This power results in a wide range of hydrogen SOC (red line on the right axis in Figure 9b) arriving at 21% and consuming about 64 g of hydrogen; in contrast, the battery SOC (green line on the right axis in Figure 9b) starts at 100% and arrives at 70%. During this drive, the fuel cell delivers a larger fraction of energy, about 965 Wh, than the battery due to the smallest power fluctuations compared to the other drive cycles. In other words, the fuel cell works as a prime mover, covering the energy demand and operating quasi-steady-state conditions, while the battery covers the power peaks when the energy demand is high and is charged by the fuel cell when the energy demand is low.

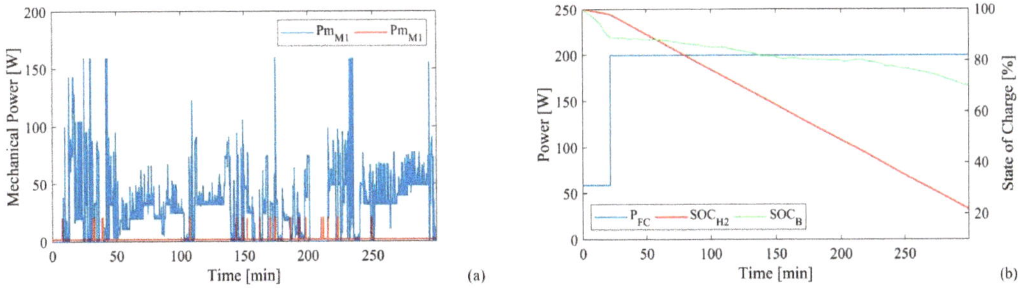

Figure 9. AMV performance in Journey 1: (**a**) Motor mechanical power trends; (**b**) Fuel cell power, hydrogen tank and battery state of charge trends.

3.2.2. Journey 2

The second Journey is the shortest among the four drive-cycles, being more or less a quarter compared to Journey 3, around 85 min, and also the one with the lowest energy demand, consuming about 425 Wh, which is also almost a quarter of the energy required for the Journey 3.

The performance of the AMV in Journey 2 is shown in Figure 10. Given the low value of the required energy, the vehicle can complete the journey with satisfactory performance. The mechanical power of the two motors follows the trend of the energy demand, excluding the consumption by the auxiliary and measurement systems. The power of the propulsion motor (blue line in Figure 10a) varies in a wide interval between 0 W and 340 W, while the power of the azimuth motor (red line), which is characterized by a sequence of numerous impulses, has 20 W as the maximum mechanical power.

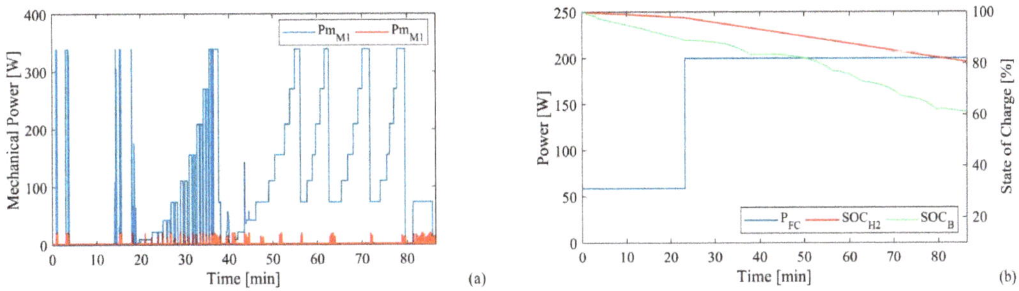

Figure 10. AMV performance in Journey 2: (**a**) Motor mechanical power trends; (**b**) Fuel cell power, hydrogen tank and battery state of charge trends.

As for the power system performance, the power of FC goes from the minimum value to the optimal value after 25 min and remains the same until the end of the cycle. According to this trend, the hydrogen SOC reaches a value of 80% at the end of the cycle, with a hydrogen consumption of 16 g. On the other hand, the battery SOC is very variable, with successive charging and discharging phases; the final value is about 61%, and, consequently, the energy provided by the battery is about 195 Wh.

3.2.3. Journey 3

The third Journey is the most critical because it takes the longest, about 6 h, and has the highest energy consumption, about 1.6 kWh, compared to the four drive-cycles tested. It is the only Journey where two motors are used. As shown in Figure 11a, the mechanical power of the drive motors reaches a maximum value of 170 W. The power of the azimuth motor, on the other hand, follows the same trend as that of the propulsion motor, with a narrow interval between 2 W and 12 W. Although the power interval is not very large, the long travel time results in huge energy consumption that the drive system cannot provide. The AMV stops driving after 340 min because the fuel cell reaches the imposed limits for the hydrogen SOC (11%), as shown in Figure 11b, and consequently, the power-sharing strategy aborts the simulations. As for the fuel cell performance, it adopts the three fixed rates given by the wide interval of the battery SOC. In fact, the trend of the battery SOC is oscillating due to the numerous charging and discharging phases and its minimum value is 24%, reached at the end of the Journey.

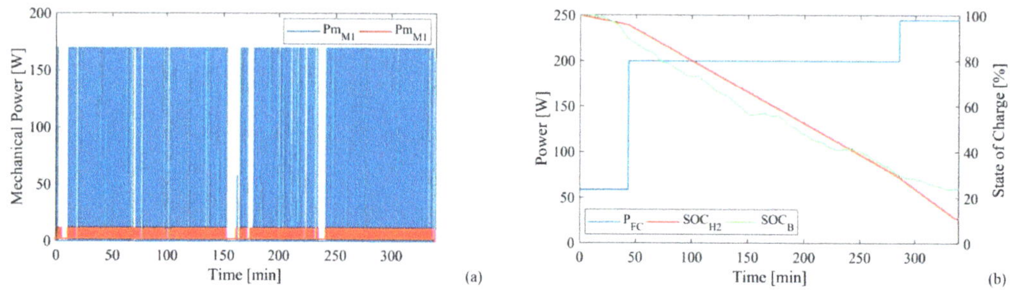

Figure 11. AMV performance in Journey 3: (**a**) Motor mechanical power trends; (**b**) Fuel cell power, hydrogen tank and battery state of charge trends.

3.2.4. Journey 4

The last Journey shows medium values compared to the other cycles, with an energy consumption of 980 Wh. As in the first two cycles, the energy stored on board is sufficient to complete the drive cycle, but a correction of the control system is required for the AMV to complete the journey. In fact, Figure 12 shows the performance of the AMV without corrections and testifies that one of the two SOC limits is reached. This is due to the characteristics of the control strategy and the large difference between the two SOCs; in other words, the hydrogen SOC is much higher than the battery SOC at all times, and the limit conditions are easily reached. Although the hydrogen SOC remains at 55%, the battery SOC reaches the lower limit of 20%, causing the simulation to terminate. The main reason for this behavior is related to the journey features, namely the pronounced and contiguous power variations between 176 W and 585 W. Therefore, the battery cannot deliver the energy variations of the journey if the fuel cell maintains the optimal power output between 80% and 30% of the trend of the battery SOC.

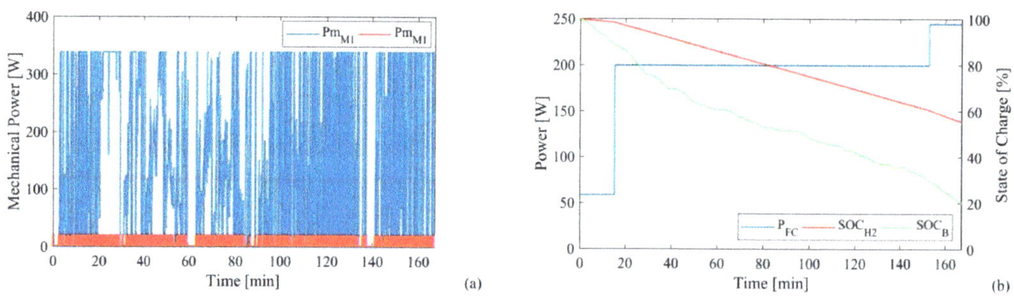

Figure 12. AMV performance in Journey 4, without control system correction: (**a**) Motor mechanical power trends; (**b**) Fuel cell power, hydrogen tank and battery state of charge trends.

To address this issue, the SOCL and SOCML values of the control strategy are changed from 30% and 40% to 50% and 60% for Journey 4 only. The updated performance with the adjusted control strategy is shown in Figure 13. The propulsion motor reaches high values, as in Journey 2 and more than the other Journeys, fluctuating in an interval between 0 W and 340 W; its trend shows sudden fluctuations, as in Journey 3. The azimuth motor, on the other hand, since four motors are active, fluctuating in the standard range between 2 W and 20 W; it follows the operating conditions of the propulsion motor and assumes its maximum power when the propulsion motor is on. Figure 13b illustrates the power system performance. As with Journey 3, unlike the other journeys, the fuel cell power assumes all three values specified by the power-sharing strategy, even if its energy demand falls in between. The maximum fuel cell power (250 W) is required from about 100 min, about halfway through the trip, until the end. Under these circumstances, the hydrogen SOC has a final value of 47%, eight points less than in the case without adjustments to the control strategy, requiring 43 g of hydrogen. At the same time, the battery SOC is at 25% at the end of the journey, which means that the lowest threshold of the control system is not exceeded, and the journey can be ended.

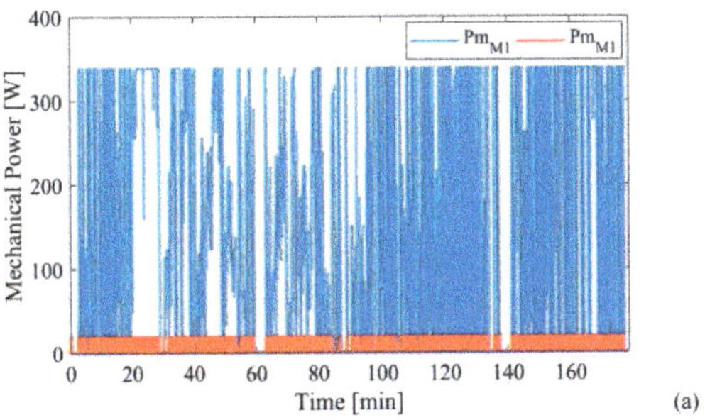

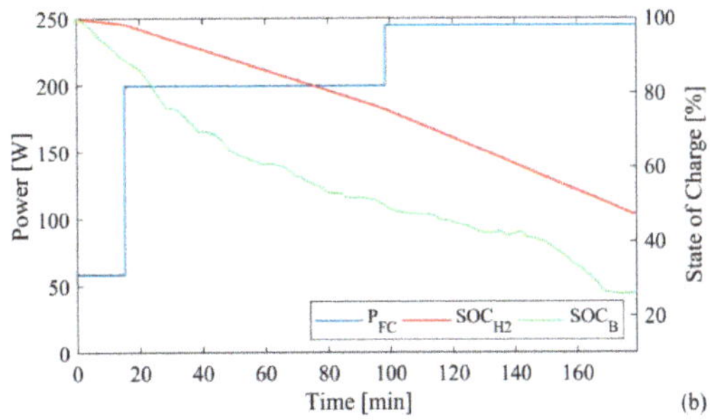

Figure 13. AMV performance in Journey 4, with control system correction: (**a**) Motor mechanical power trends; (**b**) Fuel cell power, hydrogen tank and battery state of charge trends.

3.2.5. Summary Considerations

In the result analysis, the AMV powertrain model has been numerically tested on four journeys, with different time intervals, power, and energy demands (shown in Table 6), and its performance is summarized in Table 7. During the first Journey, the motor system mechanical power assumes a maximum value of 180 W with variable trends; the hydrogen consumption is approximately 64 g, given by the fuel cell energy consumption (1 kWh) while the battery provides a moderate energy amount, remaining with a final SOC value of 70%. On the contrary, Journey 2 requires a maximum motor power of approximately 360 W; being the journey with the lowest energy demand than the other ones, the hydrogen consumption is only 16 g, but the battery provides almost 195 Wh, reaching a minimum SOC value of 61%. The third Journey instead requires a higher energy amount, about 1.60 kWh, more than the one needed in the other cycles; nevertheless, the maximum mechanical motor system power is only 180 W. The energy stored on board, as the hydrogen in the tank and the battery capacity, is not enough to complete the journey, stopping at 340 min (96% of the journey). The fuel cell supplies about 1087 Wh, consuming the whole hydrogen amount usable, while the remaining energy part is provided by the battery, about 375 Wh. In the end, the last Journey requires 980 Wh, with the same level of maximum motor power for Journey 2. Owing to the high and consecutive power variations, a modified control strategy is required to complete the journey. With these new features, the powertrain completely fulfills the vehicle demands; the fuel cell provides 632 Wh, consuming 43 g of hydrogen, while the battery supplies 368 Wh, concluding the journey with an SOC level of 25%.

Table 6. Performance features of the four analyzed journeys.

	Journey 1	Journey 2	Journey 3	Journey 4
Power [W]	(176–370)	(176–585)	(176–385)	(176–585)
Energy [Wh]	1100	425	1615	980
Time [min]	300	85	355	180

Table 7. AMV achieved performance in the four analyzed journeys.

	Journey 1	Journey 2	Journey 3	Journey 4
Journey conclusion rate	100%	100%	96%	100%
Max mechanical power [W]	180	360	180	360
Fuel cell Energy [Wh]	965	235	1087	632
Battery Energy [Wh]	157	195	375	368
Hydrogen consumption [g]	64	16	71	43

4. Conclusions

This paper presents an innovative fuel cell hybrid powertrain for an autonomous marine vehicle, whose performance is analyzed using multiphysics and dynamic numerical modeling of the main vehicle components, namely fuel cell, battery, inverter, electric motors, and control systems.

The implemented model was tested to simulate the fuel cell hybrid powertrain in four journeys, with different characteristics to evaluate the performance and range of the vehicle and to discuss the SOC of the fuel cell and battery. In the tests, hydrogen consumption varied from 16 g for Journey 2 to 71 g for Journey 3, which is closely related to the energy delivered by the fuel cell. However, each component exhibited different peculiarities in the four journeys discussed. For example, Journey 3 required a higher amount of energy, about 1.60 kWh, than the other journeys. Nevertheless, the maximum motor power was only 210 W because only 2 of 4 motors were on during the journey. The energy stored on board was not enough to complete the trip. The trip was stopped after 340 min (96% of the cycle). The fuel cell supplied about 1087 Wh, consuming all of the hydrogen stored on board (about 71 g), while the remaining energy portion (about 375 Wh) was supplied by the battery, which completed the trip at 25% SOC.

Journey 4, on the other hand, requires an intermediate amount of energy (980 Wh) with a maximum mechanical motor power of 360 W. Due to the significant and continuous power fluctuations between 176 W and 585 W, the journey could not be completed with the implemented simplified control system. Therefore, a modified control strategy was implemented in which the thresholds were changed to terminate the run. With these new features, the powertrain fully met the requirements of the vehicle; the fuel cell delivered 632 Wh and consumed 43 g of hydrogen, while the battery delivered 368 Wh and completed the trip with an SOC level of 25%.

Given these limitations, further work could consider evaluating new control systems capable of increasing the vehicle's range by exploiting the characteristics of the propulsion system.

In other words, in this paper:

1. A fuel cell hybrid powertrain for an autonomous marine vehicle was developed by means of dynamic and multi-physics relations;
2. Four journeys were considered to test the innovative vehicle on different conditions;
3. The main parameter of the powertrain, concerning power, consumption and SOC were monitored by means of a control system ad-hoc implemented;
4. The feasibility of each work journeys was verified, highlighting criticality.

The application tested proves that the use of hydrogen in marine applications can improve the performance of the electric powertrain and increase the range of the vehicle with less environmental impact compared to battery-based powertrains. The energy transition requires radical choices in all sectors, especially in the mobility sector, which produces a substantial amount of emissions. In this scenario, hydrogen-based solutions could be an important support, also in maritime applications, guaranteeing zero emissions in seawater.

Author Contributions: Conceptualization, G.D.L., F.P., F.L., G.T., V.B., N.P., M.C. and P.F.; Data curation, G.D.L., F.P. and P.F.; Formal analysis, G.D.L., F.P. and P.F.; Investigation, G.D.L., F.P. and P.F.; Methodology, G.D.L., F.P. and P.F.; Project administration, F.L., M.C. and P.F.; Resources, G.T., V.B., N.P. and M.C.; Software, G.D.L. and F.P.; Supervision, P.F.; Validation, G.D.L., F.P. and P.F.; Visualization, F.P. and P.F.; Writing—original draft, G.D.L., F.P. and P.F.; Writing—review and editing, G.D.L., F.P. and P.F. All authors have read and agreed to the published version of the manuscript.

Funding: This research received no external funding.

Institutional Review Board Statement: Not applicable.

Informed Consent Statement: Not applicable.

Acknowledgments: The authors gratefully acknowledge the Italian Ministry of University and Research for the financial support to Project No. ARS01 00682 ARES—Autonomous Robotics for the Extended Ship—CUP H56C18000160005, under the Italian National Research Program PON "Ricerca e Innovazione" 2014–2020, Action II.2, Specialization area: "Blue Growth".

Conflicts of Interest: The authors declare no conflict of interest.

Nomenclature

Symbols	Parameter	Units
A	Exponential Voltage	V
B	Exponential zone time constant	A^{-1} h^{-1}
C	Capacity	Ah
C	Capacitance	F
D	Duty ratio	-
E	Constant Voltage	V
η	Efficiency	-
φ	Electromotive forces in per-unit	-
HHV	high heating value	J/kg
$\imath$	Current	A
K	Polarization constant or polarization resistance	V A^{-1} h^{-1} or Ω
k	conversion parameter	C
L	Inductance	H
λ	Flux amplitude induced by the rotor magnets in the stator phases	V s
mol	moles	mol
n	number	-
OCV	Open Circuit Voltage	V
ω	Angular velocity	rad/s
P	Power	W
p	Number of pole pairs	-
R	Resistance	Ω
SOC	State of charge	-
t	time	s
V	Voltage	V
Subscripts	**Parameter**	
a	Phase a of the three-phase parameter	
AC	Alternative current	
an	anodes	
aux	referred to the ancillaries	
b	Phase b of the three-phase parameter	
B	Referred to the battery	
$batt$	battery	
c	Phase c of the three-phase parameter	
cat	cathodes	

el	electrolyte
exp	exponential
ext	extracted
fc	Fuel cell
full	fully charged state
H	High level
H_2	hydrogen
in	At the inlet
L	Low level
loss	Owing to losses
m	Motor
max	Maximum level
min	Minimum level
ML	Medium-low level
MH	Medium-high level
nom	nominal
opt	Optimum level
out	At the outlet
s	stack
sw	Stator winding
user	referred to the electric user
1	referred to the resistive branch
2	referred to the capacitive branch
Superscripts	**Parameter**
*	Low frequency filtered

References

1. Yan, R.; Pang, S.; Sun, H.; Pang, Y. Development and Missions of Unmanned Surface Vehicle. *J. Mar. Sci. Appl.* **2010**, *9*, 451–457. [CrossRef]
2. Naval Studies Board; National Research Council. *Autonomous Vehicles in Support of Naval Operations*; National Academies Press: Cambridge, MA, USA, 2005; ISBN 978-0-309-09676-8.
3. Carlson, D.F.; Fürsterling, A.; Vesterled, L.; Skovby, M.; Pedersen, S.S.; Melvad, C.; Rysgaard, S. An Affordable and Portable Autonomous Surface Vehicle with Obstacle Avoidance for Coastal Ocean Monitoring. *HardwareX* **2019**, *5*, e00059. [CrossRef]
4. Bauk, S.; Kapidani, N.; Lukšic, Ž.; Rodrigues, F.; Sousa, L. Autonomous Marine Vehicles in Sea Surveillance as One of the COMPASS2020 Project Concerns. *J. Phys. Conf. Ser.* **2019**, *1357*, 012045. [CrossRef]
5. Thompson, F.; Guihen, D. Review of Mission Planning for Autonomous Marine Vehicle Fleets. *J. Field Robot.* **2019**, *36*, 333–354. [CrossRef]
6. Benjamin, M.R.; Schmidt, H.; Newman, P.M.; Leonard, J.J. Nested Autonomy for Unmanned Marine Vehicles with MOOS-IvP. *J. Field Robot.* **2010**, *27*, 834–875. [CrossRef]
7. Bruzzone, G.; Odetti, A.; Caccia, M.; Ferretti, R. Monitoring of Sea-Ice-Atmosphere Interface in the Proximity of Arctic Tidewater Glaciers: The Contribution of Marine Robotics. *Remote Sens.* **2020**, *12*, 1707. [CrossRef]
8. Caccia, M.; Bibuli, M.; Bono, R.; Bruzzone, G.; Bruzzone, G.; Spirandelli, E. Unmanned Surface Vehicle for Coastal and Protected Waters Applications: The Charlie Project. *Mar. Technol. Soc. J.* **2007**, *41*, 62–71. [CrossRef]
9. Odetti, A.; Bruzzone, G.; Altosole, M.; Viviani, M.; Caccia, M. SWAMP, an Autonomous Surface Vehicle Expressly Designed for Extremely Shallow Waters. *Ocean. Eng.* **2020**, *216*, 108205. [CrossRef]
10. Bradley, T.H.; Moffitt, B.A.; Fuller, T.F.; Mavris, D.N.; Parekh, D.E. Comparison of Design Methods for Fuel-Cell-Powered Unmanned Aerial Vehicles. *J. Aircr.* **2009**, *46*, 1945–1956. [CrossRef]
11. Thomas, J.P.; Qidwai, M.A.; Kellogg, J.C. Energy Scavenging for Small-Scale Unmanned Systems. *J. Power Sources* **2006**, *159*, 1494–1509. [CrossRef]
12. Verstraete, D.; Lehmkuehler, K.; Wong, K.C. *Design of a Fuel Cell Powered Blended Wing Body UAV*; American Society of Mechanical Engineers Digital Collection: New York, NY, USA, 2013; pp. 621–629.
13. Choi, S.; Wang, G. Advanced Lithium-Ion Batteries for Practical Applications: Technology, Development, and Future Perspectives. *Adv. Mater. Technol.* **2018**, *3*, 1700376. [CrossRef]
14. Xia, S.; Wu, X.; Zhang, Z.; Cui, Y.; Liu, W. Practical Challenges and Future Perspectives of All-Solid-State Lithium-Metal Batteries. *Chem* **2019**, *5*, 753–785. [CrossRef]
15. Sorlei, I.-S.; Bizon, N.; Thounthong, P.; Varlam, M.; Carcadea, E.; Culcer, M.; Iliescu, M.; Raceanu, M. Fuel Cell Electric Vehicles—A Brief Review of Current Topologies and Energy Management Strategies. *Energies* **2021**, *14*, 252. [CrossRef]
16. Palencia, J.C.G.; Nguyen, V.T.; Araki, M.; Shiga, S. The Role of Powertrain Electrification in Achieving Deep Decarbonization in Road Freight Transport. *Energies* **2020**, *13*, 2459. [CrossRef]

17. De Lorenzo, G.; Andaloro, L.; Sergi, F.; Napoli, G.; Ferraro, M.; Antonucci, V. Numerical Simulation Model for the Preliminary Design of Hybrid Electric City Bus Power Train with Polymer Electrolyte Fuel Cell. *Int. J. Hydrogen Energy* **2014**, *39*, 12934–12947. [CrossRef]
18. De Luca, D.; Fragiacomo, P.; De Lorenzo, G.; Czarnetzki, W.T.; Schneider, W. Strategies for Dimensioning Two-Wheeled Fuel Cell Hybrid Electric Vehicles Using Numerical Analysis Software. *Fuel Cells* **2016**, *16*, 628–639. [CrossRef]
19. Piraino, F.; Genovese, M.; Fragiacomo, P. Towards a New Mobility Concept for Regional Trains and Hydrogen Infrastructure. *Energy Convers. Manag.* **2021**, *228*, 113650. [CrossRef]
20. Piraino, F.; Fragiacomo, P. A Multi-Method Control Strategy for Numerically Testing a Fuel Cell-Battery-Supercapacitor Tramway. *Energy Convers. Manag.* **2020**, *225*, 113481. [CrossRef]
21. Sezgin, B.; Devrim, Y.; Ozturk, T.; Eroglu, I. Hydrogen Energy Systems for Underwater Applications. *Int. J. Hydrogen Energy* **2022**, *47*, 19780–19796. [CrossRef]
22. González, E.L.; Cuesta, J.S.; Fernandez, F.J.V.; Llerena, F.I.; Carlini, M.A.R.; Bordons, C.; Hernandez, E.; Elfes, A. Experimental Evaluation of a Passive Fuel Cell/Battery Hybrid Power System for an Unmanned Ground Vehicle. *Int. J. Hydrogen Energy* **2019**, *44*, 12772–12782. [CrossRef]
23. Shih, N.-C.; Weng, B.-J.; Lee, J.-Y.; Hsiao, Y.-C. Development of a 20 KW Generic Hybrid Fuel Cell Power System for Small Ships and Underwater Vehicles. *Int. J. Hydrogen Energy* **2014**, *39*, 13894–13901. [CrossRef]
24. Electrocraft Power Innovation (EPI). Available online: https://www.electrocraft.com/motors-for/marine-shipbuilding/unmanned-vehicles/#product-cats (accessed on 31 July 2022).
25. Hacker, V.; Mitsushima, S. *Fuel Cells and Hydrogen: From Fundamentals to Applied Research*; Elsevier: Amsterdam, The Netherlands, 2018; ISBN 978-0-12-811537-4.
26. Samsun, R.C.; Rex, M.; Antoni, L.; Stolten, D. Deployment of Fuel Cell Vehicles and Hydrogen Refueling Station Infrastructure: A Global Overview and Perspectives. *Energies* **2022**, *15*, 4975. [CrossRef]
27. Alaswad, A.; Omran, A.; Sodre, J.R.; Wilberforce, T.; Pignatelli, G.; Dassisti, M.; Baroutaji, A.; Olabi, A.G. Technical and Commercial Challenges of Proton-Exchange Membrane (PEM) Fuel Cells. *Energies* **2021**, *14*, 144. [CrossRef]
28. Fragiacomo, P.; De Lorenzo, G.; Corigliano, O. Performance Analysis of an Intermediate Temperature Solid Oxide Electrolyzer Test Bench under a CO_2-H_2O Feed Stream. *Energies* **2018**, *11*, 2276. [CrossRef]
29. Cigolotti, V.; Genovese, M.; Fragiacomo, P. Comprehensive Review on Fuel Cell Technology for Stationary Applications as Sustainable and Efficient Poly-Generation Energy Systems. *Energies* **2021**, *14*, 4963. [CrossRef]
30. Tremblay, O.; Dessaint, L.-A. Experimental Validation of a Battery Dynamic Model for EV Applications. *World Electr. Veh. J.* **2009**, *3*, 289–298. [CrossRef]
31. Zhu, C.; Li, X.; Song, L.; Xiang, L. Development of a Theoretically Based Thermal Model for Lithium Ion Battery Pack. *J. Power Sources* **2013**, *223*, 155–164. [CrossRef]
32. Saw, L.H.; Somasundaram, K.; Ye, Y.; Tay, A.A.O. Electro-Thermal Analysis of Lithium Iron Phosphate Battery for Electric Vehicles. *J. Power Sources* **2014**, *249*, 231–238. [CrossRef]
33. Shepherd, C.M. Design of Primary and Secondary Cells: II. An Equation Describing Battery Discharge. *J. Electrochem. Soc.* **1965**, *112*, 657. [CrossRef]
34. Plett, G.L. Extended Kalman Filtering for Battery Management Systems of LiPB-Based HEV Battery Packs: Part 2. Modeling and Identification. *J. Power Sources* **2004**, *134*, 262–276. [CrossRef]
35. Map Batteries. Available online: https://www.mapbatterie.it (accessed on 5 June 2022).
36. Piraino, F.; Fragiacomo, P. Design of an Equivalent Consumption Minimization Strategy-Based Control in Relation to the Passenger Number for a Fuel Cell Tram Propulsion. *Energies* **2020**, *13*, 4010. [CrossRef]
37. Zhou, J.; Feng, C.; Su, Q.; Jiang, S.; Fan, Z.; Ruan, J.; Sun, S.; Hu, L. The Multi-Objective Optimization of Powertrain Design and Energy Management Strategy for Fuel Cell–Battery Electric Vehicle. *Sustainability* **2022**, *14*, 6320. [CrossRef]
38. Li, Q.; Yang, H.; Han, Y.; Li, M.; Chen, W. A State Machine Strategy Based on Droop Control for an Energy Management System of PEMFC-Battery-Supercapacitor Hybrid Tramway. *Int. J. Hydrogen Energy* **2016**, *41*, 16148–16159. [CrossRef]
39. Singh, B.; Singh, S. State of the Art on Permanent Magnet Brushless DC Motor Drives. *J. Power Electron.* **2009**, *9*, 1–17.
40. He, C.; Wu, T. Permanent Magnet Brushless DC Motor and Mechanical Structure Design for the Electric Impact Wrench System. *Energies* **2018**, *11*, 1360. [CrossRef]
41. Toliyat, H.A. Analysis and Simulation of Multi-Phase Variable Speed Induction Motor Drives under Asymmetrical Connections. In Proceedings of the Applied Power Electronics Conference, APEC'96, San Jose, CA, USA, 3–7 March 1996; Volume 2, pp. 586–592.
42. Bose, B.K. *Modern Power Electronics and AC Drives*; Prentice Hall PTR: Old Tappan, NJ, USA, 2002; ISBN 978-0-13-016743-9.
43. Erickson, R.W.; Maksimovic, D. *Fundamentals of Power Electronics*; Springer Science & Business Media: Berlin/Heidelberg, Germany, 2007; ISBN 978-0-306-48048-5.
44. Usha, S.; Dubey, P.M.; Ramya, R.; Suganyadevi, M.V. Performance Enhancement of BLDC Motor Using PID Controller. *Int. J. Power Electron. Drive Syst.* **2021**, *12*, 1335–1344. [CrossRef]
45. Jin, C.-S.; Kim, C.-M.; Kim, I.-J.; Jang, I. Proposed Commutation Method for Performance Improvement of Brushless DC Motor. *Energies* **2021**, *14*, 6023. [CrossRef]

46. Maxon Electric Motors. Available online: https://www.maxongroup.com/maxon/view/product/motor/ecmotor/ec4pole/305013 (accessed on 31 July 2022).
47. Faulhaber Electric Motors. Available online: https://www.faulhaber.com/it/prodotti/serie/2232bx4/ (accessed on 31 July 2022).

energies

Article

Numerical 3D Model of a Novel Photoelectrolysis Tandem Cell with Solid Electrolyte for Green Hydrogen Production

Giosuè Giacoppo [1], Stefano Trocino [1], Carmelo Lo Vecchio [1], Vincenzo Baglio [1], María I. Díez-García [2], Antonino Salvatore Aricò [1] and Orazio Barbera [1,*]

1 CNR-ITAE, Via Salita S. Lucia Sopra Contesse 5, 98126 Messina, Italy
2 Departament de Química Física i Institut Universitari d'Electroquímica, Universitat d'Alacant, Apartat 99, E-03080 Alicante, Spain
* Correspondence: orazio.barbera@itae.cnr.it

Abstract: The only strategy for reducing fossil fuel-based energy sources is to increase the use of sustainable ones. Among renewable energy sources, solar energy can significantly contribute to a sustainable energy future, but its discontinuous nature requires a large storage capacity. Due to its ability to be produced from primary energy sources and transformed, without greenhouse gas emissions, into mechanical, thermal, and electrical energy, emitting only water as a by-product, hydrogen is an effective carrier and means of energy storage. Technologies for hydrogen production from methane, methanol, hydrocarbons, and water electrolysis using non-renewable electrical power generate CO_2. Conversely, employing photoelectrochemistry to harvest hydrogen is a sustainable technique for sunlight-direct energy storage. Research on photoelectrolysis is addressed to materials, prototypes, and simulation studies. From the latter point of view, models have mainly been implemented for aqueous-electrolyte cells, with only one semiconductor-based electrode and a metal-based counter electrode. In this study, a novel cell architecture was numerically modelled. A numerical model of a tandem cell with anode and cathode based on metal oxide semiconductors and a polymeric membrane as an electrolyte was implemented and investigated. Numerical results of 11% solar to hydrogen conversion demonstrate the feasibility of the proposed novel concept.

Keywords: photoelectrochemical tandem cell; hydrogen production; hematite photoanode; numerical model; solid electrolyte membrane; metal oxide semiconductor

Citation: Giacoppo, G.; Trocino, S.; Lo Vecchio, C.; Baglio, V.; Díez-García, M.I.; Aricò, A.S.; Barbera, O. Numerical 3D Model of a Novel Photoelectrolysis Tandem Cell with Solid Electrolyte for Green Hydrogen Production. *Energies* **2023**, *16*, 1953. https://doi.org/10.3390/en16041953

Academic Editor: Attilio Converti

Received: 8 December 2022
Revised: 31 January 2023
Accepted: 6 February 2023
Published: 16 February 2023

1. Introduction

Increases in world population and human activities have led to a considerable growth in global energy demand, most of which is fulfilled using non-renewable sources. With a view to a sustainable future and to address global energy challenges, current research is shifting toward environmentally neutral technologies. Indeed, consuming sustainable energy sources such as hydropower, wind, solar, ocean, and geothermal seems to be the only way to reduce usage of fossil fuel-based sources. However, using renewable energy sources (RES) for large-scale, cost-effective energy production is still a difficult challenge. Continuous energy provision is incompatible with the intermittent nature of RES, so an energy storage medium is essential for implementing a carbon-neutral economy [1,2]. Of the renewable energy sources mentioned above, the potential of solar energy, to date, has not been sufficiently exploited and is likely to contribute significantly to a green future. Various approaches exist to directly convert solar energy into chemical energy, as in the case of batteries or by producing a storable, high-energy density fuel, such as hydrogen [3]. Although batteries have long been manufactured and used with energy efficiency approaching 90%, this technology suffers from energy losses and capacity degradation over time. In contrast, hydrogen has the potential to be an excellent energy carrier because of its high energy density (about 140 MJ/kg) and the possibility of being a fuel for internal combustion engines, turbines, and fuel cells.

It does not release any greenhouse gases, only water as a by-product; if generated using RES, it is environmentally friendly [1,2]. In the gaseous state, hydrogen is not found in nature; it can be extracted from plants, methane, methanol, and higher hydrocarbons. In particular, it can be obtained from water. Currently, it is produced through several thermochemical, greenhouse gas-generating methods. Production by conventional water electrolysis requires massive external electrical energy sources. This process is environmentally friendly only if the electricity comes from RES. Other water electrolysis technologies include bio-aided, carbon- or hydrocarbon-aided, and photoelectrochemical [2,4]. Water electrolysis can be obtained in a photoelectrochemical cell (PEC), a device capable of implementing this splitting using electricity directly converted from sunlight. It is composed of two electrodes immersed in an aqueous electrolyte or separated by a hydrated solid electrolyte, at least one of which consists of a semiconductor that is exposed to sunlight and capable of absorbing it. The two electrodes of a PEC can be configured as an n-type semiconductor photoanode and a working in-the-dark cathode, two semiconductors (an n-type photoanode and a p-type photocathode), or a p-type semiconductor photocathode and a working in-the-dark anode [4,5]. Using only one semiconductor electrode is a challenging, and, to date, no materials based on earth-abundant and cheap elements offer a reasonable solution. One solution to this is to use two semiconductors in a tandem configuration, where the electrodes are excited by the same light beam, with the second electrode receiving the radiation filtered by the first. This configuration optimizes sunlight and combines the photovoltage generated by the two electrodes. Several n-type semiconductors, including TiO_2, ZnO, $BiVO_4$, Ta_3N_5, WO_3, and Si, have been studied as anodes for water photoelectrolysis. They show good potential but, conversely, are characterized by a large bandgap and/or instability in solution. Among the n-type semiconductors, hematite (α-Fe_2O_3) is characterized by low valence band position, abundance, stability in aqueous solution, nontoxicity, and bandgap of about 1.9–2.2 eV. It can utilize up to 40% of solar radiation and achieve a solar-to-hydrogen conversion efficiency of up to 15.5% [6]. From an experimental point of view, Du et al. [7] studied the water oxidation performance of a hematite photoanode in a PEC, enhanced via Co doping and surface modification. Yang et al. [8] altered the electrode surface with catalysts to improve semiconductor efficiency using atomic deposition-growth of MnO_x on hematite. Wang et al. [9] used quick high-temperature annealing to enhance the performance of thin hematite layers on a TiO_2 nano-sheet. Deshmukh et al. [10] investigated the electrochemical behavior of ZnO nanorods in 1 M Na_2SO_4 solution, employing various electrochemical methods. Wang et al. [11] considered hematite nanorod films deposited on titanium foils and modified with phosphorous, to improve the photoelectrochemical process. Syrek et al. [12] deposited electrochemically Cu_2O thin films on FTO-coated glass (Flourine Thin Oxide) at three different temperatures, finding that the FTO/Cu_2O cathode synthesized at the highest temperature showed superior performance. Meda et al. [13] studied different techniques to produce nanostructured WO_3 photoanodes on transparent conducting oxides, considering the Santato-Augustynski formula. Finally, Kwon [14] proposed a simple and fast method for producing hierarchical nanostructure hematite using an ultraviolet-assisted process. Similar studies on materials were carried out in [15,16]. PEC prototyping with hematite photoanodes was performed by Lopes et al. [17]; they designed a PEC for testing different photoelectrode configurations; they investigated two semiconductors: tungsten trioxide and undoped hematite. Brinkert et al. [18] demonstrated that, under microgravity conditions, it is possible to produce hydrogen efficiently in a photoelectrochemical manner; this represents a possible alternative to the current space mission life-support technologies. Hogerwaard et al. [19] developed an integrated concentrated photovoltaic and photoelectrochemical hydrogen reactor. Vilanova et al. [20] developed a 50 cm^2 tandem PEC-PV cell named "CoolPEC" (compact, optimized, open light PEC cell). Recently, important progress has been made in the modelling and simulation-supported development of PECs. Three-dimensional multiphysics models capable of simulating photoelectrochemical processes have been used for this purpose, providing design criteria for semiconductors, electrocatalysts, and electrolytes.

Moreover, modelling highlighted the most feasible device architectures regarding efficiency, stability, and safety. Nevertheless, the need remains to model the more complex physical aspects (e.g., thermal effects and bubble formation) that can induce malfunctions [21]. In this context, Berger et al. [22] developed a one-dimensional model of PEC that is valid for "wired" and "wireless" architectures. Stevens et al. [23] presented a computational method to evaluate the hourly temperature profiles and solar-hydrogen efficiency to achieve an optical concentration of PECs during a typical year. Hankin et al. [24] presented a model with a hematite photoanode and platinized titanium cathode, reproducing the experimental photoelectrolysis values obtained with a device equipped with 0.1×0.1 m^2 electrodes of various geometries and architectures. Njoka et al. [25] developed a two-dimensional numerical model of a hematite photoelectrode in a two-chamber reactor separated by a proton-permeable membrane. By considering the light absorption, charge transport, and redox reactions, the model investigated the effects of materials, height, and electrolyte velocity on hydrogen and oxygen production. Xiang et al. [26] investigated a detailed geometric parameter in two cell designs for a photoelectrolysis system sustained by a water vapor feed. Walczak et al. [27] simulated an integrated photoelectrolysis system equipped with a WO_3/FTO/p+n Si anode, a Pt/TiO_2/Ti/n+p Si cathode, and a Nafion membrane separator. Harmon et al. [28] considering charge transport in semiconductor-electrolyte-based PEC solar cells, modelled using the drift-diffusion-Poisson equations, proposed numerical methods for their resolution. Other models with liquid electrolytes can also be found in the literature [29–33]. As seen in the above literature review, models have generally considered liquid electrolytes with a polymeric separator. This work studied a novel concept of tandem cells with non-precious materials and hydrated solid electrolytes. A solid polymer electrolyte has many advantages; indeed, a PEC works with only pure water, and no corrosive liquid electrolytes with recirculation/purification systems are needed. Moreover, active materials can be easily sandwiched to form a solid-state and simplified device. This model aims to simulate water splitting into H_2 and O_2 through photoelectrolysis, in view of subsequent prototyping of the PEC. The system consists of a PEC composed of four main components: the photoelectrodes, the water channels, the solid electrolyte alkaline membrane, and a cathodic diffusion layer. The numerical model calculates the 3D distribution of the fluid dynamic and electrochemical quantities, the polarization curve (I-V), and their correlation with the cell design. The local current source was analytically simulated using the Gartner formulation at the photoanode [34] and the Butler–Volmer at the photocathode. This tandem configuration makes good use of the incident sunlight and combines the photovoltage generated from the two electrodes. Notably, the simulation covers fluid-dynamic and chemical aspects (conservation and transport) that a 1D model does not consider.

2. Model Definition

The novel concept of a PEC in a tandem configuration was based on a sandwich of FTO-coated glass supporting a layer of Fe-based semiconductor (anode), a solid polymer alkaline membrane, and a CuO-based photocathode supported on a hydrophobic carbonaceous layer.

Once the photoelectrochemical cell is assembled and fed with distilled water, the incident sunlight is transmitted across the transparent FTO-coated glass layer and promotes the water-splitting reaction. The gaseous products, H_2 and O_2, are collected by the channels formed over the membrane surface at the cell outlet. As dehydrated hydrogen production represents one of the main goals of this research activity, a highly hydrophobic porous layer was considered, in conjunction with the photocathode, to accomplish this task. Indeed, a hydrophobic porous layer allows the collection of hydrogen gas only, which passes through the pores, while keeping the water inside the cell. This is usually constituted by carbonaceous-based materials (Figure 1).

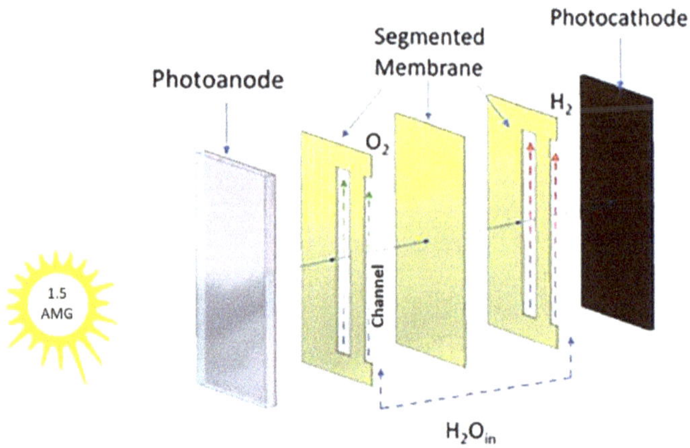

Figure 1. Sketch of the PEC novel concept.

Figure 2a shows the 3D single PEC model depicted in the simulation software. It consists of an anode collector (the transparent FTO-coated glass), a photoanode layer, a segmented membrane, a photocathode, a GDL (gas diffusion layer) for hydrogen collection, and a cathode collector. The segmented membrane performs three functions, it operates as a solid electrolyte, retains the water in the channels, and allows the escape of gaseous species. The PEC's elements are layered on each other, forming the geometric domain. They were parametrized in height and width in the x–z plane, according to Figure 2b; the length of the domain was defined along the y direction (Figure 2a).

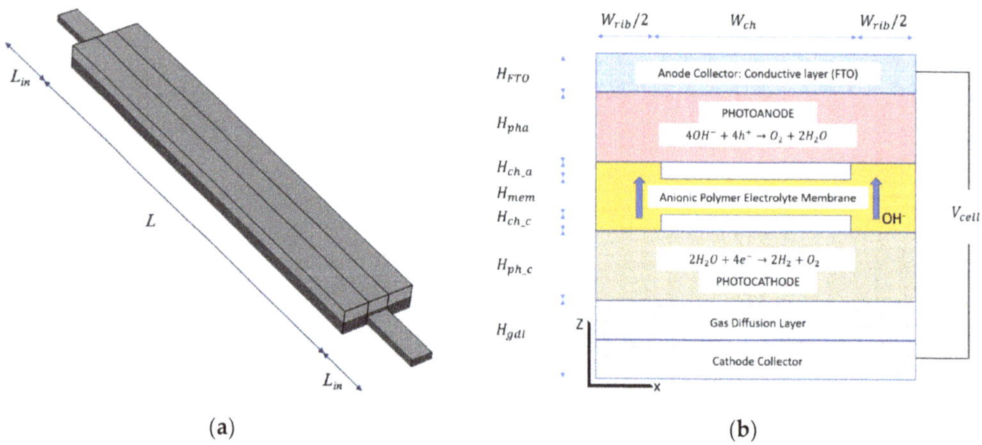

Figure 2. (**a**) Drawing of the numerical 3D domain (**b**) PEC layers with their dimensional parameters (z–x plane).

Table 1 lists the geometric and reference physical characteristics of the numerical model.

Table 1. Geometrical parameters adopted for the model.

Name	Value	Description
L	10.0×10^{-3} m	Cell length
L_{in}	1.5×10^{-3} m	In/out length
H_{FTO}	4.0×10^{-3} m	Conductive layer height (FTO)
H_{pha}	4.0×10^{-8} m	Photoanode thickness
H_{ch_a}	5.0×10^{-5} m	Channel height
H_{mem}	1.0×10^{-4} m	Membrane thickness
H_{ch_c}	5.0×10^{-5} m	Channel height
H_{ph_c}	4.0×10^{-6} m	Photocathode thickness
H_{gdl}	3.0×10^{-4} m	GDL width
W_{ch}	5.0×10^{-4} m	Channel width
W_{rib}	1.2×10^{-3} m	Rib width
A_{inlet}	2.5×10^{-8} m^2	Inlet Area
A_{phele}	1.7×10^{-5} m^2	Photoelectrode Area
A_{cell}	1.7×10^{-5} m^2	Cell Area

A structured mesh (Figure 3) was implemented for the model discretization; the mesh was refined at the inlet, outlet, and central part of the active area, in correspondence with the channel considered for the escape of the reactants. The adopted mesh consisted of about 300,000 rectangular elements.

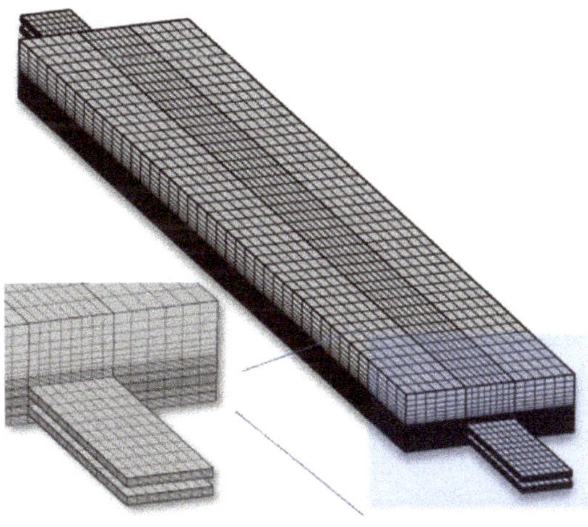

Figure 3. A picture of the mesh adopted for the model, the zoomed part (on the left) represents the inlet area with the refinement of the mesh.

Water is supplied through the anode and cathode sides of the simulated PEC and divided into H_2 and O_2 by the energy provided by the solar radiation (considered constant over the active surface) and the eventually applied bias E, with E < 1.23 V, under the semi-redox reactions below.

$$4OH^- + 4h^+ \rightarrow 2H_2O + O_2 \uparrow \text{(anodic)}$$

$$2H_2O + 2e^- \rightarrow 2OH^- + H_2 \uparrow \text{(cathodic)}$$

Produced and consumed species are transported through the anodic and cathodic channels and photoelectrodes. As the nanopillar structure characterizes the hematite photoanode, this layer should be porous (eventually with a low permeability), so the oxygen formed at the interface can pass through the photoanode and escape the PEC through the channels.

The main adopted model assumption was:

- Electrochemical reactions include the oxidation and reduction of water at the photoanode and cathode.
- Charge conservation within the electrolyte was assumed in the membrane, under the hypothesis that it is a homogeneous and continuous media, and no bubble generation effect was considered.
- Photocurrent densities were limited by the incident photon flux.
- Temperature effects were neglected. All the parameter values adopted here were evaluated at room temperature.

Comsol Multiphysics was used to implement the numerical simulations.

The model's inputs were the condition of the water flow entering the anode and the cathode, the operating pressure at the channel outlet (set as ambient pressure), and the voltage applied across the cell as the difference between the electrode potentials. The model calculates the ionic and electronic potential distribution, current density, velocity, pressure, and species distribution within the domain volume.

Table 2 reports the operating conditions for the base case, whilst Table 3 lists the model's physical parameters.

Table 2. Operating conditions of the model.

Name	Value	Description
p_{ref}	1.01×10^5 Pa	Reference pressure
T	308.15 K	Cell temperature
u_{in_anode}	5×10^{-4} m/s	Anode inlet flow velocity (Equations (1) and (2))
$u_{in\_cathode}$	5×10^{-4} m/s	Cathode inlet flow velocity (Equations (1) and (2))
μ_{anode}	1.19×10^{-5} Pa·s	Anode viscosity (Equations (2) and (3))
$\mu_{cathode}$	2.46×10^{-5} Pa·s	Cathode viscosity (Equations (2) and (3))
μ_{water}	7.972×10^{-4} Pa·s	Mixture viscosity (Equations (2) and (3))
ρ_{water}	1000 kg/m^3	Mixture density

Table 3. Model physical parameters.

Name	Value	Description
σ_{gdl}	222 S/m	GDL electric conductivity
σ_m	9.825 S/m	Membrane conductivity
w_{H2_in}	0.743	Inlet H_2 mass fraction (anode)
w_{H2O_in}	0.023	Inlet H_2O mass fraction (cathode)
w_{O2_in}	0.228	Inlet oxygen mass fraction (cathode)
M_{H2}	0.002 kg/mol	Hydrogen molar mass
M_{N2}	0.028 kg/mol	Nitrogen molar mass
M_{H2O}	0.018 kg/mol	Water molar mass
M_{O2}	0.032 kg/mol	Oxygen molar mass
c_{O2ref}	40.88 mol/m^3	Oxygen reference concentration (Equation (1))
c_{H2ref}	40.88 mol/m^3	Hydrogen reference concentration (Equations (1) and (11))

Table 3. *Cont.*

Name	Value	Description
D_{H2_H2O}	9.2048×10^{-5} m^2/s	H_2-H_2O binary diffusion coefficient ($9.15 \times 10^{-5} \cdot (T/307.19)^{1.75}$) (Equation (1))
D_{O2_H2O}	2.8208×10^{-5} m^2/s	O_2-H_2O binary diffusion coefficient ($2.82 \times 10^{-5} \cdot (T/308.1)^{1.75}$) (Equation (1))
V_{cell}	1.6 V	Cell voltage
ε_{gdl}	0.4	GDL porosity (Equation (3))
ε_{l}	0.3	Electrolyte phase volume fraction (Equation (3))
ε_{cl}	0.3	Open volume fraction for gas diffusion in porous electrodes ($1 - \varepsilon_l - \varepsilon_{GDL}$) (Equation (3))
k_{cl}	2.36×10^{-12} m^2	Permeability (porous electrode, $k_{\mathrm{GDL}}/5$, Equation (3))
k_{GDL}	1.18×10^{-11} m^2	GDL permeability (Equation (3))
i_{0_cat}	0.03 A/m^2	Cathode exchange current density (Equation (11))
V_{chan}	2.5×10^{-10} m^3	Channel volume
L_{in}	0.0015 m	Channel inlet length
α_{cat}	0.5	Cathode transfer coefficient (Equation (11))
q	1.61×10^{-19} C	Elemental charge (Equation (9))
E_{eq_a}	1.185 V	Anode equilibrium potential
$E_{fb,an}$	−0.7 V	Hematite flat band potential (Equation (9))
$E_{bp,cat}$	0	CuO flat band potential
Φ_0	3.8×10^{-18} m^{-2}·s^{-1} nm	Incident photon flux at 550 nm (Equation (9))
λ	5.5×10^{-7} m	Wavenumber length (Equation (10))
ε_0	8.854×10^{-12} F/m	Vacuum permittivity (Equation (9))
ε_{hema}	80	Hematite permittivity (Equation (9))
N_D	1×10^{25} 1/m^3	Donor concentration (Equation (9))
L_p	5×10^{-9} m	Diffusion length hematite
α_λ	5.879×10^{6}	Hematite extinction coefficient at 550 nm (Equation (10))
k	1	Kinetic constant (Equation (11))

2.1. Species Transport and Conservation

In the computational domains, conservation equations for mass, chemical species, momentum, and charge are solved. Diffusion and convection result in the transportation of species, which takes place in the channels, anodic, and cathodic photoelectrodes, and the GDL. In liquid water streams, the generated hydrogen and oxygen are considered diluted species. The species transportation was implemented using the molar balance equation:

$$\nabla \cdot (-D_i \nabla c_i) + \boldsymbol{u} \cdot \nabla c_i = 0 \quad (1)$$

In (1), D is the binary diffusion coefficient, c is the concentration, and $\boldsymbol{u}$ is the velocity vector. The first addend in (1) indicates the transport by diffusion according to Fick's law; the second is the convective transport caused by water flow. The anodic and cathodic fluxes are mixtures of two species (H_2O-O_2, H_2O-H_2); therefore, to consider the actual volume available for the diffusion of the species in the porous media, D is modified with Bruggeman's condition in the porous electrode domain. Porous materials can be considered a volume characterized by a solid matrix and voids, in which one or more fluid phases can occur. In a porous medium, the theoretical and real diffusion coefficients are different. This occurs because the fluid has less transverse diffusion area available to it than the free volume. Since the fluid must move between the interstices of the solid matrix, it must travel a greater distance than to move geometrically between two points in the material itself, which makes the real concentration gradient lower than the theoretical one. By correlating the tortuosity of the path within the porous medium with the porosity, Bruggeman's theory allows the determination of the real diffusion coefficient to be used in Fick's law.

2.2. Flow Channels and Porous Electrodes

The Navier–Stokes Equation (2) describes the flow in the channels (anode and cathode):

$$\rho(\boldsymbol{u}\cdot\nabla\boldsymbol{u}) = \nabla\cdot\left[-p\boldsymbol{I} + \mu\left(\nabla\boldsymbol{u} + (\nabla\boldsymbol{u})^T\right)\right] \tag{2}$$

and Darcy Brinkman's Equation (3) gives the flow in the porous domains (photo-electrodes and GDL):

$$\frac{\rho}{\varepsilon}\left((\boldsymbol{u}\cdot\nabla)\frac{\boldsymbol{u}}{\varepsilon}\right) = \nabla\cdot\left[-p\boldsymbol{I} + \frac{\mu}{\varepsilon}\left(\nabla\boldsymbol{u} + (\nabla\boldsymbol{u})^T\right) - \frac{2}{3}\frac{\mu}{\varepsilon}(\nabla\cdot\boldsymbol{u})\boldsymbol{I}\right] - \left(\frac{\mu}{k} + \frac{Q}{\varepsilon^2}\right)\boldsymbol{u} \tag{3}$$

The momentum Equations (2) and (3) are solved together with the continuity Equation (4):

$$\rho\nabla\cdot\boldsymbol{u} = Q \tag{4}$$

In Equations (2)–(4), $\boldsymbol{u}$ represents the velocity, ρ is the fluid density, μ is the viscosity, p is the pressure, ε is the porosity, k is the permeability, Q is the mass flux, $\boldsymbol{I}$ is the identity tensor

2.3. Transportation and Conservation of Charge

Ohm's law, in conjunction with the charge balance for electrons and ions, is used to set the charge transport and conservation in photoelectrodes and electrolyte interfaces:

$$\nabla\cdot\boldsymbol{i_s} = 0 \tag{5}$$

$$\boldsymbol{i_s} = -\sigma_s\nabla\varphi_s \tag{6}$$

$$\nabla\cdot\boldsymbol{i_l} = 0 \tag{7}$$

$$\boldsymbol{i_l} = -\sigma_l\nabla\varphi_l \tag{8}$$

where $\boldsymbol{i_s}$ and $\boldsymbol{i_l}$ are the electronic and ionic current densities, respectively, and φ_s and φ_l are the electronic and ionic potentials, σ_s and σ_l the electronic and ionic conductivities. The flow of charged species is assumed to be dominated by the electric potential gradient instead of the concentration gradients. Considering the presence of liquid water in the channels, the membrane can be regarded as fully hydrated, and independent of external water flow, within certain conditions. Locally, at the photoelectrode–membrane interface, the electronic and ionic currents have the same order of magnitude, with the same number of electrons and ions exchanged in the redox half-reactions.

2.4. Photoelectrode Kinetics

Local current densities from sunlight were modelled using an analytical approach. The Gartner–Butler equation, a simplified equation arising from the Gartner model, was applied to the photoanode [35]:

$$\varphi_s - \varphi_l - E_{fb,an} = \left(\frac{N_D}{2\cdot q\cdot\varepsilon_{hema}\cdot\varepsilon_0}\right)\left(\frac{i_{loc,an}}{\alpha\Phi_0}\right)^2 \tag{9}$$

where φ_s and φ_l are the electronic and ionic potentials, $E_{fb,an}$ is the anode flat band potential, q is the elemental charge, N_D is the donor concentration, $i_{loc,an}$ is the local current density, Φ_0

is the incident photon flux, α is the absorption coefficient, which is related to the wavelength λ and extinction coefficient α_λ by:

$$\alpha = \frac{4\pi\alpha_\lambda}{\lambda} \tag{10}$$

For the photocathode case, a parametric equation with a form similar to that of the diffusion-limited Butler–Volmer equation was used as an approximation, so the local current $i_{loc,cat}$ is

$$i_{loc,cat} = i_{0,cat}\cdot\left(\frac{c_{H_2}}{c_{H_{2,ref}}}\right)^{k_{H_2}}\cdot\left[exp\left(\frac{\alpha_{cat}\cdot F\cdot\eta_{act,cat}}{RT}\right) - exp\left(\frac{-\alpha_{cat}\cdot F\cdot\eta_{act,cat}}{RT}\right)\right] \tag{11}$$

where $i_{0,cat}$ is the cathode exchange current density, c_{H_2} is the H_2 concentration, $c_{H_{2,ref}}$ is the H_2 reference concentration, F is the Faraday constant, R is the universal Gas constant, T is the temperature, $\eta_{act,cat}$ is the activation overpotential, α_{cat} is the charge transfer coefficient, and k_{H2} is the kinetic constant

3. Results and Discussion

Figure 4 shows the I–V curve obtained from the continuum model, varying the voltage applied at the cell leads. The figure reports the current density at two different values of the incident photon flux, because the model solves the equations specifically for monochromatic light. In this case, the incident photon flux corresponds to $\Phi_0 = 3.8 \times 10^{18}\ [m^{-2}\ s^{-1}\ nm]$ and corresponds to the energy content limited to just one wavelength number (λ = 550 nm). By increasing the incident photon flux up to $\Phi_0 = 5 \times 10^{21}\ [m^{-2}\ s^{-1}\ nm]$ (that corresponds to the total spectra), the current density magnitude increases accordingly. In the case of a monochromatic incident photon flux, the maximum value of current density was about 2.3 mA/cm^2; as expected, the full spectra generated a maximum current density of about 9.0 mA/cm^2

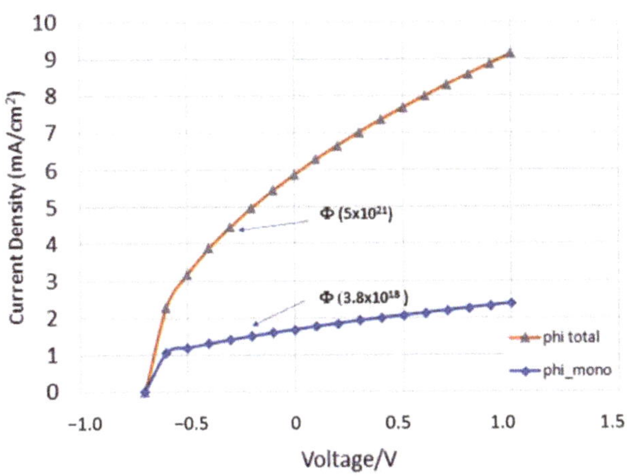

Figure 4. I–V curves derived from the continuum models and related to the total spectra and a monochromatic case.

The current density distribution in the electrodes (electronic current) and electrolyte (ionic current) is shown in Figure 5, which is illustrated as a cross-section of the PEC model.

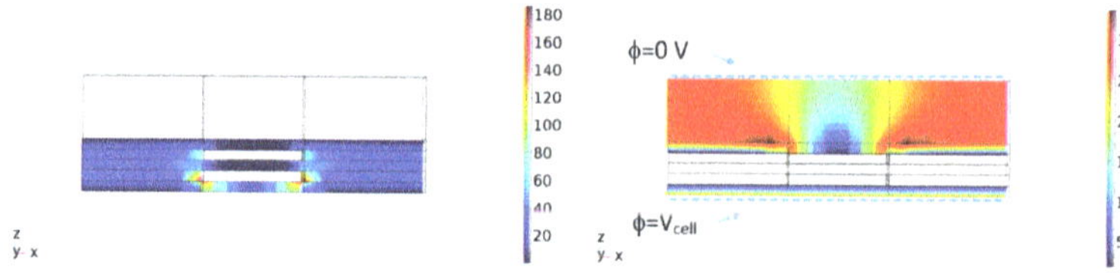

Figure 5. Current density ionic in the electrolyte (**left**) electronic in the photoelectrode (**right**) values in mA/cm^2.

The top and lower bound edges (on the anode side) show the highest achievable electrolyte current density. The three-phase boundary between the electrode, reactant, and the electrolyte is realized at this site. The reduced interface availability for the photo electrolysis process might be an issue for a scaled-up prototype. Therefore, the channel width and membrane rib width must be chosen appropriately, to maximize the surface for the reaction and minimize parasitic losses of the water and products entering/leaving the photoelectrochemical cell.

According to this trend, the photo-generated electronic current was characterized by the complementary tendency in the proximity of the current collection surfaces.

The current density field, as distributed, causes oxygen and hydrogen to form at the interface between the photoelectrodes and the water channels, as shown in Figure 6a.

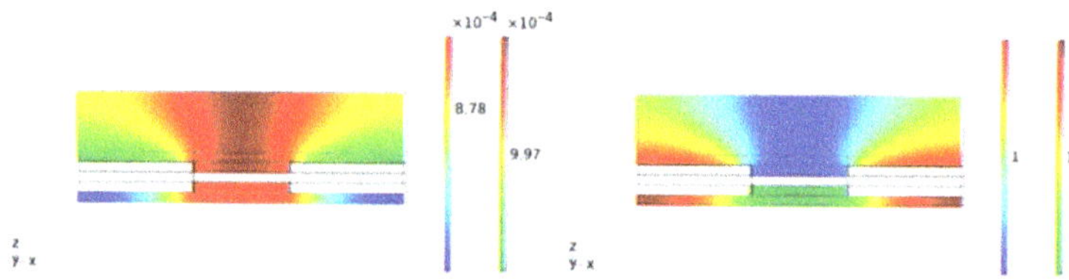

(**a**) (**b**)

Figure 6. (**a**) Mass fraction distributions of oxygen/hydrogen, (**b**) water consumed at both the anode and cathode side @ 1 V potential.

Specularly, water is consumed on both the anode and cathode sides (Figure 6b).

The results demonstrate how the presence of outlet channels for the gases produced and for the permanence of the water is essential for the proper functioning of the cell, and how modelling helps visualize this behavior. Using a continuous membrane, as often occurs in small laboratory cells as the cell area increases, implies the impossibility of escape for the oxygen and the difficulty, if not impossibility, of supplying of water for the solid electrolyte. This may negatively affect or even impede the device's proper functioning after a short time.

4. Conclusions

A new concept of PEC was numerically modelled using 3D multiphysics software, covering the physical (species and mass transport, heat transport, and charge transport) and electrochemical (electrode reactions) processes that occur as a result of sunlight. The numerical results showed that the cell could reach current densities in the 10 mA/cm^2 @ 1 V bias. Simulation results are helpful for cell design because they allow one to "visualize", albeit virtually, the nonuniform distribution of species, potentials, and currents that often cause significant performance reductions, especially in prototypes with a sizeable active surface area. In addition, they can be used as a tool to address the enhancement of the surface reactions and the cell configuration.

Further improvements of the continuum model are possible, for instance, with the explicit consideration of the losses caused by recombination.

Notwithstanding that Gartner's equation provides a good description of the photoelectrode–electrolyte interface in PEC cells, its limitations are well known, especially in not accounting for recombination effects, and thus are most evident in photoelectrolysis cells, where the slowness of redox processes, e.g., oxygen evolution, can become a determining factor in the rate of water molecule separation.

With the assumptions of Gartner's model, the polarization curve of the photoelectrolysis device can vary significantly. A simple way to eliminate these deviations might involve using explicit, potential-dependent formulations for the kinetic constants that appear in Equation (9).

Author Contributions: Conceptualization, G.G. and O.B.; methodology, G.G., O.B. and M.I.D.-G.; data curation, S.T., C.L.V., V.B. and A.S.A.; writing—original draft, G.G. and O.B.; writing—review and editing, G.G. and O.B. All authors have read and agreed to the published version of the manuscript.

Funding: This research was funded by the European Union's Horizon 2020 research and innovation program under grant agreement ID 760930 (FotoH2 project).

Data Availability Statement: Not applicable.

Conflicts of Interest: The authors declare no conflict of interest.

References

1. Burton, N.A.; Padilla, R.V.; Rose, A.; Habibullah, H. Increasing the efficiency of hydrogen production from solar powered water electrolysis. *Renew. Sustain. Energy Rev.* **2021**, *135*, 110255. [CrossRef]
2. Saravanan, P.; Khan, M.R.; Yee, C.S.; Vo, D.V.N. An overview of water electrolysis technologies for the production of hydrogen. In *New Dimensions in Production and Utilization of Hydrogen*; Elsevier: Amsterdam, The Netherlands, 2020; pp. 161–190, ISBN 9780128195536.
3. Modestino, M.A.; Haussener, S. An Integrated Device View on Photo-Electrochemical Solar-Hydrogen Generation. *Annu. Rev. Chem. Biomol. Eng.* **2015**, *6*, 13–34. [CrossRef]
4. Chatterjee, P.; Ambati, M.S.K.; Chakraborty, A.K.; Chakrabortty, S.; Biring, S.; Ramakrishna, S.; Wong, T.K.S.; Kumar, A.; Lawaniya, R.; Dalapati, G.K. Photovoltaic/photo-electrocatalysis integration for green hydrogen: A review. *Energy Convers. Manag.* **2022**, *261*, 115648. [CrossRef]
5. Bak, T.; Nowotny, J.; Rekas, M.; Sorrell, C.C. Photo-electrochemical hydrogen generation from water using solar energy. Materials-related aspects. *Int. J. Hydrogen Energy* **2002**, *27*, 991–1022. [CrossRef]
6. Li, C.; Luo, Z.; Wang, T.; Gong, J. Surface, Bulk, and Interface: Rational Design of Hematite Architecture toward Efficient Photo-Electrochemical Water Splitting. *Adv. Mater.* **2018**, *30*, 1707502. [CrossRef] [PubMed]
7. Du, C.; Yang, J.; Yang, J.; Zhao, Y.; Chen, R.; Shan, B. An iron oxide -copper bismuth oxide photoelectrochemical cell for spontaneous water splitting. *Int. J. Hydrogen Energy* **2018**, *43*, 22807–22814. [CrossRef]
8. Yang, X.; Du, C.; Liu, R.; Xie, J.; Wang, D. Balancing Photovoltage Generation and Charge-Transfer Enhancement for Catalyst-Decorated Photoelectrochemical Water Splitting: A Case Study of the Hematite/MnOx Combination. *J. Catal.* **2013**, *304*, 86–91. [CrossRef]
9. Wang, D.; Chen, Y.; Zhang, Y.; Zhang, X.; Suzuki, N.; Terashima, C. Boosting photoelectrochemical performance of hematite photoanode with TiO_2 underlayer by extremely rapid high temperature annealing. *Appl. Surf. Sci.* **2017**, *422*, 913–920. [CrossRef]
10. Deshmukh, P.R.; Sohn, Y.; Shin, W.G. Chemical synthesis of ZnO nanorods: Investigations of electrochemical performance and photo-electrochemical water splitting applications. *J. Alloys Compd.* **2017**, *711*, 573–580. [CrossRef]
11. Wang, X.; Gao, W.; Zhao, Z.; Zhao, L.; Claverie, J.P.; Zhang, X.; Wang, J.; Liu, H.; Sang, Y. Efficient photo-electrochemical water splitting based on hematite nanorods doped with phosphorus. *Appl. Catal. B Environ.* **2019**, *248*, 388–393. [CrossRef]
12. Syrek, K.; Jażdżewska, M.; Kozieł, M.; Zaraska, L. Photoelectrochemical activity of Cu_2O electrochemically deposited at different temperatures. *J. Ind. Eng. Chem.* **2022**, *115*, 561–569. [CrossRef]
13. Meda, L.; Tozzola, G.; Tacca, A.; Marra, G.; Caramori, S.; Cristino, V.; Bignozzi, C.A. Photo-electrochemical properties of nanostructured WO3 prepared with different organic dispersing agents. *Sol. Energy Mater. Sol. Cells* **2010**, *94*, 788–796. [CrossRef]
14. Kwon, J.; Yeo, J.; Hong, S.; Suh, Y.D.; Lee, H.; Choi, J.H.; Lee, S.S.; Ko, S.H. Photoreduction Synthesis of Hierarchical Hematite/Silver Nanostructures for Photoelectrochemical Water Splitting. *Energy Technol.* **2016**, *4*, 271–277. [CrossRef]
15. Gao, J.; Sahli, F.; Liu, C.; Ren, D.; Guo, X.; Werner, J.; Jeangros, Q.; Zakeeruddin, S.M.; Ballif, C.; Grätzel, M.; et al. Solar Water Splitting with Perovskite/Silicon Tandem Cell and TiC-Supported Pt Nanocluster Electrocatalyst. *Joule* **2019**, *3*, 2930–2941. [CrossRef]
16. Heremans, G.; Bosserez, T.; Martens, J.A.; Rongé, J. Stability of vapor phase water electrolysis cell with anion exchange membrane. *Catal. Today* **2019**, *334*, 243–248. [CrossRef]
17. Lopes, T.; Dias, P.; Andrade, L.; Mendes, A. An innovative photoelectrochemical lab device for solar water splitting. *Sol. Energy Mater. Sol. Cells* **2014**, *128*, 399–410. [CrossRef]

18. Brinkert, K.; Richter, M.H.; Akay, Ö.; Liedtke, J.; Giersig, M.; Fountaine, K.T.; Lewerenz, H.J. Efficient solar hydrogen generation in microgravity environment. *Nat. Commun.* **2018**, *9*, 2527. [CrossRef]
19. Hogerwaard, J.; Dincer, I.; Naterer, G.F. Experimental investigation and optimization of integrated photovoltaic and photoelectrochemical hydrogen generation. *Energy Convers. Manag.* **2020**, *207*, 112541. [CrossRef]
20. Vilanova, A.; Lopes, T.; Spenke, C.; Wullenkord, M.; Mendes, A. Optimized photoelectrochemical tandem cell for solar water splitting. *Energy Storage Mater.* **2018**, *13*, 175–188. [CrossRef]
21. Xiang, C.; Weber, A.Z.; Ardo, S.; Berger, A.; Chen, Y.; Coridan, R.; Fountaine, K.T.; Haussener, S.; Hu, S.; Liu, R.; et al. Modellierung, Simulation und Implementierung von Zellen für die solargetriebene Wasserspaltung. *Angew. Chem.* **2016**, *128*, 13168–13183. [CrossRef]
22. Berger, A.; Newman, J. An Integrated 1-Dimensional Model of a Photoelectrochemical Cell for Water Splitting. *J. Electrochem. Soc.* **2014**, *161*, E3328–E3340. [CrossRef]
23. Stevens, J.C.; Weber, A.Z. A Computational Study of Optically Concentrating, Solar-Fuels Generators from Annual Thermal- and Fuel-Production Efficiency Perspectives. *J. Electrochem. Soc.* **2016**, *163*, H475–H484. [CrossRef]
24. Hankin, A.; Bedoya-Lora, F.E.; Ong, C.K.; Alexander, J.C.; Petter, F.; Kelsall, G.H. From millimetres to metres: The critical role of current density distributions in photo-electrochemical reactor design. *Energy Environ. Sci.* **2017**, *10*, 346–360. [CrossRef]
25. Njoka, F.; Ookawara, S.; Ahmed, M. Influence of design and operating conditions on the performance of tandem photoelectrochemical reactors. *Int. J. Hydrogen Energy* **2018**, *43*, 1285–1302. [CrossRef]
26. Xiang, C.; Chen, Y.; Lewis, N.S. Modeling an integrated photoelectrolysis system sustained by water vapor. *Energy Environ. Sci.* **2013**, *6*, 3713–3721. [CrossRef]
27. Walczak, K.; Chen, Y.; Karp, C.; Beeman, J.W.; Shaner, M.; Spurgeon, J.; Sharp, I.D.; Amashukeli, X.; West, W.; Jin, J.; et al. Modeling, simulation, and fabrication of a fully integrated, acidstable, scalable solar-driven water-splitting system. *ChemSusChem* **2015**, *8*, 544–551. [CrossRef]
28. Harmon, M.; Gamba, I.M.; Ren, K. Numerical algorithms based on Galerkin methods for the modeling of reactive interfaces in photoelectrochemical (PEC) solar cells. *J. Comput. Phys.* **2016**, *327*, 140–167. [CrossRef]
29. Haussener, S.; Xiang, C.; Spurgeon, J.M.; Ardo, S.; Lewis, N.S.; Weber, A.Z. Modeling, simulation, and design criteria for photoelectrochemical water-splitting systems. *Energy Environ. Sci.* **2012**, *5*, 9922–9935. [CrossRef]
30. Carver, C.; Ulissi, Z.; Ong, C.K.; Dennison, S.; Kelsall, G.H.; Hellgardt, K. Modelling and development of photoelectrochemical reactor for H_2 production. *Int. J. Hydrogen Energy* **2012**, *37*, 2911–2923. [CrossRef]
31. Qureshy, A.M.; Ahmed, M.; Dincer, I. Simulation of transport phenomena in a photo-electrochemical reactor for solar hydrogen production. *Int. J. Hydrogen Energy* **2016**, *41*, 8020–8031. [CrossRef]
32. Haussener, S.; Hu, S.; Xiang, C.; Weber, A.Z.; Lewis, N.S. Simulations of the irradiation and temperature dependence of the efficiency of tandem photoelectrochemical water-splitting systems. *Energy Environ. Sci.* **2013**, *6*, 3605. [CrossRef]
33. Andrade, L.; Lopes, T.; Ribeiro, H.A.; Mendes, A. Transient phenomenological modeling of photoelectrochemical cells for water splitting-Application to undoped hematite electrodes. *Int. J. Hydrogen Energy* **2011**, *36*, 175–188. [CrossRef]
34. Pleskov, Y.; Gurevic, Y. *Semiconductor Photoelectrochemistry*, 1st ed.; Consultants Bureau: New York, NY, USA; Springer: New York, NY, USA, 1986; ISBN 978-1468490800.
35. Butler, M.A. Photoelectrolysis and physical properties of the semiconducting electrode WO_2. *J. Appl. Phys.* **1977**, *48*, 1914–1920. [CrossRef]

MDPI AG
Grosspeteranlage 5
4052 Basel
Switzerland
Tel.: +41 61 683 77 34

Energies Editorial Office
E-mail: energies@mdpi.com
www.mdpi.com/journal/energies

www.ingramcontent.com/pod-product-compliance
Lightning Source LLC
Chambersburg PA
CBHW061503210926
51850CB00008B/1809
9783725803538